T0317576

Computational Methods for Mass Spectrometry Proteomics

Computational Methods for Mass Spectrometry Proteomics

Ingvar Eidhammer
University of Bergen, Norway

Kristian Flikka
University of Bergen, Norway

Lennart Martens
European Bioinformatics Institute, Hinxton, Cambridge, UK

Svein-Ole Mikalsen
The Norwegian Radium Hospital, Oslo, Norway

BICENTENNIAL 1807 WILEY 2007 BICENTENNIAL

John Wiley & Sons, Ltd

Copyright © 2007 John Wiley & Sons Ltd, The Atrium, Southern Gate, Chichester, West Sussex PO19 8SQ, England

Telephone (+44) 1243 779777

Email (for orders and customer service enquiries): cs-books@wiley.co.uk
Visit our Home Page on www.wiley.com

All Rights Reserved. No part of this publication may be reproduced, stored in a retrieval system or transmitted in any form or by any means, electronic, mechanical, photocopying, recording, scanning or otherwise, except under the terms of the Copyright, Designs and Patents Act 1988 or under the terms of a licence issued by the Copyright Licensing Agency Ltd, 90 Tottenham Court Road, London W1T 4LP, UK, without the permission in writing of the Publisher. Requests to the Publisher should be addressed to the Permissions Department, John Wiley & Sons Ltd, The Atrium, Southern Gate, Chichester, West Sussex PO19 8SQ, England, or emailed to permreq@wiley.co.uk, or faxed to (+44) 1243 770620.

Designations used by companies to distinguish their products are often claimed as trademarks. All brand names and product names used in this book are trade names, service marks, trademarks or registered trademarks of their respective owners. The Publisher is not associated with any product or vendor mentioned in this book.

This publication is designed to provide accurate and authoritative information in regard to the subject matter covered. It is sold on the understanding that the Publisher is not engaged in rendering professional services. If professional advice or other expert assistance is required, the services of a competent professional should be sought.

Other Wiley Editorial Offices

John Wiley & Sons Inc., 111 River Street, Hoboken, NJ 07030, USA

Jossey-Bass, 989 Market Street, San Francisco, CA 94103-1741, USA

Wiley-VCH Verlag GmbH, Boschstr. 12, D-69469 Weinheim, Germany

John Wiley & Sons Australia Ltd, 33 Park Road, Milton, Queensland 4064, Australia

John Wiley & Sons (Asia) Pte Ltd, 2 Clementi Loop #02-01, Jin Xing Distripark, Singapore 129809

John Wiley & Sons Canada Ltd, 6045 Freemont Blvd, Mississauga, Ontario L5R 4J3, Canada

Wiley also publishes its books in a variety of electronic formats. Some content that appears in print may not be available in electronic books.

Anniversary Logo Design: Richard J. Pacifico

Library of Congress Cataloging in Publication Data

Computational methods for mass spectrometry proteomics / Ingvar Eidhammer . . . [et al.].
p. ; cm.
Includes bibliographical references and index.
ISBN 978-0-470-51297-5 (alk. paper)
1. Proteomics—Data processing. 2. Mass spectrometry—Data processing.
3. Bioinformatics. I. Eidhammer, Ingvar.
[DNLM: 1. Proteomics—methods. 2. Computational Biology—methods.
3. Mass Spectrometry—methods. QU 58.5 C7385 2007]
QP551.C713 2007
572′.60285—dc22

2007040536

British Library Cataloguing in Publication Data

A catalogue record for this book is available from the British Library

ISBN: 978-0-470-51297-5

Typeset in 10/12pt Times by Integra Software Services Pvt. Ltd, Pondicherry, India

FSC
Mixed Sources
Product group from well-managed forests and other controlled sources
Cert no. SGS-COC-2953
www.fsc.org
© 1996 Forest Stewardship Council

Contents

Preface

Mass spectrometry proteomics is a relatively young field, but it is rapidly gaining importance in both large and small laboratories around the world, and a steadily increasing number of scientists and students have come into more or less intimate contact with proteomics, through the literature, through collaborations, or through their own practical work. The field relies heavily on bioinformatics to process and analyze the acquired data, and our aim has therefore been to write a book that can serve as a solid introduction to the crossroads where proteomics and bioinformatics meet.

The intention has been to give an introduction to students and beginners in both 'proteoinformatics', experimental proteomics, and researchers who are in more peripheral contact with these fields. Thus, the book includes the basic information needed for bioinformaticians who want to develop algorithms and programs for proteomics. For such work, the informaticians must have some knowledge of molecular biology. They should also understand the function of the instruments, whether it is for writing programs that control the instruments, or for further processing of the data produced. The beginner in practical proteomics, on the other hand, is often confused by the technical and biological complexity of the field, including the theoretical basis underlying proteomics. We hope that this book can contribute to a shortening of the period of confusion. However, we will underline that this book is no replacement for contact with experienced people within bioinformatics or experimental proteomics, and we hope that the book can give the beginner a better basis of knowledge, and thereby improve the reciprocal understanding between bioinformaticians and the experimentalist.

The book concentrates on the identification and characterization problem of proteomics, though quantification and sample comparison are also briefly described. The first chapter provides a fairly detailed introduction to the peptides and proteins that are studied in mass spectrometry proteomics, and their relevant properties. It also gives an overview of the two main approaches used: peptide mass fingerprinting (PMF) and tandem MS (MS/MS). The next four chapters explain techniques and methods used in both of the approaches, including informatics solutions of selected problems. Chapters 6 and 7 describe PMF, with special emphasis on the informatics challenges. Chapters 8 and 9 describe the instruments and principles for tandem MS, and the next four chapters concentrate on algorithms and programs for protein identification and characterization by tandem MS. The last five chapters deal with quantification and some more specific topics.

The application of bioinformatics in mass spectrometry proteomics is a young field, and little theoretical foundations have been developed. It is not the goal of this book to provide such a theoretical foundation, but rather to present the main problems, and extract and systematize some common principles used in solving these problems. The presentation is illustrated by figures and numerous examples.

We have tried to restrict the description of a subject to one section or chapter, but for some of the subjects it is necessary to treat them several times in different contexts. In these cases we have included the necessary cross-references.

We have also included many references and websites, although we are aware that websites can change quickly.

Terms

Many terms used in molecular biology and proteomics do not have unique or commonly accepted definitions. We have mainly tried to follow the IUPAC (International Union of Pure and Applied Chemistry) Compendium of Analytical Nomenclature. IUPAC has an ongoing revision of the terms used in mass spectrometry, and a first draft exists.

IUPAC	http://www.iupac.org/publications/analytical_compendium/
Ongoing revision	http://www.msterms.com/
First draft	http://www.sgms.ch/links/IUPAC_MS_Terms_Draft.pdf

However, for some cases where other terms seem to be more established, we have used those.

Also, new terms are introduced in some of the literature quoted. We have relied partly on these, but in some cases have resorted to defining our own terms. When applicable, we also mention synonymous terms that may be encountered.

Acknowledgements

Many articles and some websites have been used in the writing of this book, and citations have been given for these works. However, in order to enhance readability we have tried to minimize literature references in the text, and we have instead included a bibliographic section at the end of each chapter. The sources we consulted are detailed there, as well as some additional literature that we felt may be interesting or relevant to the reader. It is our sincere hope that this structure will serve to provide the authors of the source literature with the appropriate acknowledgements.

IE would like to thank Rein Aasland for valuable discussions, and for writing some paragraphs for the first chapter. Further thanks are due to Frode Berven, Harald Jensen, Einar Solheim, Pål Puntervoll, and Harald Barsnes for valuable information during several scientific discussions. Special thanks to my wife, Maria, for being so patient during the most busy writing times.

KF would like to thank Inge Jonassen, Kari Espolin Fladmark, and Trond Hellem Bø for their support and valuable scientific input. Kaja and Silje, my daughter and wife, also deserve gratitude for being patient and providing a different angle to life.

LM would like to thank Joël Vandekerckhove, Kris Gevaert, Rolf Apweiler, and Henning Hermjakob for their support over the years and for sharing their extensive knowledge. Special thanks go out to my wife, Leen, who has patiently endured the many evenings and weekends during which bookwriting took precedence over family life.

SOM would like to thank Per Seglen, Anders Øverbye, and Hamid Samari for the local discussions on LC and MS, and Véronique Cruciani for asking important questions. Harald Barsnes is thanked for trying to get some bioinformatics understanding into the brain of a biologist and experimentalist. My family and friends are thanked for their support during the busy periods of work.

1 Protein, proteome, and proteomics

Proteins are generally the executive molecules in the cells. Many of them are enzymes that catalyze myriads of biochemical reactions. Others are regulatory proteins that contribute to the correct expression of the genome. Some are structural proteins that build the many cellular components, and some proteins are carriers that move molecules around in the cells or between cells. Annotation of genomes (that is, deciding the functions and names for the individual genes that constitute the genome) is therefore primarily done at the level of the *proteome*. While the genome is essentially identical in all cells in an organism, the set of expressed proteins varies extensively through time as well as according to the specific conditions a cell finds itself in. The term *proteome* may therefore be used with several meanings. One definition is all the proteins encoded by the genome of a species. Another use of the term is the set of proteins expressed in a particular cell (or tissue or organ) at a particular time and under specific conditions. The term can also be used for the set of proteins of a subcellular structure or organelle. *Proteomics* is therefore the study of the subsets of proteins present in different parts of the organism and how they change with time and varying conditions.

1.1 Primary goals for studying proteomes

The overall goal of proteomics is to understand the function of all proteins found in an organism. Since this implies the collection of large quantities of proteomics data, frameworks are needed for the storage and presentation of such data. The Gene Ontology Consortium (GO), an international collaboration that aims to catalogue and standardize the existing knowledge about protein function, has established a large controlled vocabulary (CV) for gene and protein function. This CV is subdivided in three orthogonal types: molecular function, biological process, and cellular component. Each of these is briefly described below.

Molecular function describes activities at the molecular level, such as catalytic or binding activities. It is important to note that molecular function terms represent activities rather than the molecules (for example, proteins and molecular assemblies) that perform them. Furthermore, these terms do not specify where and when, or in which context,

Computational Methods for Mass Spectrometry Proteomics I. Eidhammer, K. Flikka, L. Martens and S.-O. Mikalsen
© 2007 John Wiley & Sons, Ltd

the activities occur. The molecular function terms describe function in the full range of detail, from the broadest terms (for example, 'catalytic activity') to the more molecular precise terms (for example, 'Toll receptor binding').

Biological process describes processes that are typically composed of a series (or collection) of molecular functions. 'Cell division' is an example of a higher level biological process, while 'regulation of Ras protein signal transduction' is an example of a lower level process. Higher level processes can also contain one or many subprocesses.

Cellular component is just that – a component of a cell that is not an individual protein. Examples are 'rough endoplasmic reticulum' and 'nucleus' at the high level and 'ribosome' and 'protein dimer' at the detailed level. The 'nucleus' further contains 'nuclear chromosomes', 'nuclear envelope', etc.

As of April 2007, GO has a total of 22 945 terms divided by 13 446 biological processes, 7563 molecular functions, and 1936 cellular components. The GO tree is organized as a directed acyclic graph (DAG) since a term can have more than one parent term.

It is important to note that while GO describes possible functions, it does not describe dynamics and chains of functions as in metabolic pathways and regulatory networks. Similarly, GO does not describe supercellular structures such as tissues, organs, and body parts. Such terms and terminologies are currently being developed in other ontologies and classification schemes which complement GO. Examples are KEGG (Kyoto Encyclopedia of Genes and Genomes) that describes metabolic and regulatory pathways and The Adult Mouse Anatomical Dictionary for anatomical structures of the mouse. A new Sequence Ontology Project (SO) is being developed for detailed descriptions of features and objects of genes and proteins. Other initiatives, such as OBO (Open Biological Ontologies), aim to establish the means for the integration of ontologies from different domains within biology.

Establishing such systems for the digitalized description of both function and properties of genes, proteins, cells, and organs as well as pathways and networks will pave the way for the development of computerized models of cells, organs, and organisms. Such static models form the basis for the dynamic modeling of processes in biological systems in the emerging field of research known as systems biology.

We can say that there exist four cornerstones of present-day proteomics: identification, characterization, quantification, and sample comparison.

Protein identification is the determination of which protein we have in our sample. This can be done by determining the sequence of the protein, or by measuring so many properties of the protein that it is statistically unlikely that it could be another protein. In this book we primarily consider determining the sequence.

Protein characterization is the determination of the various biophysical and/or biochemical properties of the protein (regardless of whether the protein has been identified). Although there are many important protein properties, in this book we mainly concentrate on the problem of determining the posttranslational modifications.

Protein quantification is the determination of the amount (abundance) of a protein in the sample, either as a relative or an absolute value. It should be noted that the determination of protein abundance is most often far from trivial.

Protein sample comparison is the determination of the similarities and the differences in the protein composition of different samples. Some aspects of sample comparison are:

- relative occurrence, the presence of proteins in some samples but not in others;
- relative abundance, the presence of proteins in different amounts in different sample s;
- differential modification, the presence of different modified forms of proteins in different samples.

In this book we present one of the most important basic techniques to achieve the goals of proteomics: *mass spectrometry*. We mainly concentrate on the first two tasks (identification and characterization), and give more briefly an introduction to the last two (quantification and sample comparison). Before we proceed to a more thorough treatment of the tasks, we will look at the protein itself in more detail.

1.2 Defining the protein

A protein is usually defined as consisting of one or more polypeptides, a polypeptide being a macromolecular chain of amino acids. More elaborate assemblies of polypeptides are often referred to as protein complexes. As many protein complexes are dynamic in composition, a subunit can spend a considerable amount of its lifetime in the cell as a single chain protein. This is one of the reasons why it has become commonplace to refer to polypeptide subunits of protein complexes as proteins. In this book, we will therefore use the term *protein* to denote a polypeptide, while assemblies of polypeptides will be called protein dimers, trimers, etc., or protein complexes as appropriate. Unless otherwise stated, the term protein thus refers to a single polypeptide chain.

Some proteins can have bonds to non-protein molecules such as small bio-organic molecules (for example, fatty acids, co-enzymes, nucleotides) or, as in the case of ribosomes and chromatin, they can be stably associated with RNA or DNA. Derivation of a comprehensive terminology for proteins (and parts of proteins) is non-trivial and beyond the scope of this book. It is, however, treated briefly in Chapter 1.4.

1.2.1 *Protein identity*

The objective of proteomics is to assign experiment-derived information on protein function to a particular protein. To do this, it is necessary to identify the protein. Ideally, a protein should have a unique amino acid sequence and a single source of origin (a species name). While this may appear trivial at first sight, there are several reasons why fulfilling these requirements is not always straightforward.

First, a single organism may contain several genes that encode proteins of identical sequence, violating the first of the above constraints (sequence uniqueness). This is for instance the case for several histone genes. Second, two proteins from two different organisms may also have identical sequences, violating the single source of origin constraint. A third complication is that the species concept itself is far from trivial and sometimes controversial. Finally, it is worth noticing that the emerging field of metagenomics generates DNA (and protein) sequence data for which a particular species name simply cannot be provided. Yet for the purposes of this book we need not consider these issues further. There are, however, other levels of complexity, variation, and ambiguity that proteomics must consider. These issues are outlined in the following sections.

1.2.2 Splice variants

The primary transcript of a gene (transcription unit) can often be spliced in different ways, giving rise to different mRNAs. The different proteins translated from such differentially spliced mRNAs are often called splice variants and can have different molecular functions, localize to different cellular components, and be involved in different biochemical processes. A detailed proteomic analysis must therefore be able to assign functions correctly to different splice variants. A similar situation occurs for genes that can be transcribed from different initiation sites (and/or end at different termination sites).

1.2.3 Allelic variants – polymorphisms

In natural populations, many genes are polymorphic (estimates indicate that this is the case for 20 % to 30 % of human genes), some of which result in protein variants with different sequences. Although the majority of such variants are functionally indistinguishable from each other, some have altered function, and in rare cases, allelic variants are associated with disease. Although a large number of polymorphisms have been identified and recorded in protein databases, many remain to be discovered. Algorithms for the identification of proteins should therefore ideally be able to deal with allelic variants.

1.2.4 Posttranslational modifications

Another source of variation in proteins (and their masses) is the many *posttranslational modifications*. These include both simple modifications such as phosphorylation and methylation as well as more complex modifications such as glycosylation and sumoylation. Many modifications can significantly alter the function of the protein and the investigator may therefore like to distinguish the modified forms from the unmodified ones. In other applications, however, it could be desirable to identify the proteins regardless of their modification status. It should also be considered that some modifications are transient and/or chemically vulnerable and may not survive until detection. Although many software applications have been designed to deal with posttranslational modifications, it is often far from trivial to detect and identify them, in particular for proteins which carry multiple modifications. Posttranslational modifications are considered in more detail in Section 1.4.

Proteolytic cleavage

Proteolytic cleavage is a particular kind of posttranslational modification that will result in variants. In this case, the variants may appear as if they are the result of alternative splicing (splice variants). Cleavage results in N- or C-terminal truncation of the primary protein product as well as removal of internal segments, creating two polypeptide chains from the original single polypeptide chain. N-terminal truncation is for example a common mechanism for the removal of the N-terminal signal sequence of proteins targeted to the secretory system, while insulin is a classic case of the proteolytic excision of an internal segment.

1.2.5 *Protein isoforms*

The term *protein isoforms* has been frequently used in the literature. Unfortunately, the term has not been used consistently. It can refer to any of the variants of modified forms discussed above. In some cases it is also used to describe two distinct, but closely related proteins encoded by different genes. In this book we therefore avoid the use of this term, but *forms* of proteins are used in some places. Also, we occasionally use *protein variants* for clarification.

1.3 Protein properties – attributes and values

Proteins are essentially chains of linked amino acids. There are 20 different naturally occurring amino acids, all with an amino group,[1] a carboxyl group, a central carbon atom (often denoted as the α (alpha) carbon), and a side chain. The generalized structure of an amino acid is depicted in Figure 1.1(a). The different amino acids are distinguished by the structure of their side chain (Figure 1.1(b)). In writing, each amino acid can be represented in one of three ways: by its full name, its three-letter abbreviation, or its one-letter notation (see the first column of Table 1.1). When amino acids are part of a chain, they are commonly referred to as *residues*. The bonds between two consecutive amino acid residues in the chain are amide bonds, and are denoted as *peptide bonds*. Correspondingly, a short chain of amino acids (smaller than a few tens of residues) is usually referred to as a peptide. From the structure shown in Figure 1.1(c), it is evident that an amino acid loses a water molecule (H_2O) when it is incorporated into a chain. Structurally speaking, a residue is therefore what remains of the amino acid after this water molecule is released. The chain along the atoms $N - C_\alpha - C - N\ldots$ is called the protein *backbone*. The end that carries the free amino group ($-NH_2$ or $-NH_3^+$) is called the *amino-terminal* (or *N-terminal*) end of the chain, while the other end, carrying the free carboxyl group ($-COOH$ or $-COO^-$), is referred to as the *carboxy-terminal* (or *C-*terminal) end. By convention, the amino acid sequence[2] is written as a string of residues (usually in one-letter notation) from the N-terminus to the C-terminus.

The side chains of the different amino acids are structurally distinct and carry characteristic physical and chemical properties, some of which are listed in Table 1.1. For some of the properties in the table the values vary slightly with the experimental conditions used for measuring them, conditions such as temperature, salt concentration in solutions, etc. A lot of experiments suitable for measuring amino acid properties are collected in

[1] Strictly speaking, proline contains an imino group rather than an amino group.
[2] We will use amino acid sequence and protein sequence interchangeably.

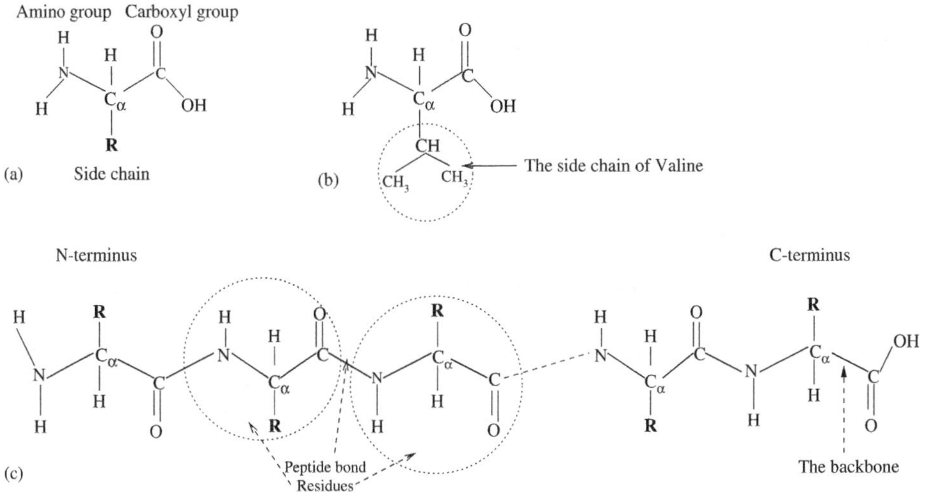

Figure 1.1 (a) All 20 amino acids have the same basic structure, but differ in their side chains. (b) The amino acid valine with its side chain. (c) A polypeptide is formed by chaining amino acids. The bonds between consecutive amino acids are called peptide bonds. When two amino acids are bound, a water molecule is released

The Amino Acid Repository database for which the web address can be found in the bibliographic notes.

A protein can be considered as having a set of properties. A common way of describing a property of an object is to use an *(attribute,value)* pair. For example, a protein can have a 'mass' attribute with a value of '8500 Da', and a 'cellular component' attribute with value 'Endoplasmic reticulum'.

Since we mainly concentrate on protein identification, characterization of posttranslational modifications, and comparison of proteins, we are primarily interested in attributes that can be of help in solving these tasks. For such attributes there exist some desirable requirements:

1. The value of the attribute should be (easily) measurable.
2. The attribute should have a (high) degree of specificity, meaning that different proteins should ideally have different attribute values.
3. The value should be constant (minimal variation with time and other external factors).
4. It should be possible to calculate or predict the value from the sequence of the protein, if the sequence is known (this is useful for protein identification).

We can distinguish between properties that are *intrinsic* to the protein sequence (given standard buffer conditions) and *contextual* properties that depend on the molecular or cellular context of the protein. From point 4 above, it is clear that we are mainly interested in the intrinsic properties.

Table 1.1 Some amino acid properties. The residue masses are from the i-mass guides (http://i-mass.com/guide/aamass.html), pI values are from Nelson and Cox (2004), pK values are from DTASelect (http://fields.scripps.edu/DTASelect/), the hydrophobicity index is from Kyte and Doolittle (1982), average occurrences are from the composition of the Swiss-Prot database release 52.1 of 20. March-07

Amino acid			Residue mass		pI values	Hydro-phobicity index	pK values	Occurrence (%)
			Mono-isotopic	Average				
Ala	A	alanine	71.04	71.08	6.01	1.8		7.87
Arg	R	arginine	156.10	156.19	10.76	−4.5	12.0	5.42
Asn	N	asparagine	114.04	114.10	5.41	−3.5		4.13
Asp	D	aspartic acid	115.03	115.09	2.77	−3.5	4.4	5.34
Cys	C	cysteine	103.01	103.15	5.07	2.5	8.5	1.50
Gln	Q	glutamine	128.06	128.13	5.65	−3.5		3.96
Glu	E	glutamic acid	129.04	129.12	3.22	−3.5	4.4	6.66
Gly	G	glycine	57.02	57.05	5.97	−0.4		6.95
His	H	histidine	137.06	137.14	7.59	−3.2	6.5	2.29
Ile	I	isoleucine	113.08	113.16	6.02	4.5		5.91
Leu	L	leucine	113.08	113.16	5.98	3.8		9.65
Lys	K	lysine	128.10	128.17	9.74	−3.9	10.0	5.92
Met	M	methionine	131.04	131.20	5.74	1.9		2.39
Phe	F	phenylalanine	147.07	147.18	5.48	2.8		3.95
Pro	P	proline	97.05	97.12	6.48	−1.6		4.82
Ser	S	serine	87.03	87.08	5.68	−0.8		6.84
Thr	T	threonine	101.05	101.11	5.87	−0.7		5.41
Trp	W	tryptophan	186.08	186.21	5.89	−0.9		1.13
Tyr	Y	tyrosine	163.06	163.18	5.66	−1.3	10.0	3.02
Val	V	valine	99.07	99.13	5.97	4.2		6.73
N-terminus			1.01[a]				8.0	
C-terminus			17.01[a]				3.1	

[a] Note that the terminus masses are the masses of the atoms added to the end residues, normally H at the N-terminus and OH at the C-terminus.

There exist a large number of protein attributes, and many of them can be used in protein analysis in some way. We will, however, concentrate on the physicochemical ones that are used in the analytic methods described later in the book. We will also briefly mention how they can be measured, whether the values can be calculated or estimated from the sequence, and, if possible, how this can be done.

1.3.1 The amino acid sequence

The amino acid sequence is the most fundamental attribute of a protein. Correspondingly, the sequence is also referred to as the primary structure of the protein. If the sequence is known, the protein is considered identified. When we consider a protein we should therefore first try to determine its complete sequence. This is, however, not a straightforward process.

Protein sequencing

The traditional technique for protein sequencing is *Edman degradation*. The first step of Edman degradation relies on the ability to remove the N-terminal amino acid from a polypeptide while leaving the rest of the chain intact. In the next step the identity of the removed amino acid is determined. Repeated cycles of these two steps ultimately result in a readout of the polypeptide sequence from the N-terminus to the C-terminus. This process is restricted to chains of roughly 60 residues and is very laborious: one day of work for 40–50 amino acids.

Tandem mass spectrometry (Chapter 12) can also be used for protein sequencing and is now becoming the standard technique. This approach is aimed at sequencing peptides rather than whole proteins, and direct sequencing of intact proteins remains a very tough challenge today. Note that many of the protein sequences found in databases are derived from *in silico* translations of (predicted or annotated) open reading frames (that is, genetic sequences that are assumed to encode for proteins).

1.3.2 Molecular mass

The mass (symbol m) of a molecule or atom is expressed in *unified atomic mass units* (symbol u), defined as 1/12 the mass of carbon-12 (which is $1.660\,540\,2 \cdot 10^{-27}$ kg). A more common term for unified atomic mass unit is dalton (Da).

Molecular weight is not the same as molecular mass, and is also known as relative molecular mass (denoted as M_r). Formally, weight is a ratio relative to ^{12}C, and can vary with the location. This is unlike mass, which stays the same regardless of location. When the literature gives a mass in Da or kDa it refers to molecular mass. It is incorrect to express molecular weight (relative molecular mass) in daltons. Nevertheless you will find the term *molecular weight* used with daltons or kilodaltons in some of the literature, often using the abbreviation *MW* for molecular weight. Another term sometimes used in this context is the *atomic mass unit* (symbol *amu*). Note that this term is deprecated since it is used both as synonymous with dalton and as a relative definition to ^{16}O.

All chemical elements have naturally occurring *isotopes*. Isotopes are elements that have the same atomic number (and therefore similar chemical properties), but different molecular mass (slightly different physical properties). This is important for the mass definitions.

The following definitions of terms related to ions or molecules are mainly based on the IUPAC third draft. The reader should therefore be aware that these terms can sometimes be used with slightly different meanings in the literature.

- *Exact mass* is the *calculated mass* of an ion or molecule containing a single isotope for each atom (most frequently the lightest isotope of the element). It is calculated using an appropriate degree of accuracy.
- *Monoisotopic mass* is the exact mass of an ion or molecule calculated using the mass of the most abundant isotope of each element. For small elements the most abundant isotope is the lightest one (as for C, H, N, O, S), but this is not necessarily the case for larger elements (like B and Fe).

- *Average mass* is the mass of an ion or molecule *calculated* using the average mass of each element weighted for its natural isotopic abundance.
- *Mass number* is the sum of the number of protons and neutrons in an atom, molecule, or ion.
- *Nominal mass* is the mass of an ion or molecule *calculated* using the mass of the most abundant isotope of each element rounded to the nearest integer value. It is equivalent to the sum of the mass numbers of all constituent atoms.
- *Mass defect* is the difference between the mass number and the monoisotopic mass.
- *Mass excess* is the negative of the mass defect.
- *Accurate mass* is an *experimentally* determined mass of an ion that is used to determine an elemental formula.
- *Mole* is a measure of the amount of substance. One mole of any substance contains $6.022 \cdot 10^{23}$ (Avogadro's number) molecules or atoms.
- *Apparent mass* is not an IUPAC term. It is frequently used in the literature to indicate a molecular mass that is estimated from an inaccurate experiment (for example, gel electrophoresis, Chapter 2).

Example Consider the amino acid valine. The derived residue consists of five C, nine H, one O, and one N. We calculate the different residue masses (in daltons).

$$\begin{array}{ll} \text{Monoisotopic mass:} & 5 \cdot 12.0000 + 9 \cdot 1.0078 + 15.9949 + 14.0031 = 99.0682 \\ \text{Average mass:} & 5 \cdot 12.0107 + 9 \cdot 1.0079 + 15.9994 + 14.0067 = 99.1307 \\ \text{Nominal mass:} & 5 \cdot 12 + 9 \cdot 1 + 16 + 14 = 99 \end{array}$$

△

Measuring the mass of a protein

In the laboratory, the mass of a protein is commonly estimated by SDS-PAGE (sodium dodecyl sulfate polyacrylamide gel electrophoresis), as explained in Chapter 2. This method has limited accuracy and the inaccuracy increases with increasing mass. Other methods like gel filtration and analytical ultracentrifugation are also used to estimate the masses, in particular of larger proteins and protein complexes. The mass of (small) proteins can, however, be accurately determined by mass spectrometry, as described in Chapter 6.

Experimentally measured masses necessarily carry errors or uncertainties. The total error of a general measurement is often given as either a relative or an absolute value, and the two scales used for these are the *part per million* (symbol *ppm*) scale for relative error, and the *milli unit* (symbol *mu*) scale for absolute error. For mass measuring, the absolute error is commonly given in *milli-mass unit* (*mmu*) or daltons: 1 mmu is 0.001 Da. For ions less than 200 Da, a measurement with 5 ppm accuracy is considered sufficient to determine the elemental composition.

Example Suppose we have a measured value of 5 kDa, and an error of 1 Da. The relative error is $\frac{1}{5000} = \frac{200}{1\,000\,000} = 200\,\text{ppm}$. For our value of 5 kDa, a relative error of 200 ppm is therefore equivalent to an absolute error of 1 Da.

△

Theoretical calculation

Theoretical calculation of the mass of a polypeptide is done simply by adding up the masses (monoisotopic or average) of all the residues, and adding the masses of the extra atoms at the N- and C-terminus (normally H and HO).

Distribution of the protein masses

If we want to use protein mass for identification purposes, we first need to know about the protein mass distribution. Release 52.1 of the Swiss-Prot sequence database contains 261 513 sequence entries, with lengths in the interval [2; 34 350], and an average length of 365 residues; 195 202 of the sequences lie in the length interval [51; 500]. If we assume an average residue mass of 110 Da, this means that the masses of 195 202 proteins lie in the interval [5.5; 55] kDa. If the proteins are sorted by increasing mass, the average difference between two neighboring masses in this interval is 0.28 Da.

1.3.3 *Isoelectric point*

The isoelectric point of a protein is defined using the term *pH*, which is a measure of the acidity of a solution. pH is an abbreviation for *potential of Hydrogen*, or more precisely 'The negative of the base 10 logarithm of the concentration of hydrogen ions in a solution.' It is measured using the number of H^+ and OH^- ions in the solution.

Water dissociates (splits into simpler fragments) as

$$H_2O \rightarrow H^+ + OH^-$$

Using the chemical convention that concentrations are indicated by square brackets [], the equilibrium of dissociation can be expressed as

$$\frac{([H^+]\cdot[OH^-])}{[H_2O]} = K$$

The concentrations in such formulas are expressed in moles per liter, or molar (M). The equation above can be rearranged as

$$K[H_2O] = K_w = [H^+][OH^-]$$

K_w is called the *dissociation constant*, or the ion product, of water, and at 25 °C, $K_w = 1.0 \cdot 10^{-14}\,M^2$. A neutral solution of pure water (at 25 °C) implies equal amounts of H^+ and OH^- and correspondingly means $[H^+] = [OH^-]$. Thus,

$$K_w = [H^+][OH^-] = [H^+]^2 = 1.0 \cdot 10^{-14}$$

and $[H^+] = 1.0 \cdot 10^{-7}$.

By definition, pH is the negative logarithm of the concentration of H^+. Thus, for neutral solutions $pH = -\log_{10} 10^{-7} = 7$. When the concentration of H^+ is high (due to the negative logarithm, this corresponds to a low pH), the solution is considered acidic.

According to the Brønsted – Lowry definition, molecules that readily donate protons (H^+) to the solution are acids, and molecules that accept protons from the solution are bases (note that the hydroxyl ion (OH^-) is the quintessential base, as it readily combines with a proton to form water).

Proteins (like amino acids and many other molecules) have both acidic and basic characteristics, and thus carry chemical groups that act as either proton acceptors or proton donors. Accepting a proton results in a positive charge, while loss of a proton yields a negative charge. These charges are distributed along the chain of the molecule, according to the location of the specific basic or acidic groups. Depending on the availability of protons in the solution (the pH) in which the protein is dissolved, the number of positive and negative charges of the protein will vary. At a low pH, there is an abundance of protons available that will quickly populate the available acceptor sites while inhibiting dissociation of the donor sites. The result is a net positive charge for the protein. At a high pH, very few protons are available to bind to the acceptor sites, and the donor sites will readily lose their protons, leading to a net negative charge for the protein. At a certain pH, however, the protein will have an equal number of positive and negative charges. The net charge is then zero, and the protein is electrically neutral. This pH value is called the isoelectric point (*pI*) of the protein. Different proteins will have different pIs, which depends on their amino acid composition.

Measuring the pI

The pI of a protein can experimentally be determined by use of isoelectric focusing *pH gradients*, and is explained in Chapter 2.

Theoretical calculation

Protein ionization is determined by the properties of the amino acids that the proteins consist of. Each amino acid carries two ionizable groups: the amino group (proton acceptor, basic) and the carboxyl group (proton donor, acidic). Additionally, some amino acids also carry an ionizable group in their side chain. The amino group already ionizes at a relatively high pH and will carry a positive charge at neutral and acidic pHs. The carboxylic group ionizes at a relatively low pH and carries a negative charge at neutral and high pHs. However, since these two groups form the peptide bond between two consecutive amino acids in a protein, it is only the N-terminal amino acid that has a free amino group, and only the C-terminal amino acid that has a free carboxyl group. If these two groups were the only ionizable groups in the proteins, all proteins would obviously have very similar pIs. Yet proteins actually have widely varying pIs, and this is due to the influence of the ionizable groups in the side chains of certain amino acids. Histidine (`H`), lysine (`K`), and arginine (`R`) carry a proton acceptor group in their side chains and can therefore become positively charged, while cysteine (`C`), aspartate (`D`), glutamate (`E`), and tyrosine (`Y`) have side chains that contain a proton donor group and can become negatively charged. The dissociation constants can be determined for the ionizable groups of these amino acids. The negative logarithm of the dissociation constant is defined as the *pK value*, analogous to the definition of pH. Simply put, the pK value corresponds to the pH where 50 % of the particular ionizable group is ionized, and 50 % is not ionized. The pK value thus represents a measure of the acidity or basicity of the ionizable group.

The pK values are obtained experimentally, and some values for pK are shown in Table 1.1. Note, however, that the values vary slightly with the environment and the experimental conditions in which the values are measured. Also bear in mind that the values reported for amino acids are experimentally determined for free amino acids. When the residues are bound in a complex macromolecule, their pK values can vary significantly (something that is exploited to great advantage by many enzymes). The pK values for the termini also vary with the actual amino acids that form the corresponding ends.

Note also that pK is a generic dissociation constant, and pK_a is the acid dissociation constant, thus more specific than pK. For simplicity we use pK here.

Theoretical calculation (or rather estimation) of the pI of a protein is performed by using these dissociation constants. Indeed, the pK values can be used to calculate the partial charge of the ionizable amino acids b at any given pH according to standard equations:

$$C_b = \frac{1}{1+10^{pH-pK_b}}$$

if amino acid b is positively charged, and

$$C_b = \frac{-1}{1+10^{pK_b-pH}}$$

if amino acid b is negatively charged. Note that $|C_b| < 1$. From these, we can define a function $C_P(pH)$ that estimates the net charge of a protein P for a given pH. Let B be the set of charged amino acids including the termini, and n_b the number of occurrences of amino acid b in the protein. Then $C_P(pH)$ is calculated as

$$C_P(pH) = \sum_{b \in B} n_b C_b$$

The protein's pI is then obtained as the pH where $C_P(pH)$ is zero. This value can be found by numerical methods as described in Algorithm 1.3.1, which implements a binary search procedure. For simplicity, the function $C_P(pH)$ is denoted as $C(x)$ in the algorithm.

Algorithm 1.3.1 Theoretical calculation of the isoelectric point of a polypeptide
Let the function be $C(x)$. We search for the value where $C(x) = 0$

const
ϵ limit for accuracy
var
x, x_0, x_1
begin
find values x_0, x_1 *such that* $sign(C(x_0)) \neq sign(C(x_1))$
while $|x_1 - x_0| \geqslant \epsilon$ **and** *max number of iterations not reached* ***do***
$x := (x_0 + x_1)/2$
if $sign(C(x_0)) \neq sign(C(x))$ then $x_1 := x$ **else** $x_0 := x$ **end**
end
end the isoelectric point is in $[\min(x_0, x_1), \max(x_0, x_1)]$

Example Consider the (small) polypeptide P=QEAYEGK. Then we calculate

$$C(5)(=C_P(pH=5)) = -1.6, \quad C(4) = 0.6$$

This means that the isoelectric point for P is between 4 and 5. The average of these limits (4.5) is calculated and the function $C(4.5)$ is solved. If this value is less than zero, then the isoelectric point must lie between 4 and 4.5, otherwise it is between 4.5 and 5. In the next iteration, the average of these limits (4.25) is again used and $C(4.25)$ calculated. The process continues until a sufficiently small interval is obtained (the absolute difference of the limits must be less than the predefined value ϵ).

△

The distribution of pI values

Several analyses have been done on the distribution of the pI values for proteins. Almost all proteins have values in the interval [4, 12], but the actual distribution varies with the organism under study (deviating values are for instance obtained for organisms that have adapted to highly basic or acidic environments).

1.3.4 Hydrophobicity

Compounds that tend to repel water are termed hydrophobic, as opposed to hydrophilic compounds that dissolve easily in water. Though hydrophobicity is one of the most important physicochemical properties of amino acids and proteins, it is still poorly defined. The main characteristic is that when there are many consecutive hydrophobic amino acids in a certain part of the sequence, that sequence part will tend to avoid water. This can be achieved in one of several ways. The sequence may be packed in the interior of the 3D structure of the protein, the sequence may integrate into one of the many membranes of the cell (that is, the protein becomes a membrane protein), or the sequence may appear as a hydrophobic patch on the surface of the protein, serving to attract a similar hydrophobic patch on another protein. In all cases the hydrophobic sequence avoids the entropically unfavorable contact with water.

Hydrophobicity scales

The hydrophobicity of the amino acids has been determined through either calculation or measurement. There are different ways to determine hydrophobicity, relying on different properties of the amino acids. Because of this, many scales exist for the hydrophobicity values of amino acids. Fortunately most of them are fairly similar, and for most applications it does not matter which of them is used. The most often used scale is the Kyte–Doolittle scale from 1982 (Table 1.1), in which the most hydrophobic amino acids have the highest positive values. Hydrophobicity scales are often used for finding hydrophobic regions in proteins.

Theoretical calculation

A hydrophobicity value for a protein can be calculated simply by adding the hydrophobicity values of all the residues and dividing the sum by the number of residues. This is called a GRAVY score (GRand AVerages of hYdrophobicity). For proteins, however, it is generally more common to calculate a hydrophobicity plot (or hydrophobicity profile). A sliding window is used in this approach, and the average hydrophobicity of the residues inside the window is plotted at the residue in the middle of the window. If the size of the window is nine, the first average value will be plotted at residue 5, the next at residue 6, and so on. This technique can be used to find hydrophobic regions along the sequence.

1.3.5 *Amino acid composition*

The amino acid composition of a protein can be determined by first cleaving (hydrolyzing) all the peptide bonds in the protein to release the constituent amino acids. The amino acids are subsequently separated, and after staining the intensity of each amino acid is determined. From this the prevalence of each amino acid is calculated.

Theoretical calculation

Theoretical calculation of the amino acid composition from a protein sequence is performed simply by counting the number of occurrences of each amino acid in the protein.

1.4 Posttranslational modifications

A posttranslational modification (PTM) can be defined as any alteration to the chemical structure of the protein effected by the cellular machinery after the formation of the protein. PTMs thus occur strictly *in vivo* and are correspondingly sometimes called *in vivo* modifications. Note that this implies that the chemically induced modifications that often occur in the lab during sample preparation are (strictly speaking) not PTMs. These are usually described as *chemical*, *artefactual*, or *in vitro* modifications and should be taken into consideration when performing proteomics. A very specific type of chemical modification is the introduction of *isotopic labels*. These modifications exploit the different masses of different isotopes of the same element to distinguish between different samples. The distinction is achieved by introducing a different isotope (and thereby a different mass) to each sample, as explained in Chapter 15.

PTMs are very important to the survival of the cell; they contribute to the correct localization of proteins, act as controllers and mediators of protein function, and play crucial roles in cellular communication and signal transduction. Some examples of common PTMs are:

1. cleavage of the protein, for example the removal of inhibitory sequences or sorting signals;
2. ligand binding;
3. covalent addition of further groups, for example glycosylation, acetylation, etc.;
4. phosphorylation and dephosphorylation.

In this book we will only consider modifications that affect the structure of one residue. The modifications will change some of the attributes of the proteins, most notably the mass but also the pI and the hydrophobicity.

Several databases of modifications exist. Some focus explicitly on PTMs, while others also include a range of chemical modifications. The RESID database at the European Bioinformatics Institute (EBI) is an example of the former category, while UniMod and DeltaMass are examples of the latter category. UniMod and DeltaMass have been developed for use in mass spectrometry applications and this explains their inclusion of chemical modifications. Recently, these three databases have started to collaborate on the HUPO PSI Modifications ontology. Because of our focus on mass spectrometry, we here describe UniMod in some more detail.

Each modification has one entry in the UniMod database. Each entry can have several *specifications*, and each specification refers to the modification of a specific amino acid. A specification consists of a *site*, a *position*, and a *mass*:

Site is the residue where the modification can take place. There are 22 possibilities: the 20 amino acids, the N-terminus, and the C-terminus.

Position is where the modification can occur, and here there are five possibilities.

Anywhere, the modification is position independent.

Any N-term, the modification occurs only at a *peptide* N-terminus. For definition of peptides, see Section 1.6.2. Note that this option is primarily used for chemical modifications as peptides are usually artificial molecules obtained in the laboratory.

Any C-term.

Protein N-term, the modification occurs only at the N-terminus of the intact protein.

Protein C-term.

Mass is the mass difference effected by the modification and this varies from −128 to 1769 Da, with only nine over 1000. The mass difference is specified as both monoisotopic mass and average mass. In addition a neutral loss may be specified, since some modifications exhibit a neutral loss on fragmentation in a mass spectrometer. This loss has no biological meaning by itself and is simply a useful artifact uniquely associated with mass spectrometry experiments.

Example The entry phosphorylation has seven specifications, one for each of the amino acids S,T,Y,D,H,C,R. The addition is HO_3P, with monoisotopic mass 79.966 331. △

Modifications are mainly discovered by observing changes in the mass and/or pI. However, since there are so many possible modifications, it is often not easy to uniquely determine which modification led to the observed parameter shift. Additionally, if there are several possible positions for the modification, it can be difficult to determine the exact residue where the protein modification took place. Another method for detecting protein modifications relies on the specificity of antibodies to bind a particular residue carrying a particular modification. Antibodies can be spotted on arrays and can then analyze many modifications in parallel. Some modification-specific antibodies may be specific for the modification and a certain protein, others recognize the modification regardless of the protein. However, reliable antibodies are difficult to obtain.

Theoretical calculation of the mass of a modified protein is simply done by adding the mass of the modification. The change in pI is similarly calculated by changing the pK value of the modified residues in accordance with the modifications. It is important to note, however, that for many modifications the pK values are either unknown or very rough estimates, making accurate pI predictions practically impossible.

1.5 Protein sequence databases

Sequence databases build upon the most basic source of information about proteins, their sequence. In the next three subsections the most commonly used sequence databases for protein identification will be briefly discussed and in the fourth subsection a note is given on the instability in time of sequence databases.

1.5.1 UniProt KnowledgeBase (Swiss-Prot/TrEMBL, PIR)

The central UniProt database builds on three well-established pillars in the sequence database world: the *Swiss-Prot*, *TrEMBL* (Translated EMBL), and *PIR* (Protein Information Resource) databases. The key difference between Swiss-Prot and PIR on the one hand and TrEMBL on the other hand lies in the manual curation effort that underlies the former two databases. We will focus here on the Swiss-Prot database since we mostly refer to and take examples from Swiss-Prot. All entries in Swiss-Prot have passed through a rigorous manual control by human curators. During this curation process diverse information sources are consulted and cross-verified in order to establish which annotations are clearly supported by trustworthy evidence. The result of these efforts is an extremely high-quality and stable[3] protein sequence database with heavily cross-linked annotations. Obviously, the curation process is labor intensive and this limits the reach of the Swiss-Prot database. The TrEMBL database complements this nicely, however. It contains automatically annotated proteins, including predicted protein sequences. UniProt is more than a list of sequences and their annotations. Apart from the UniProt KnowledgeBase, the system encompasses a complete sequence archive called UniParc (UniProt archive), as well as UniRef, a set of reference clusters at different sequence identity levels.

As stated above, Swiss-Prot contains much curated protein information and annotations, but in our context it is enough to consider the following points:

- the accession number, which supplies a unique identifier for the entry;
- the sequence;

[3] The number of sequences curated into Swiss-Prot continues to grow, so 'stable' here simply means that it is unlikely that an entry will be deleted from Swiss-Prot.

- molecular mass (denoted as molecular weight);
- observed and predicted modifications.

The sequence is the sequence after translation, yet before any modifications that may cleave off parts of it. This is often explicitly annotated in the entry by affixing *[Precursor]* to the protein name.

A *feature table* contains reported differences in the sequence for a protein. They are of different types:

Conflict, positions where different (bibliography) sources report different amino acids;

Variants, where the sources report that sequence variants (alleles) exist;

Mutagen, positions where alterations have experimentally been performed;

Varsplic, the description of sequence variants produced by alternative splicing. Note that following our definition this would be different proteins, and that a derived database called *Swiss-Prot Varsplic* exists in which all splice variants are included as separate protein entries. The accession numbers for these entries are composed of the accession number of the parent entry, followed by a dash (-) and a number, for example `P12345-3` for the third splice variant of entry `P12345`.

The feature table also includes observed PTMs. The type of modification, and the position, is specified for each observed modification.

Finally, it is important to keep in mind that even Swiss-Prot may contain errors, although the curation process is specifically designed to minimize potential errors.

1.5.2 The NCBI non-redundant database

The National Center for Bioinformatics Information provides a non-redundant sequence database that is usually referred to as the *NCBI nr* database. This database groups sequence information from a variety of sources, including Swiss-Prot, TrEMBL, and RefSeq. The last one consists of two distinct types of entries, NP and XP, which are readily identifiable by their accession numbers (they start with 'NP_' or 'XP_', respectively). The NP sequences have corroborating evidence such as cDNA to back up their validity, while the XP sequences are based purely on predictions.

It is clear that the level of annotation and available cross-links in the database varies considerably between entries and is largely dependent on the source database of the sequence.

The NCBI nr database is non-redundant at the absolute protein sequence level, meaning that no two sequences are completely identical in the database. History management is also provided via the Entrez web interface.

1.5.3 The International Protein Index (IPI)

This database has its origins in the human genome project and was originally conceived as a non-redundant view on all known human proteins. Over the years the reach and scope of the IPI database has grown, yet the basic premise of providing an intelligently designed non-redundant protein sequence database has remained. IPI is now available for a variety of model organisms, including human, mouse, rat, and Arabidopsis. IPI presents an automatically curated view on the total contents of a large collection of sequence databases (including UniProt, RefSeq, and Ensembl) by a rather complex algorithm to remove sequence redundancy in a thorough way. Simply put, the algorithm can be described as follows: instead of simply requiring each protein sequence to be unique, sequences are also collapsed into clusters when they show more than 95 % overlap over the full match length. When a cluster consists of sequences of different lengths (which often happens due to the presence of protein fragments as separate entries in several source sequence databases), the longest sequence is chosen as the master sequence. Notable exceptions to the clustering rule are the annotated UniProt splice variants, which are retained as separate entries. Each cluster is finally assigned an IPI accession number and all source references aggregated in the cluster are expressly reported in the IPI entry. IPI also provides complete history files, which trace the history of every entry that has ever carried an IPI identifier.

1.5.4 *Time instability of sequence databases*

In the above sections it has been mentioned that each of the sequence databases maintains sequence history in a specific way. The ability to trace a sequence (or its (versioned) accession number) through time is a much overlooked, yet highly important characteristic of a sequence database. Indeed, sequences in any database are subject, to a greater or lesser extent, to changes over time. The primary event of any sequence is of course its first inclusion in a database. This can be traced for each entry in the above-mentioned sequence databases. All sequence databases continue to grow, so the creation of novel entries occurs frequently. Once a sequence has been recorded in a database, it can undergo alterations as well. Depending on the database, these alteration events can trigger a change in the version number or even cause the assignment of a new accession number.

In the latter case, tracing is usually provided in such a way that the original accession number will still link up to the sequence. Certain sequences can be subjected to removal from the database, invalidating the accession number without providing a replacement. This effect is most pronounced for purely theoretical predictions (such as the above-mentioned RefSeq XP entries). Indeed, a change in the prediction algorithm usually renders a number of previous predictions obsolete. Also note that databases such as NCBI nr or IPI, which combine information from different sources (including RefSeq XP), necessarily suffer the same fluctuations as their source databases.

1.6 Identification and characterization of proteins

Protein identification can be performed by collecting properties of the proteins under consideration, and explore whether we can recognize these properties among the known proteins. This is generally done by comparing the measured properties to the calculated or documented properties of the proteins in a protein database, and could be performed by the following procedure:

1. Choose a protein database, or a subset of a protein database, with the candidate proteins.
2. Compare the properties of an unknown protein to the properties of the database proteins, and select the proteins that best fit the observed properties.
3. Compute a probability for the hypothesis that the best fit database protein and the unknown protein really are the same protein. Note that we do not always know whether the unknown protein is in the database in the first place. The probability calculation mentioned earlier will be different depending on whether we know the protein to be present in the database, or not.

We have seen that, especially due to the modifications, there does not exist a single, or even a set of, protein attributes that make this procedure workable in the general case. It has therefore been necessary to develop other techniques and procedures for identifying proteins. Mass measurement has proved to be the most useful technique, with *mass spectrometers* as the instruments of choice. Mass spectrometers are general instruments for measuring the masses of the molecules in a sample. However, mass spectrometers are only able to recognize charged molecules, therefore the molecules must be ionized. In proteomics the ionization is commonly achieved by the addition of protons, and more rarely by loss of protons. Hence the mass of the peptide or protein is increased by the nominal mass of 1 Da times the number of charges (protons) in the case of addition, and decreased by the nominal mass of 1 Da times the number of charges in the case of loss of protons. By convention the number of added (or lost) protons is denoted by z. It is then important to recognize that the mass is only indirectly determined; it is the *mass-to-charge* ratio, denoted m/z, that is measured. Figure 5.4 shows examples of *mass spectra*, with m/z values along the horizontal axis, and intensities along the vertical axis. The masses are then only obtained after processing of the m/z values. Depending on the instrument and the experimental conditions, the processing of m/z data to masses can be anything from trivial to exceedingly difficult. This will be described later.

1.6.1 Top-down and bottom-up proteomics

Although proteomics can be performed in a number of different ways, it may be useful to divide the existing approaches into two types of paradigms: top-down and bottom-up. In the top-down paradigm, intact proteins are directly used for the analysis. In the bottom-up paradigm, the proteins are first cleaved into smaller parts, and these parts are then used for identification, characterization, and quantification. These smaller parts are called *peptides*. Methods using a combination of both paradigms (hybrid methods) are also in use.

The bottom-up paradigm is most often used, and there are several reasons for doing mass spectrometry on peptides, and not (solely) on intact proteins, as follows:

- The absolute error in the measurement increases with the measured m/z.
- A protein does not have one defined mass, but rather a mass distribution. This is due to the occurrence of stable isotopes. Due to their long sequence, proteins will

have a much more complex mass distribution than the short peptides. As a result, it is a lot easier to obtain a single, monoisotopic mass for a peptide than for a protein.

- It is not possible to measure the mass of all proteins, especially (very) large and hydrophobic proteins.
- Sensitivity of measurement of intact protein masses is not nearly as good as sensitivity for peptide mass measurement.
- The existence of modifications complicates the analysis. This effect is much more noticeable for whole proteins as their long sequence can be modified several times with several different modifications. For the much shorter peptides, the combinatorial possibilities are more manageable.

Also, it is difficult to perform large-scale analysis of intact proteins with top-down proteomics techniques. The bottom-up paradigm is therefore used as the basis for the presentation in this book. For the sake of completeness, an introduction to the top-down paradigm is given in Chapter 17.

1.6.2 Protein digestion into peptides

As we have seen, a peptide means a contiguous stretch of only a few tens of residues (in practice, the desired length lies between 6 and 20 residues). Protein cleavage into such peptides is generally done by enzymes called *proteases*, and the cleaving operation is called *digestion*. The measured masses (*experimental masses*) of the resulting peptides can be used for comparison against the predicted masses (*theoretical masses*) of the peptides obtained after an *in silico* digestion of candidate protein sequences in a database. *In silico* digestion is done by simulating the cleavage on a computer using the specificity of the protease (for example, that it cleaves only after arginine residues) and the peptide masses can theoretically be calculated as the sum of the residue masses with the addition of the masses of the intact termini: a hydrogen (H) at the N-terminus and a hydroxyl group (OH) at the C-terminus.

Example Let a (short) protein sequence be `MALSTRVATSKLICDVTRASDT`. Using trypsin as a protease (cleaving after each arginine and lysine, if not followed by proline), the protein should ideally be cleaved into the peptides `MALSTR`, `VATSK`, `LICDVTR`, `ASDT`, resulting in the (nominal) masses {677, 504, 818, 392}. △

1.7 Two approaches for bottom-up protein analysis by mass spectrometry

The foundation for bottom-up proteomics is to digest the proteins into peptides, and compare peptide properties to the properties of theoretical peptides from a protein sequence database. Two main approaches have evolved to perform such analysis, differing in the peptide properties used, the mass spectrometry instruments used, and the method for separation of molecules to reduce the complexity.

Peptide masses or peptide sequences Peptide masses and peptide sequences are the main peptide properties used in bottom-up analysis. The process of comparing experimental masses to theoretical masses has been named *peptide mass fingerprinting* (PMF) and is sometimes also referred to as mass profile fingerprinting and peptide mass maps. If our unknown protein corresponds to a database protein, each of the theoretical peptide masses of the database protein should coincide with an experimental peptide mass, and vice versa. However, in reality full experimental coverage of the protein sequence is never achieved; 20–40 % *sequence coverage* is more usually obtained. The search tools for peptide mass fingerprinting therefore calculate statistics to indicate the confidence of the suggested identifications, where one of the components is the coverage.

For the peptide mass fingerprinting approach, we can usually determine only one property of the peptide, namely its mass, and this only to within a certain accuracy, for example 50 ppm. There are probably hundreds, or even thousands, of peptides in a database that have a mass within this accuracy limit. Even if we know the exact amino acid content of the peptide, we do not know the positions for each of the amino acids (except that its C-terminal amino acid is probably K or R if the peptides have been generated by the action of trypsin). The following peptides have for example exactly identical mass: WGAR, WAGR, GWAR, WAGR, AWGR, and AGWR. Additionally, this mass is very similar to the mass of the peptide ENAR (there is a 31 ppm difference). Thus deriving the sequence (or part of the sequence) of a peptide gives much higher confidence in the subsequent identification of the protein of origin.

Mass spectrometers Different mass spectrometers are used for measuring the mass, and for deriving the sequence. One mass spectrometry analysis is sufficient for measuring the masses, using an MS instrument. For deriving the sequence two mass analyses are used in the MS/MS-capable instruments. The first mass analysis is used to specifically select ionized molecules (in our context peptide ions) from a particular m/z interval. The selected ions are then often subjected to a fragmentation step, yielding a number of fragments *per* selected ion. The second mass analysis finally measures the m/z of the fragment ions, from which sequence data might be derived. More mass analyses can be chained to yield, generically speaking, *MS*^{n} analysis. Instruments capable of performing such analysis are generally called MS^n instruments. Since sequence information can be derived, the MS/MS studies have become the most commonly used type of mass spectrometry analysis. The selected peptide ion in MS/MS experiments is usually called the *precursor* and the isolation interval is often small enough to admit only a single precursor to the following sequencing steps.

Separation When presented with the problem of analyzing a mixture of proteins, the capacities of mass spectrometers are easily overcome by a too complex mixture, resulting in the analysis of only a minor part of the total protein complement of the sample. By *fractionating* the initial sample into *fractions*, the mass spectrometer can be used to analyze each obtained fraction separately. This will mean that more of the proteins in the sample are analyzed; the *sample coverage* is increased. Fractionation is usually achieved by different methods of *separation*. When proteins or peptides are separated, they are split into fractions (groups), in which the members share those properties that are used by the separation process. The main separation steps may be performed on either

Table 1.2 Illustration of two different approaches for mass spectrometry proteomics

	MS (peptide mass fingerprinting)	MS/MS (tandem MS)
Peptide property	Mass	Sequence
Separation	Protein separation before digestion	Peptide separation after digestion
Number of MS analyses	1	2

the proteins or the peptides, and can therefore be done either before or after proteolytic digestion.

The two different main approaches are summarized in Table 1.2. They are briefly described in the following two subsections, and further detailed in the following chapters.

1.7.1 MS – peptide mass fingerprinting

This approach uses separation of intact proteins, and the ideal separation should yield all the molecules of a single protein into a single fraction. In practice, protein separation techniques rarely achieve this ideal resolution, especially if the starting material is a complex mixture containing many proteins. For example, the variants of a single protein will usually fall into different fractions. Protein separation is described in more detail in Chapter 2. However, if each fraction contains only a small number of proteins (one to three), the separated fractions may still be manageable for further analysis by mass spectrometry.

The approach is sketched in Figure 1.2.

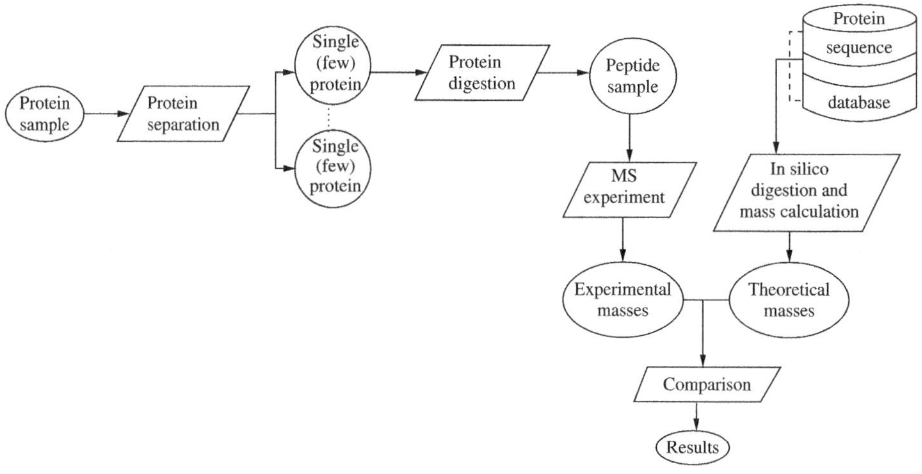

Figure 1.2 MS protein analysis approach, using protein separation

Example We have a sample of proteins, which we want to identify. Assume there are five proteins in the sample, A, B, C, D, E, and that we separate on isoelectric point, resulting in the separated fractions $\{A, D\}, \{B, E\}, \{C\}$. Note that this means that proteins A and D (and proteins B and E) have similar isoelectric points.

We then digest each fraction separately. Assume that ideal digestion (digestion at all possible places that the protease can cleave) would result in the following peptides:

Protein	Peptides
A	A_1, A_2, A_3, A_4
B	B_1, B_2, B_3
C	C_1, C_2, C_3, C_4, C_5
D	D_1, D_2, D_3, D_4
E	E_1, E_2, E_3

Suppose we now analyze the fraction that consists of protein C. We can expect to find the following peptides: C_1, C_2, C_3, C_4, and C_5. Even if the C_2 and C_5 peptides are missed by the instrument, we can still use the other three peptide masses to identify protein C from the database. When we analyze the fraction consisting of proteins A and D, however, we are faced with a mixture of peptides. Suppose the instrument presents the masses for five peptides: A_1, D_2, D_3, D_4, and A_4. The database search (which assumes a single protein to have been isolated) obtains a higher coverage of protein D and will list this identification as the most probable candidate. Without additional analysis (such as the removal of all identified peptides and a subsequent re-search with only the masses of peptides A_1 and A_4), protein A would be missed. Note that missed cleavage sites and similar masses of different peptides usually confound the identification process in practice. Additionally, the mass spectrometer will also report the masses of any ionized impurities, further complicating the analysis.

$\triangle$

1.7.2 MS/MS – tandem MS

This approach uses separation of peptides, and is also termed *peptide-centric proteomics*. A relatively simple protein sample, such as an *Escherichia coli* cell lysate, will contain somewhere between 2500 and 5000 proteins. With an average of 25 peptides per protein in *E. coli*, this yields 62 500 peptides if 2500 proteins are expressed. If the separation step is only performed after the proteolytic digestion, one thus ends up with a large number of peptides. In order to obtain a reasonable sample coverage, the peptide separation step therefore must be able to handle this increased complexity. The approach is illustrated in Figure 1.3.

Example We have the same data as in the example above, and assume that a (non-ideal) proteolytic digestion results in $A_1, A_2, A_{3,4}, B_1, B_2, B_3, C_{1,2,3}, C_4, C_5, D_1, D_2, D_3, D_4, E_1, E_{2,3}$. Note that we have introduced some missed cleavage sites here, resulting in the unexpected peptides $A_{3,4}$, $C_{1,2,3}$, and $E_{2,3}$. Assume separation of these

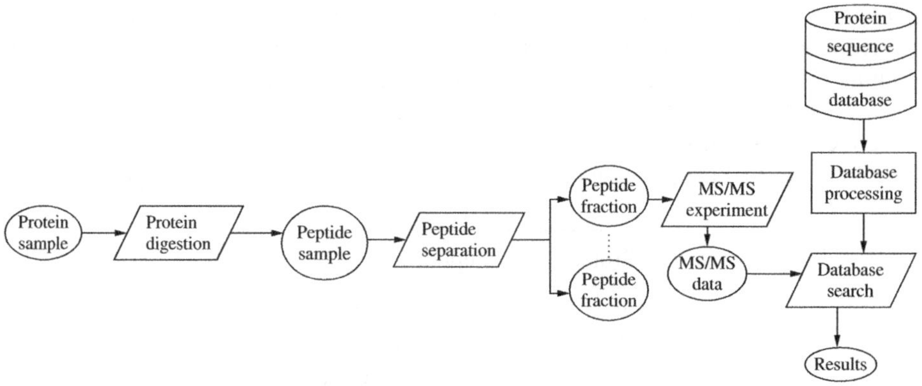

Figure 1.3 MS/MS protein analysis approach, using peptide separation

peptides by hydrophobicity, which results in three fractions, where the content of the first is $\{A_2, B_1, B_2, C_{1,2,3}, E_{2,3}\}$, the content of the second is $\{B_3, D_1, D_4\}$, and the content of the third fraction is $\{C_5, D_2, E_1\}$. (Note that not all peptides are observed.) Remember that all the peptides within each fraction have similar hydrophobicities. Suppose now that MS is performed for the first fraction, and that all peptides have different masses. Assume that the masses of $\{A_2, B_1, E_{2,3}\}$ each are selected for an MS/MS experiment, and that the resulting fragmentation spectra from these experiments are used to search the database. Depending on the outcomes of these searches, none or up to all three of the proteins A, B, E can be identified, with each protein supported by a single peptide. The same procedure can be performed on the other two fractions. Again, this example is a simplification, but it serves to illustrate the approach.

△

1.7.3 Combination approaches

It is also possible to first perform a separation at the protein level, proteolytically digest the isolated protein fractions, and finally separate the resulting peptides. By using the separations in series, the sample complexity can be greatly reduced. On the other hand, the extra level of separation results in a significant increase in the amount of work.

In general, if a step n yields F_n fractions after separation, the number of times the $(n+1)$th step needs to be applied is given by the following multiplicative series:

$$\prod_{i=0}^{n} F_i = F_0 \cdot F_1 \cdot F_2 \cdot \cdots \cdot F_{n-1} \cdot F_n$$

It is obvious that the increase in the number of sample processing steps reflects negatively on the speed of analysis. Combined approaches quickly become very laborious and are therefore not often used in high-throughput experiments.

1.7.4 *Reducing the search space*

Often some overall knowledge about the sample is available. Usually the species of origin is known, and it is therefore possible to select as potential candidates only those proteins from the sequence database that are derived from this organism. For experiments that apply some sort of protein separation, the approximate values for pI and the mass are sometimes known. Again, this information can be used to filter candidate proteins.

It is important to bear in mind that restricting the number of candidate peptides can influence the performance of the search algorithm employed, and should always be used judiciously.

1.8 Instrument calibration and measuring errors

Different types of instruments exist for measuring the values of different protein or peptide attributes. They are all similar in that they must be calibrated, however, and the result may contain errors.

1.8.1 *Calibration*

The numerical value given by a measuring instrument can depend on a variety of external factors such as the chemicals used, the temperature, voltage fluctuations, etc.

This means that the actual measured values may be incorrect, and have to be calibrated (corrected). Calibration is performed by measuring the attributes of interest of some molecules for which the correct values are known. These molecules are called *calibrants* or *standards*. The measured values for the standards are compared to the known values, and possible deviations are calculated. The measured values of any other molecules are then corrected in accordance with these observed deviations. Note that this process implicitly assumes that the deviations occurring during calibrant measurement will be the same as those occurring during sample measurement.

The correction of the measurements is done by calculating a function $f(v)$, such that $v* = f(v)$, where v is a measured value, and $v*$ the corrected value. Such a function can be found by *least squares approximation*. The correction function is often assumed to be linear, which corresponds to a function of the type $v* = a + bv$. Note that we will need to calculate a and b before we can apply this function. Assume that n standards are used, with correct values $\{c_i\}$, and measured values $\{e_i\}$. a and b are then found such that

$$\sum_{i=1}^{n} (a + be_i - c_i)^2$$

is minimized. Calculation of a and b from this expression is documented in textbooks on numerical analysis and optimization. The overall procedure is illustrated in Figure 1.4.

The calibration can be done by either *external* or *internal* standards. The values for internal standards are measured in the same experiment (measurement) as the unknown molecules. External standards are measured in a separate experiment. Internal standards come closest to satisfying the requirement introduced above that standard and unknown molecules should be measured with similar deviations. A disadvantage of

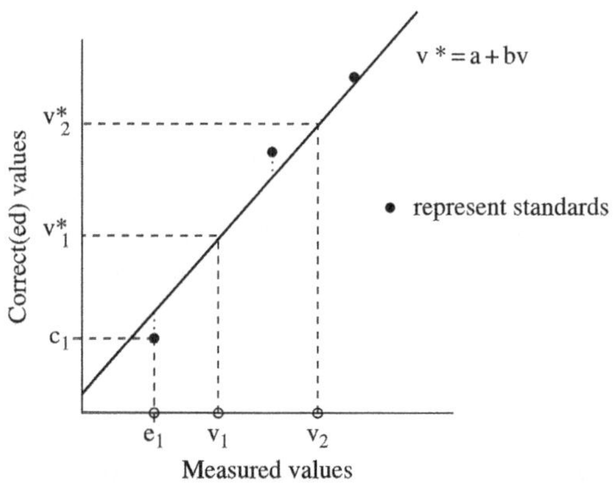

Figure 1.4 Figure showing calibration. The three full dots represent the standards, used to calculate the line. The calibrated value of the experimental value v_i is v_i^*

internal standards is that they can affect the detection of the unknown molecules in some cases. External standards suffer from the possibility that the environment (chemical/biological/physical) in which measurement takes place is not identical in both experiments. As such, the factors that affect the measurement might not be the same in the two experiments, violating the assumption that the standards suffer from the same kind of deviation as the unknown molecules.

1.8.2 Accuracy and precision

The quality of a protein identification or characterization increases with more exact measurements of the relevant attribute values.

Two terms are used for specifying this exactness (and thereby the uncertainty in the measured values): *accuracy* and *precision*. The description of these terms that we give here is based on IUPAC.

Accuracy

Accuracy is a measure of the closeness of a result to the true value. It involves a combination of a random error component and a common systematic error or *bias* component.

Precision

Precision is the closeness of agreement between independent measurements obtained by applying the experimental procedure under stipulated conditions. The smaller the random part of the experimental errors which affect the results, the more precise the procedure. A measure of precision is the standard deviation, or a formula based on the standard deviation. When a more precise definition is needed, *repeatability* or *reproducibility* is used. Repeatability considers the same method with the same test material under the same conditions (same operator, same apparatus, same laboratory, and after short intervals of time) but reproducibility means under different conditions.

The value of a measurement can be precise but not accurate, neither of them, or both of them. However, it is not possible to achieve high accuracy without high precision; if the values are not similar to one another, they cannot all be similar to the correct value. If the value from a measurement is both accurate and precise, it is often said to be *valid*.

Example Let the correct mass of a peptide be 950.455. Let two instruments perform four measurements each, resulting in the following value sets: {950.650, 950.654, 950.652, 950.653} and {950.465, 950.485, 950.445, 950.425}. The first instrument has high precision, but low accuracy; the other has higher accuracy, but lower precision.

△

Accuracy and precision are illustrated in Figure 1.5. The bias (or the systematic error) of the measurement depends on several 'fixed' factors, and is usually defined (for the instrument and method used) as the mean of the deviation from the correct value. It is often used as a measure of the accuracy of the instrument, and can also be used for calibration purposes. In fact, usage of the mean value of multiple measurements of a standard for calibration yields a two-step procedure that attempts to minimize the random fluctuations in the instrument measurements while compensating for the systematic error.

The precision is often determined by taking a sufficient number of measurements of the same sample, and performing a statistical analysis on the data. This analysis most often assumes a normal distribution for the measurement errors and relies on the 95 % confidence interval, calculated as 1.96 times the standard deviation (for a two-sided test against a normal distribution). In many cases, however, it is not convenient or even not possible to perform many measurements, and precision then needs to be estimated indirectly, with the exact mechanism depending on the type of attribute measured as well as the instrument used.

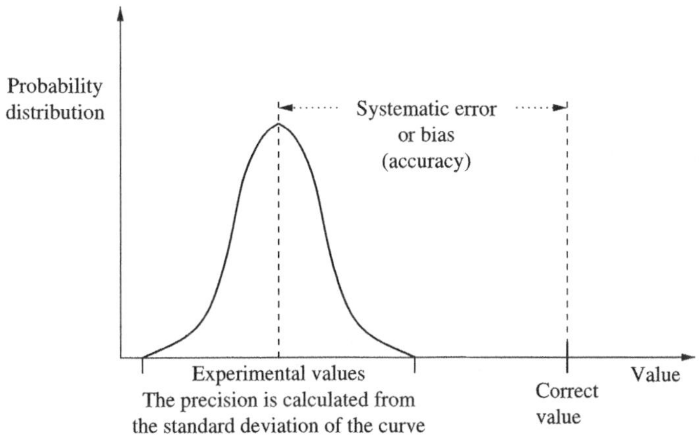

Figure 1.5 Illustration of accuracy and precision

Exercises

1.1 Calculate the monoisotopic, average, and nominal mass of the residue of serine. The chemical formula of the residue serine is $C_3H_5NO_2$, and use the atom masses as in the example in Section 1.3.2.

1.2 Suppose the mass of a protein is measured to 18 500 Da, with an uncertainty of 10 Da. What is the uncertainty in mmu and ppm?

1.3 Consider the peptide `EVEDLQVR`.

(a) Show that the theoretical pI value is between 4 and 4.5.

(b) Use the programs 'ProtParam' and 'Compute pI/Mw' (http://expasy.org/tools/) to calculate the pI value for the peptide.

1.4 Consider the peptide `ARTWLCDYEVTKRSEPTVGR`.

(a) Calculate the GRAVY score for the first windows of length five.

(b) Use the program 'Kyte-Doolittle Hydropathy Plots' (http://gcat.davidson.edu/rakarnik/kyte-doolittle-background.htm) to show a hydropathy (hydrophobicity) plot of the whole peptide.

1.5 Write a program that given the (nominal) mass of a peptide calculates the possible amino acid composition of the peptide. Remember that the mass of a peptide is the sum of the residue masses plus the mass of a water molecule (H_2O). Run the program on the peptide mass 677.

1.6 Use the search program Mascot (http://expasy.org/tools/) to search in Swiss-Prot with the peptide masses 611.3, 794.4, 1052.5, 818.4. Try with some and all of the masses, and with different accuracy values. Use trypsin as the protease, and try with zero and one missed cleavages. Observe the score of the highest scoring sequences, and compare to the significance level. (The masses are calculated from *in silico* digestion of the Swiss-Prot entry GLNB_PORPU.)

Bibliographic notes

Biological classifications

The Gene Ontology	http://www.geneontology.org
Reactome; Pathways	http://www.reactome.org/
KEGG	http://www.genome.jp/kegg/
SO	http://song.sourceforge.net/
OBO	http://obo.sourceforge.net/

Protein definition

HUPO	http://www.hupo.org/
PSI	http://psidev.sourceforge.net/

Amino acid properties

Repository	http://www.imb-jena.de/IMAGE_AA.html
Indices and similarity	http://www.genome.ad.jp/dbget/aaindex.html

Hydrophobicity scales

Kyte and Doolittle (1982)	http://garlic.mefos.hr/garlic/commands/sca.html

Isoelectric point

Bjellqvist (1993); Halligan *et al.* (2004); Nelson and Cox (2004); Schwartz *et al.* (2001)

DTASelect	http://fields.scripps.edu/DTASelect/20010710-pI-Algorithm. pdf

Protein modifications

RESID database	http://www.ebi.ac.uk/RESID
UniMod database	http://www.unimod.org/, Ceasy and Cottrell (2004)

Protein databases and tools

IPI	http://www.ebi.ac.uk/IPI
UniProt	http://www.uniprot.org/
ExPasy	http://www.expasy.org/
PIR	http://pir.georgetown.edu/
EBI	http://www.ebi.ac.uk/
NCBI	http://www.ncbi.nlm.nih.gov/
NCBI Entrez	http://www.ncbi.nlm.nih.gov/entrez/
MPSS	http://www.scbit.org/mpss/, Hao *et al.* (2005)
Sequence databases	Magrane *et al.* (2005)

Precision for mass spectrometers

Blom (2001); Wolff *et al.* (2003)

2 Protein separation – 2D gel electrophoresis

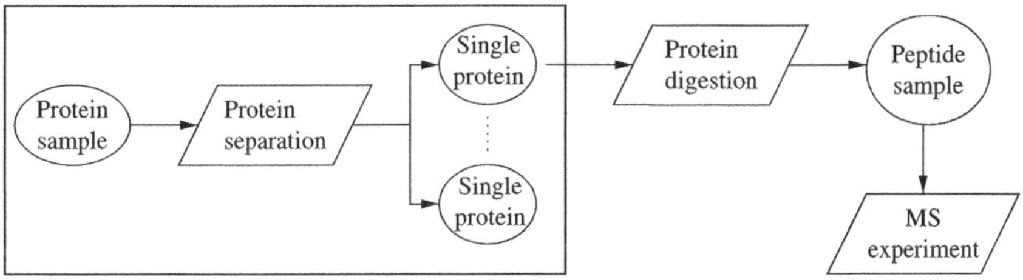

In Chapter 1 we explained why separation of proteins or peptides is important, and that different techniques or methods exist for performing this separation. Common to most of them is that the analyte[1] mixture is migrated through a substance. This substance is chosen to influence the speed of a molecule's migration depending on some property of the molecule (for example, size) and the molecules are therefore separated over time based on this property. Most of the different separation methods can thus briefly be described by:

- the substance through which the molecules migrate;
- the external force that causes the molecules to migrate;
- the preprocessing of the molecules, making them able to migrate through the substance.

The most commonly used technique for protein separation is gel electrophoresis, described in this chapter. Liquid chromatography (LC) is also used to a lesser extent. LC is frequently used for peptide separation, however, and is described in more detail in Chapter 4.

[1] The proteins (or peptides or any other molecules) under analysis are often called *analytes*.

Computational Methods for Mass Spectrometry Proteomics I. Eidhammer, K. Flikka, L. Martens and S.-O. Mikalsen
© 2007 John Wiley & Sons, Ltd

Gel electrophoresis is the only technique that can simultaneously separate thousands of unknown proteins. The protein attributes on which separation relies are commonly molecular mass and isoelectric point. Since modifications and sequence variations change these attribute values, protein variants will often be separated. The proteins can be separated on mass alone, on pI alone, or on both. It is important to note that protein mass and pI are *orthogonal* attributes. This means that they are not directly related and therefore should give optimal separation efficiency. For instance, two proteins with similar mass will most probably not have similar pIs. As shown in Section 1.3 these parameters can be calculated based on the sequence, which makes them appropriate for identification.

To achieve high-quality results, care has to be taken during sample preparation, especially to prevent other components (such as salts, lipids, polysaccharides, and nucleic acids) from interfering with the protein separation. Furthermore, one should try to avoid protein aggregation (particularly troublesome for hydrophobic proteins) and undesirable chemical modifications due to the chemicals used during sample preparation. Fortunately, a number of protocols have been developed for performing the sample preparation and the subsequent separation experiments.

2.1 Separation on molecular mass – SDS-PAGE

Strictly speaking, SDS-PAGE is a technique for separation on molecular size, since the speed of migration is inversely proportional to the size of the molecule. However, size is usually considered to be sufficiently proportional to mass.

The substance used in this separation technique is polyacrylamide. When this polymer is formed under the right circumstances, it turns into a porous gel with a lot of small holes or tunnels. It is this network of tunnels that impedes the transit of large molecules while small molecules move much more readily through them. The force applied to move the molecules through the gel is supplied by an electric field. PAGE thus stands for *PolyAcrylamide Gel Electrophoresis*.

The preparation of the proteins should therefore generate molecules in which the size is more or less proportional to the mass, while at the same time charging the analytes with a constant charge-over-mass ratio. Since proteins exist in natural form as 3D structures, a form which is inappropriate for mass-proportional size separation, they must first be denatured into a linear form. In natural form proteins also carry both negative and positive charges, depending on the charge state of the amino acids along their sequence. This is not an ideal situation with regard to the separation procedure. The necessary preprocessing of the proteins is achieved by treating them with *sodium dodecyl sulfate (SDS)*, a detergent molecule with a long hydrophobic tail and a negatively charged head. SDS will then:

- denature the proteins; and
- impart negative charges to the proteins (roughly) proportional to the size of the protein.

This is illustrated in Figure 2.1.

After being prepared, the resulting negatively charged and denatured proteins are loaded onto the polyacrylamide gel. When a gel is loaded with multiple, discrete samples, each

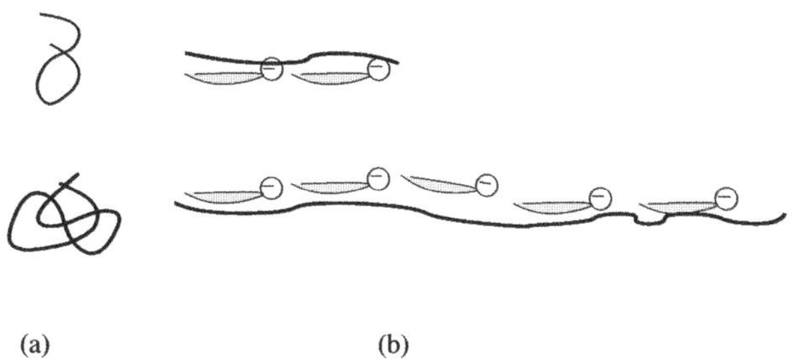

Figure 2.1 Sample preparation for SDS-PAGE. (a) The native proteins exist in 3D form. (b) They are treated by SDS, denaturing them to linear form and negatively charging them in constant charge/size ratio

loading spot is called a *well*. The loaded gel is then placed in an electric field, and the negatively charged proteins start migrating towards the positive electrode with velocities depending on their sizes, small proteins migrating fastest. When the gel is loaded by multiple samples, each protein separation track derived from a single well is called a *lane*.

In any sample there will be many copies of the same protein. They may take different paths through the substance, but will still migrate with approximately the same speed. All copies of proteins of the same mass will therefore turn up as one band on the gel, as illustrated in Figure 2.2. A standard dye is also loaded on the gel and is of sufficiently low

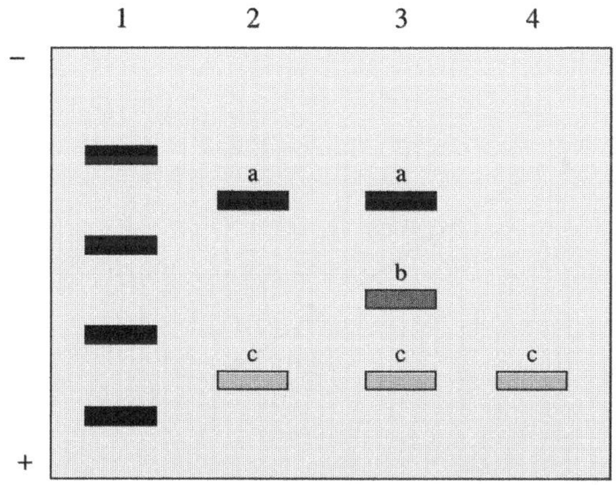

Figure 2.2 A gel with four numbered loading spaces or wells. A voltage difference is applied to the gel, with the negatively charged electrode at the top, and the positively charged electrode at the bottom. The proteins carry negative charge because of the attached SDS molecules and will move towards the bottom, positive electrode when they are loaded at the top. Well 1 is typically loaded with molecules with known masses (standards or markers), and the resulting gel bands can be used to calibrate the masses of the unknown proteins. Well 2 contains a two-protein mixture with proteins a and c. Well 3 contains a sample of three proteins a, b, and c, and well 4 contains protein c alone. In this case a is the heaviest protein. Note that a certain protein species (for example, protein c) travels equally far through the gel, independently of how many other proteins were present in the samples loaded in the wells

molecular mass to most probably be the fastest migrating molecule in the gel. Usually, the dye is colored (hence the name), which allows tracking of its progress through the gel. This progress is visible as a thin, colored line and this line is called the *dye-front*. The migration is stopped by turning off the electric field before this dye-front reaches the bottom of the gel. Biological protein samples normally contain a number of proteins with approximately the same mass, yielding bands that can contain several different proteins. The bands can also merge into each other, making the interpretation of real gels a difficult task.

2.1.1 Estimating the protein mass

Estimating the mass of a protein[2] can be done (as illustrated in Figure 2.2) by loading standard molecules of known masses in a parallel well on the gel, and after separation comparing the migration distances. An Rf value is first calculated for the proteins, which is the ratio of the distance that the protein has migrated to that migrated by the dye-front. Consequently, the Rf values of the proteins are less than or equal to one. There is a linear relationship between the logarithm of the protein mass M and its Rf value:

$$\log_{10} M = aRf + b$$

A set of standards are migrated through the gel, and a linear line is interpolated (by calculating the values for a and b). The log of the protein mass is then determined by using the Rf value on this line, as explained in the example below.

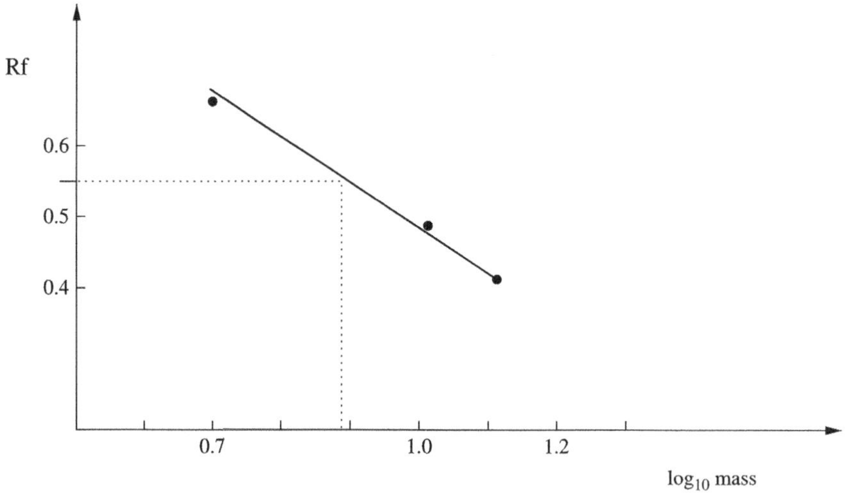

Figure 2.3 The estimation of protein mass. The full dots illustrate the standards used, and a linear relationship is interpolated. A protein with a relative migration of 0.55 will then have a $\log_{10}$ value of its mass equal to 0.88, and therefore a mass of 7.6 kDa

[2] The mass estimate as derived from an SDS-PAGE experiment is usually referred to as *apparent mass*.

Example Assume we have three standards with masses (5, 10.8, 14.5) kDa, and the relative migration distances (Rf values) are (0.66, 0.48, 0.42) respectively. The logs to base 10 of the masses are (0.70, 1.02, 1.16). Figure 2.3 illustrates how the mass of an unknown protein with Rf value 0.55 is estimated from these data points.

△

2.2 Separation on isoelectric point – IEF

Separation on isoelectric point is performed by *isoelectric focusing (IEF)*, where a *pH gradient* is used for the separation. The pH gradient used for separation these days is typically immobilized on a plastic backing. The gradient itself is generated from certain polymerized chemicals (polyaminocarboxylic acids and similar compounds) that are imbedded into agarose or polyacrylamide gel, arranged by (continuously) increasing pH values. The proteins are entered into the gradient by absorption from a buffer with the sample, often strengthened by applying an electric current. After sample transfer, an electric field is applied across the pH gradient. The proteins will move in this electric field, initially towards the electrode with the opposite charge. When the protein reaches the point in the pH gradient that is equal to its pI, the net charge will become zero, and the migration will end. The procedure is illustrated in Figure 2.4.

The pH value at both ends of the gradient is known, and the gradient itself can be linear or nonlinear. If it is linear, the pI of the proteins can be calculated directly by measuring the distance to one of the ends. If it is nonlinear, standards are used for determining the pI of the proteins.

The most popular type of pH gradient is the *immobilized pH gradient* or *IPG strips.* Immobilized means that the chemicals on the strip are not moving, which used to be a problem with older systems. Gradients producing reproducible pI measurements can be bought for different pH ranges, and vary in length from 7 to 24 cm. The pI of proteins is usually in the range of 3 to 12, but the overall majority are between 4 and 7. The strips typically have either a very broad range (for example, pH 3–10), or a very narrow range (for example, pH 4–5).

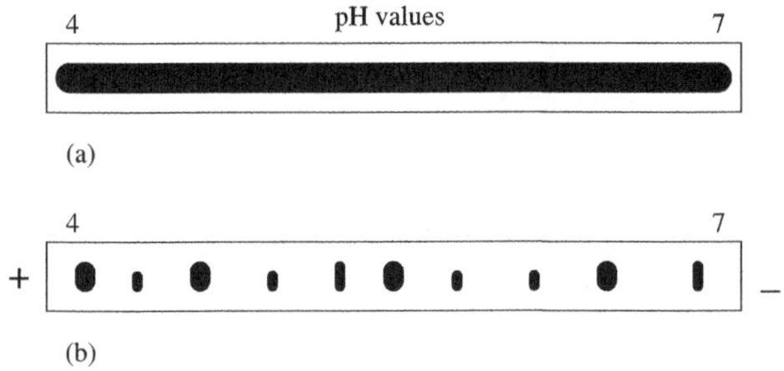

Figure 2.4 (a) A pH gradient filled initially with the protein sample. Note that the sample may also be applied at one end of the pH gradient. (b) When an electric field is established, the proteins move to the pH equal to their pI values. Each spot therefore contains a set of proteins with similar pI value

2.3 Separation on mass and isoelectric point – 2D SDS-PAGE

The SDS-PAGE or IEF separation methods alone do not have high enough separating power for sufficient resolution of complex protein samples. However, because these two methods are orthogonal, they can be combined in a method called *two-dimensional SDS-PAGE (2D SDS-PAGE)*. The proteins are separated on pI in the first dimension and on mass in the second. This method, first published in 1975, remains the most widely used protein separation technique today.

2.3.1 Transferring the proteins from the first to the second dimension

This is done manually by attaching the pH gradient to the top of the gel, after it has been treated by SDS.

2.3.2 Visualizing the proteins after separation

Normally the bands/spots (of proteins) are invisible. To make them visible on the gel they must therefore be stained, commonly with dyes or metals (most notably silver).

Which dyes or metals that are best suited depends on the types of the proteins. Desired properties for a staining agent are reproducibility, sensitivity (can detect low-abundance proteins), and linearity, so that it can be used for quantification. No staining technique will stain all types of proteins in proportion to the amount of protein present, implying that they are unsuited for accurate quantification inside a gel. Figure 2.5 shows a 2D gel.

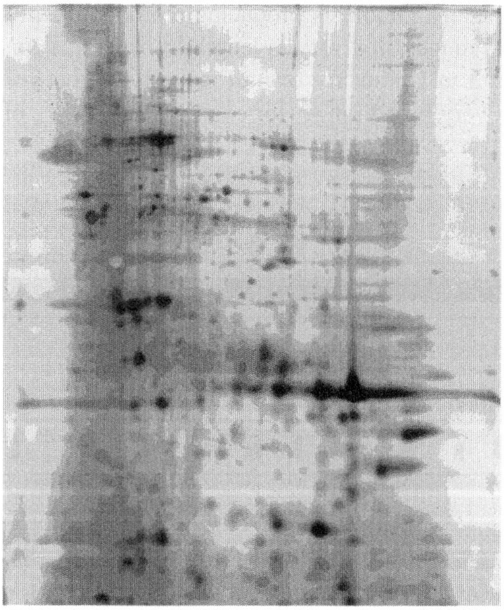

Figure 2.5 A typical 2D gel with pI in the horizontal direction, and mass in the vertical direction

2.3.3 Problems

There are several problems with 2D SDS-PAGE, as follows:

- Not all proteins will appear on the gel. These are typically:
  - hydrophobic proteins, with GRAVY values above 0.4;
  - proteins with an isoelectric point below 3 or above 10;
  - proteins with a mass below 8 kDa, or above 150–200 kDa;
  - proteins of low abundance.

  The hydrophobic proteins tend to form complexes and precipitate before they can be loaded on the pH gradient, while the proteins with extreme pI usually fall off the edges of the pH strip. The very small or very large proteins tend to be lost during gel electrophoresis and the low-abundance proteins will fail to be stained and will therefore escape detection.
- The gels may contain contaminating components.
- Protein complexes may stay intact and will therefore behave as a single, very large molecule.
- It is difficult to observe membrane proteins, which are usually quite hydrophobic (see above).
- 2D SPS-PAGE has mediocre reproducibility. It is difficult to obtain the same patterns for repeated experiments. There might also be local variations inside a single gel. Thus, proteins with a high pI may not have exactly the same migration rate as proteins with the same masses, but low pI.

2.3.4 Excising the proteins

Proteins with similar mass and isoelectric point migrate into spots on the gel. The different spots can then be cut out by a scalpel, or can be picked up by a robot, for further analysis (for example, by a mass spectrometer).

2.4 2D SDS-PAGE for (complete) proteomics

2D SDS-PAGE is mainly used as a separation technique, enabling further analysis in other instruments. However, despite the above-mentioned serious drawbacks, it has been used for identification, and was for some years considered synonymous to proteomics. High-quality sample preparation and very precise experimentation are then required. Below we briefly describe how 2D SDS-PAGE is used for further processing.

2.4.1 Identifying the proteins

The values for isoelectric point and mass can be determined by use of standards, and further used as support for identification. The accuracy and precision obtained are usually not sufficient to lead to unambiguous identification, however. As we have seen above, usually the separation is followed by mass spectrometry analysis for identification.

Another approach for identification is to use *antibodies*. This approach is for instance used in *Western blotting*. A disadvantage of this approach is that we need to know which proteins are present (in order to obtain the correct antibodies). The proteins are transferred from the gel onto a membrane by applying an electric current or a liquid flow. This procedure transfers the pattern from the gel onto the membrane. The membrane is then probed for a specific protein by adding a specific antibody, which also contains a detectable label (usually an enzyme, a fluorescent dye, or a radioisotope). The membrane is washed in a series of steps, and the bound probe is detected using photographic film, a highly light-sensitive camera, or a camera-like device.

2.4.2 Quantification

In general, 2D SDS-PAGE gel quantification is based on the signal intensity of the spot in which the protein has been found. Obviously, spots containing more than one protein present challenges to any such quantitative study as the unexpected proteins add staining intensity that will be incorrectly assumed to be derived from the identified protein. And even if the spot is known to hold more than one protein, it is no longer possible to trace back which protein contributed which amount of staining to the spot. On the other hand, if different forms (for example, truncated or modified forms) of the protein are spread over different spots, the (total) quantification of the protein may become problematic as it relies on adding the spot intensities of all the different forms. Any missed spot (that could not be identified for whatever reason) will thus result in an underestimation of the quantification. The fact that different forms often localize in different spots can, however, be beneficial if one wants to compare one form with the other (for example, the phosphorylated form compared to the unphosphorylated protein). Another problem focuses on the usable interval of these staining procedures for quantification. Several staining procedures do not stain at all below a certain amount of protein present, and their staining intensity levels off quickly above a certain amount of protein present as well. The interval between these amounts can be considered 'covered' by the staining procedure and is called the *dynamic range*. It is desirable to have a linear response in staining intensity versus amount of protein present for the dynamic range, in addition to a high sensitivity. Most non-fluorescent staining agents have relatively poor dynamic range, and even the best non-fluorescent technique (silver staining) only has a dynamic range of about one order of magnitude. The fluorescent stains can attain the sensitivity of silver staining when used under optimal circumstances, yet can yield a much better dynamic range, up to five orders of magnitude. The fluorescent detection does have some caveats, however, including fading signals if the fluorophore is (slowly) decomposed by the influx of light, and problems with the interference of fluorescence of flourophores that have become associated with detergent micelles.

2.4.3 Programs for treating and comparing gels

In most cases, the patterns obtained after 2D SDS-PAGE are analyzed and compared using computer-based tools.

In a first step, image capture devices are used to generate digital images of the actual gels. Such images consist of *pixels*, and each pixel has a corresponding value for the brightness or signal intensity. The images are then analyzed by special programs. Such programs are typically semi-automatic, in that the analysis is performed in an interactive way. The user gives parameters for calibration, spot detection, quantification, spot matching, and result presentation. The sequence of typical operations performed is listed below:

1. *Removal of noise.* Several filtering procedures are performed, depending on the types of noise that exist. The challenge is to remove the noise while retaining the spots corresponding to the molecules of interest. There are different types of filters. Generally they use a window of a certain size, usually (3×3) to (9×9) pixels, which is slid across the image. A new value for the pixel in the middle of the window is calculated from the values of all the pixels in the window by a *filtering function*. An example of a filtering function can be a weighted average of the intensities of the pixels in the window, with the pixels lying nearest the one in the middle carrying the greatest weight. The programs may offer some help in identifying what types of noise exist, and can thus suggest appropriate filters to use.
2. *Removal of the background.* The background intensity should be subtracted from the image. In more advanced noise filtering steps, a complete background image is constructed which is then subtracted in its entirety.
3. *Detection of the spots.* Cut-off parameters such as minimum peak intensity, spot size in pixels, and other parameters, depending on the program, are used. The software should detect the spots, and try to separate protein spots from any other coloring of the gel (due to artifacts or left-over noise). Fortunately, protein spots typically have special forms that can be used to recognize them.
4. *Quantification of the spots.* The amount of protein present in a spot is calculated from the pixel intensities and the spot surface area. A Gaussian filter of the pixels around the detected spot center can be used, for example, after which the quantification is based on the output of the Gaussian filter.
5. *Matching spots from different gels.* This is the most important analysis when performing gel-to-gel comparisons. Due to the low reproducibility typically encountered in 2D SDS-PAGE, it can be a difficult process. Usually, gels are aligned by using a set of internal standards that constitute reference spots. Since local deviations occur, the standards are spread over the gels. The reference spots in the two gels are recognized, and overlaid. The rest of the spots are then moved and matched in correlation with the alignment of the standards. Matching several gels is typically done by defining a *master gel*, which can be either one of the real gels, or a new gel constructed from several of the real ones. Spot matching over several gels is then performed using pairwise matchings to the master gel. Programs that attempt to directly perform multiple matchings also exist.
6. *Analysis and presentation.* The result can be statistically analyzed and presented in a variety of ways.

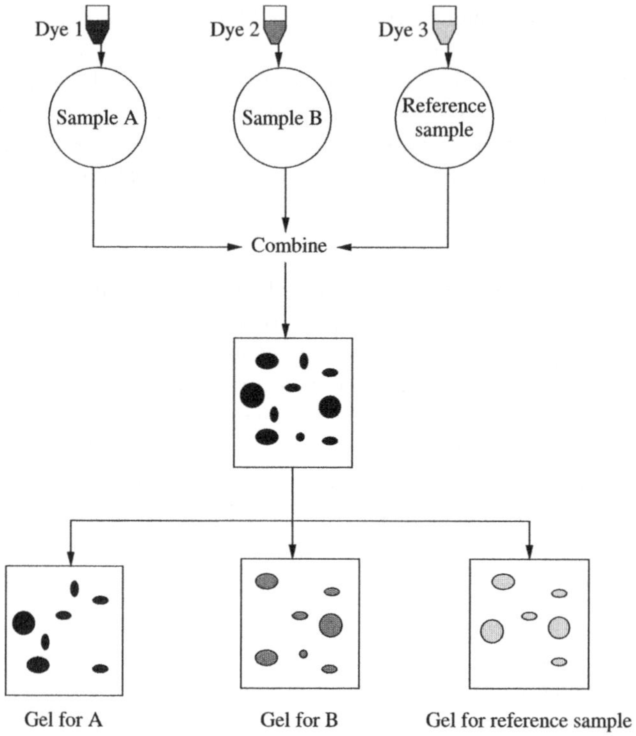

Figure 2.6 Illustration of the DIGE procedure. Two samples and a reference sample are each labeled with different dyes and then mixed. One 2D gel is used for separating the proteins. Using different wavelengths of light, three different gel images can be obtained (one for each sample) from the same gel

2.4.4 Comparing results from different experiments – DIGE

As described earlier, the poor reproducibility implies that comparing gels is not a straightforward process. A widely used technique to avoid the typical low reproducibility between different gels is the *fluorescence-based differential in-gel electrophoresis* (*DIGE*) approach. In this technique, the proteins in the different samples are labeled with different fluorescent dyes having different excitation wavelengths. The samples are then mixed, and the proteins separated on the same 2-D PAGE gel. Because of the different excitation wavelengths of the different labels, separate gel patterns can be obtained for each sample. This effectively allows the samples to be run under identical circumstances, greatly limiting the factors that can cause variation. The spots from two samples are therefore much more directly comparable. Figure 2.6 illustrates the process.

Exercises

2.1 Assume we have calculated the equation between the log of the mass (in kDa) and the relative migration as $\log_{10} M = -0.6Rf + 1.08$. What is the mass of a protein with relative migration equal to 0.8?

2.2 As indicated in Chapter 1, proteins and their amino acids may become modified post-translationally, intentionally by chemical treatment, or unintentionally during sample handling. Discuss how the following modifications could change the migration of a protein during 2D electrophoresis.

(a) Phosphorylation of the side chain of serine ($-OH \rightarrow -OPO_3^{2-}$).

(b) Methylation of the side chain of glutamic acid ($-COO^- \rightarrow -COOCH_3$).

(c) Assume we have a protein with pI of 5.6 and mass 40 000 Da, and that the N-terminal part of the protein (`MGKLLSRVHWKKLI`) is cleaved off.

Bibliographic notes

Protocols
Walker (2005)

DIGE
Marouga *et al.* (2005); Tonge *et al.* (2001)

Programs for 2D gel analysis

Melanie	http://au.expasy.org/melanie/
Bio-rad	http://www.bio-rad.com/
GELLAB-II	http://www.lecb.ncifcrf.gov/gellab/index.html
nonlinear	http://www.nonlinear.com/

Other system for protein separation

FFE	http://www.bd.com/proteomics/

3 Protein digestion

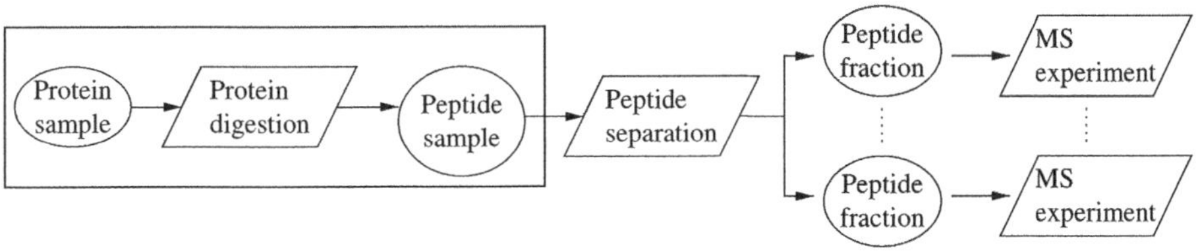

An important part of the identification process is protein digestion: the cleaving of proteins into peptides. Protein cleavage can be performed chemically or enzymatically, with enzymatic cleavage the most often used approach. The enzyme(s) that perform(s) the digestion are called *proteases*, and these are found in all organisms. For example, in the intestine, proteins derived from food are cleaved into small pieces to achieve an efficient absorption, while inside individual cells proteins are continuously degraded by proteases as a part of regulatory mechanisms. Numerous proteases from numerous species, ranging from man to bacteria, are known and characterized. Because proteases in proteomics research need to fulfill certain specific requirements, only a few proteases are routinely used.

As explained earlier, the peptides are analyzed by mass spectrometry, producing a mass spectrum. This spectrum is then compared to the theoretical peptide masses from an *in silico* digestion of database sequences. In the following we will look more deeply into this comparison.

Let P_E be the set of masses in a spectrum, which are assumed to come from a single protein. Furthermore, let P_T be the set of theoretical masses from an *in silico* digestion (simulating the protease) for that same protein. Then, in an ideal situation without modifications, $P_E = P_T$. In reality, however, this situation is rarely encountered, for the following reasons:

1. Some of the masses in P_E come from other proteins or molecules contaminating the sample. These masses are called P_E^C.
2. Not all of the peptides obtained after digestion are detected by the mass spectrometer, and these unrecorded masses will not be in P_E. This set of masses is denoted P_T^U.

Computational Methods for Mass Spectrometry Proteomics I. Eidhammer, K. Flikka, L. Martens and S.-O. Mikalsen
© 2007 John Wiley & Sons, Ltd

3. There may be a disagreement between the experimental digestion and the model used for the *in silico* digestion. An example of this is the occurrence of so-called *missed cleavages* in which not all expected cleavages are performed in the experimental digestion, hence the name. This results in a set of experimental masses, P_E^D, which are not in P_T, and, if the missed cleavages occur consistently, in a set of theoretical masses P_T^D, which are not in P_E. Note that P_T^D is a subset of P_T.

4. Some peptides may contain modified residues, P_E^M. The mass set from the corresponding unmodified peptides is denoted P_T^M.

The shared set of experimental and theoretical peptide masses based on pure mass similarities then is $P_E - (P_E^C \cup P_E^D \cup P_E^M) = P_T - (P_T^U \cup P_T^D \cup P_T^M)$. Most comparison procedures will usually take possible modifications and missed cleavages into consideration, however, and more experimental masses corresponding to theoretical masses will therefore be found. The overall approach is sketched in Figure 3.1.

Example Consider the sequence

MAVMQPRTLLLLLLSGALAKTQTWAGSHSMRYFYTSVSRPGKGEPRFIAVGYKVDDTQFVR
FDSDAASQRMEPT

Using trypsin (which cleaves after R and K, unless followed by P) for *in silico* digestion will result in the following peptides:

MAVMQPR TLLLLLLSGALAK TQTWAGSHSMR YFYTSVSRPGK GEPR FIAVGYK
VDDTQFVR FDSDAASQR MEPT

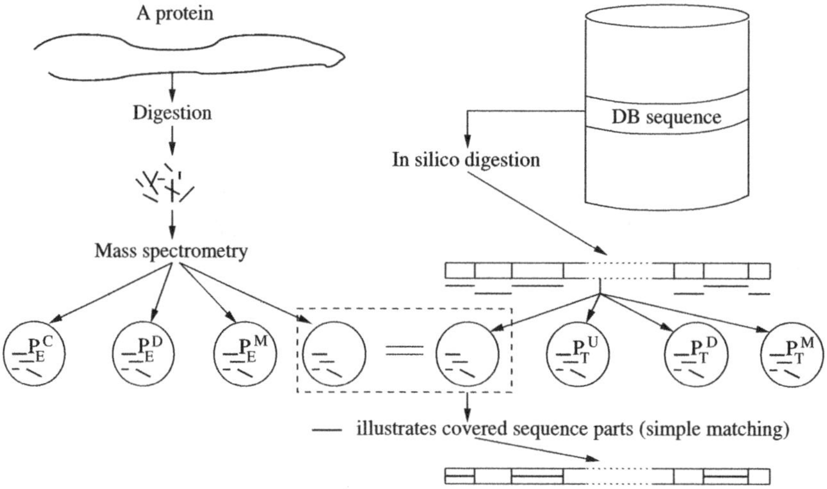

Figure 3.1 Comparing experimental and *in silico* digestion of a protein. The different set names are described in the text. The covered part of the sequence is defined by matching theoretical and experimental masses

The theoretical peptide masses are denoted $T_1, \ldots, T_9$. Assume that six experimental masses $E_1, \ldots, E_6$ are observed in the mass spectrum, and that

- E_1 corresponds to T_1 with a modification at amino acid Q;
- E_2 corresponds to T_2;
- E_3 corresponds to T_4 with a modification at amino acid S;
- E_4 corresponds to $\{T_6, T_7\}$ (one missed cleavage);
- E_5 corresponds to a contaminant;
- E_6 corresponds to T_9.

Then the different subsets are $P_E^C = \{E_5\}, P_E^D = \{E_4\}, P_E^M = \{E_3, E_1\}, P_T^U = \{T_3, T_5, T_8\}, P_T^D = \{T_6, T_7\}, P_T^M = \{T_1, T_4\}$.

A simplistic matching will then find that E_2 and T_2 have equal masses, as well as E_6 and T_9, and therefore that these parts of the sequence are *covered*. If we now revise the matching to become more robust by also allowing certain modifications, and if we assume that the modification occurring in E_3 is considered in our comparison, an additional match between E_3 and T_4 is found. Finally, if one missed cleavage is also allowed, one more match is found between E_4 and the concatenation of T_6 and T_7, thus further increasing the sequence coverage. Note that these refinements (considering modifications and missed cleavages) also substantially increase the size of P_T.

△

In peptide mass fingerprinting (PMF) the aim is to achieve as many common peptides as possible for the experimental and *in silico* digestion, hence as high a sequence coverage as possible. A very important factor in achieving high coverage is the protein separation steps performed prior to digestion. It is clearly preferable to obtain a single protein in each sample (for example, a single protein in each spot on a 2D gel) after separation, and contamination should be avoided at any step of sample handling. Human keratins (from the skin or hair) are common contaminants, and they can severely interfere with the analysis of the samples. Typical PMF-based identifications are achieved in a sequence coverage of 15 to 30 %, corresponding to 5 to 15 peptides for most proteins. There are many reasons that contribute to this relatively low overall sequence coverage. These will be discussed in some detail in Chapter 6.

As can be understood from the above discussion, the selection of the protease is a very important decision. The protease should cleave the protein in a consistent and predictable way. The protease should also not cut the protein into too many small peptides or into only a few very long peptides, and this for two reasons. First, most mass spectrometers have only a limited mass range, with a lower threshold due to interference by background noise, and an upper threshold depending on the resolution of the instrument (see Section 5.7). Second, the number of sequences that share a specific peptide mass increases with decreasing mass. For example, if we take 13 359 human proteins available in the Swiss-Prot database in January 2006 (a few restrictions were made in the selection of proteins), and digest them *in silico* with trypsin, there are 633 peptides in the *m/z* range 499.0 to 501.0, but only 195 peptides in the *m/z* range 1999.0 to 2001.0. Peptides of less than six amino acids are so difficult to distinguish by mass alone that they are less appropriate for PMF identification. It is thus clear that the protease should follow the Goldilocks doctrine: the resulting peptides should not be too small, nor should they be too big.

3.1 Experimental digestion

A protease cleaves a protein substrate by first recognizing a *cleavage site* on the protein; this results in a bond being formed between the cleavage site and the *binding site* of the protease, and subsequent cleavage. The actual cleavage reaction is performed by a subset of the residues in the binding site, and this subset is called the *catalytic site*.

The binding site of the protease consists of a set of *subsites*, by convention denoted $S_n, S_{n-1}, \ldots, S_1, S'_1, S'_2, \ldots, S'_m$. These bind to a protein substrate cleavage site, with the residues of the cleavage site denoted $P_n, P_{n-1}, \ldots, P_1, P'_1, P'_2, \ldots, P'_m$ by the same convention. The cleavage is performed between P_1 and P'_1, such that P_1 is the N-terminal to the cleavage point, and P'_1 is the C-terminal to it. The cleavage point is called the *scissile bond*. This is illustrated in Figure 3.2. Each subsite may have a preference for the properties of the corresponding amino acid, such as shape, size, and charge. The stringency of this preference is strongly variable, with some subsites requiring a specific amino acid, while others accept any amino acid.

Example The cleavage site for caspase 6 is five residues long, with four residues being recognized before the cleavage point. The requirements for the different subsites are $P_4 = \texttt{V}$, $P_3 = \texttt{E}$, $P_2 = \texttt{H}$ or $\texttt{I}$, $P_1 = \texttt{D}$, and P'_1 must be different from $\{\texttt{P,E,D,Q,K,R}\}$. P'_2 is aspecific and can accommodate any residue.

△

Of the thousands of known proteases from a wide variety of organisms, only a few cleave in a way that is appropriate for use in proteomics.

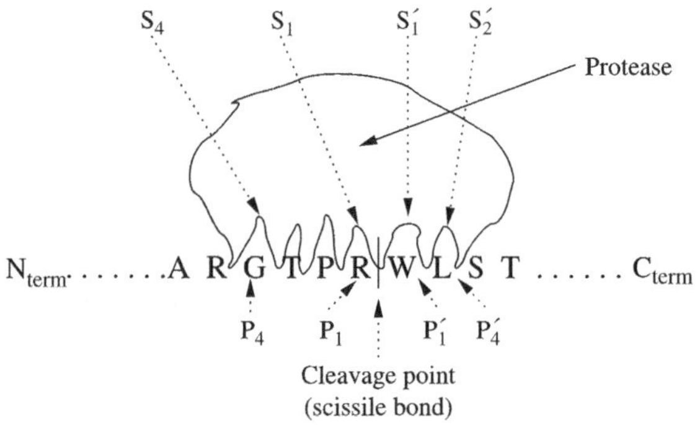

Figure 3.2 Illustration of the notation used for binding and cleavage sites. In this example the binding site has six subsites which are used to bind to a cleavage site of six residues, four before (the N-terminal to) the cleavage point, and two residues after (the C-terminal to) the cleavage point. (Redrawn from a figure in PoPS homepages (http://pops.csse.monash.edu.au/home.html), with permission)

3.1.1 Cleavage specificity

Cleavage specificity is a description of the cleavage site of a protease's substrate protein. A cleavage site can be described by the following:

cleavage activator is a set of amino acids that a subsite can bind to;

cleavage point specifies the cleavage point;

cleavage preventor is a set of amino acids that hinder the cleavage if one of them occurs at a specific position, effectively negating the occurrence of the cleavage activators.

Thus each residue of a cleavage site is part of an activator or a preventor. Note, however, that for some proteases an activator can be X, meaning that any of the amino acids can occur at that position.

A lot of research has been performed to reveal the cleavage specifications of different proteases, but many uncertainties still exist. Therefore (slightly) different specifications can be found in the literature for several proteases.

A notation for describing a cleavage site is to:

- enclose the cleavage activators in brackets, '[]';
- enclose the preventors in $\langle\rangle$;
- specify the cleavage point by a full stop, '.';
- take the length of the cleavage site as equal to the sum of the number of activators and preventors.

Table 3.1 shows the cleavage specificity for the most commonly used proteases. For some of them other specifications may exist.

Table 3.1 Cleavage specificity for the most commonly used proteases

Trypsin	$[RK].\langle P\rangle$	Cleaves C-terminal to an arginine or a lysine if not followed by a proline
Chymotrypsin	$[WYF].\langle P\rangle$	
Glu C (V8 protease)	$[ED].\langle P\rangle$	In sodium phosphate buffer,
	$[E].\langle P\rangle$	otherwise
Lys C	$[K].$	
Asp N	$.[D]$	Cleaves N-terminal to aspartic acid
Pepsin	$[WYF].$	
Pepsin (pH 1.3)	$\langle HKR\rangle\langle P\rangle\langle R\rangle.[WYFL]\langle P\rangle$	

Some important proteases have been researched in so much detail that even more specific cleavage information is known, which cannot be described in the formula notation used here. Trypsin, for example, will cleave poorly at the following compositions:

$$\langle CD\rangle[K].\langle D\rangle, \langle C\rangle[K].\langle HY\rangle, \langle C\rangle[R].\langle K\rangle, \langle R\rangle[R].\langle HR\rangle$$

`W` as activator for chymotrypsin at position P_1 is blocked by `M` at position P_1'. Note also that proteases may occasionally cleave at other positions. Trypsin is considered as a very specific protease that rarely cuts at unexpected positions, while chymotrypsin is rather unspecific and may do so quite often.

A protease with only one amino acid as activator will on average generate quite long peptides (if all 20 amino acids were equally prevalent, the protease would generate peptides with an average length of 20 residues). If there are three or more amino acids as activators, the majority of the peptides would become quite short. A general purpose protease for proteomics should thus have two amino acids as activators, no preventors, and no additional specific demands for the surrounding amino acids. Such an ideal protease does not exist. Trypsin is the protease that has properties closest to these criteria, and is by far the most commonly used proteolytic enzyme in proteomics. Other proteases may also be used depending on the properties of the protein under study and the specific purpose of the experiment (identification or characterization), in addition to the properties of the protease, and the mass spectrometer used in the analysis (MS or MS/MS; positive or negative mode). Some of these points will be elaborated upon in the following subsections.

3.1.2 Trypsin

As mentioned above, trypsin is the protease that best satisfies the desired requirements, for the following reasons:

1. It has high specificity, results in relatively few missed cleavages, and rarely or never cleaves at unexpected positions.
2. From Table 1.1 we see that arginine and lysine appear with an average distance of approximately 11 residues, and with only a small probability of being succeeded by a proline. The peptides produced are therefore of suitable length.
3. It is easily obtained and purified, and most of the 'sequencing grade' trypsin used today is obtained as a recombinant version from high-yield expression vectors.
4. It is applicable in most experimental settings and procedures, and is used to cleave proteins in solution, in gels, or even adsorbed onto surfaces.

Trypsin is particularly appropriate for positively charged mass spectrometry. If a peptide is to be observed in a mass spectrometer, the peptide must be able to become ionized by trapping and retaining a proton. Since arginine and lysine are both basic, trypsin, by cleaving after each arginine and lysine, ensures that each peptide will have a site capable of retaining a proton.

A MALDI MS (Chapter 6) analysis of missed cleavages in trypsin found that approximately two peptides with missed cleavages were observed per protein. Almost all of the missed cleavage sites had one of the amino acids K, R, D, E either before or after the activator sites K or R.

3.1.3 Chymotrypsin

In contrast to trypsin, chymotrypsin cuts the protein backbone at more than two sites, and it is less reliable in its selection of cleavage sites. It is therefore less suitable for identification purposes, but can still be valuable for protein characterization purposes. In many cases, we know the identity of the protein we want to further characterize. The purpose of the characterization is often to identify posttranslational modifications and the residue and position of the modification. To be able to do so, the modified peptide must be detected in our experiments. As explained previously, a full sequence coverage of a protein is rarely achieved, no matter the protease or instrument used. By employing an alternative protease, we may be able to extend the sequence coverage, and we will probably get an alternative sequence coverage of parts of the area detected by the first protease. Thereby we increase the chance of detecting the modified peptides and determining the modified residue.

3.1.4 Other considerations for the choice of a protease

As explained in Chapter 2, 2D gel electrophoresis is a common separation method for proteins. Proteins are found over a wide pI range from greater than 10 to less than 4. It is very likely that the highly basic proteins will contain many arginines and lysines, but few aspartic acids and glutamic acids. Conversely, a very acidic protein will probably contain many aspartic and glutamic acids, and few arginines and lysines. Thus, a very basic protein may be cleaved into a high number of small peptides by trypsin, and a very acidic protein may be cut into a few large peptides. In these cases, alternative proteases may be used, such as Lys-C, Glu-C, or Asp-N, especially if a first approach with trypsin was not successful. Similar considerations may come into play if we are interested in covering an area of a protein with an unusual amino acid sequence. In a few cases, usually for characterization purposes, a combination of two proteases may be used to achieve peptides of suitable length. For peptide-centric proteomics analyses (which do not rely on protein separation) on mass spectrometers that yield multiply charged peptide ions, larger peptides are sometimes desirable. Lys-C can then be employed to shift the average peptide length above 11. An alternative approach relies on the chemical modification of the side chain of lysine residues (for instance, by acetylation), which stops trypsin from recognizing them and therefore causes it to cleave only after arginine.

3.1.5 Random cleavage

Although not applicable for PMF, rather unspecific proteases may be used for MS/MS purposes. If such proteases are allowed to work for a long time, the result is a large number of small and useless peptides. However, by controlling time and conditions, the result can be a set of overlapping peptides that is useful for both protein identification and protein characterization. In this way, we partly achieve, in one single sample and one single experiment, the same effect as described in Section 3.1.3 by using one unspecific protease rather than two more specific ones in two parallel samples.

3.1.6 *Chemical cleavage*

Chemical cleavage is rarely used in proteomics research. The most common method of chemical cleavage employs cyanogen bromide, which cleaves the peptide backbone C-terminal to a methionine residue. The methionine is at the same time converted to homoserine, which then becomes the C-terminal amino acid in the peptide. Under acidic conditions, the hydroxyl group of the homoserine side chain may react with the carboxylic acid in a cyclization process and form homoserine lactone. Since methionine is a relatively rare amino acid, the average size of the generated peptide is much larger than for most proteases. Due to the aggressive chemicals and acids used in cyanogen-bromide-induced cleavage, more protective care is needed than when using proteases.

3.1.7 *In-gel digestion*

In-gel digestion is the most common approach to PMF. Usually a complex mixture of proteins is first separated by 2D gel electrophoresis, and the gel stained to reveal the positions of the proteins. Many, but not all, gel staining methods are compatible with subsequent mass spectrometry analysis. The spots containing the proteins of interest are excised from the gel, and, if needed, the stain is washed away. The gel piece is then dehydrated. When a small volume of buffer with protease is added, the liquid and the protease are both absorbed into the gel piece, and thereby gain access to the protein molecules. The resulting peptides are extracted from the gel by washing and dehydration.

3.2 *In silico* digestion

The *in silico* digestion of a protein sequence is performed by scanning the sequence for cleavage sites, and calculating the masses of the resulting peptides. However, one should bear in mind how the experimental data are produced. There can be:

1. missed cleavages;
2. naturally occurring modifications in some positions of the protein;
3. chemical modifications intentionally introduced;
4. unintentional chemical modifications introduced by the sample handling;
5. unsuspected cleavages during the maturation/life cycle of the protein, creating new N- or C-termini on the protein;
6. unexpected cleavages occurring during cell lysis and/or compartment mixing;
7. unexpected cleavages occurring during the experimental proteolytic treatment.

In general, the aim of the comparison between the experimental data and the theoretical data is to maximize the number of matches. Points 1 to 3 are relatively easy to take into consideration when performing the *in silico* digestion. Often (but not always) we may have some indications or suspicion of modifications of the protein, for example that the protein is a phosphoprotein. Certain unintentional modifications (point 4) happen so often that they are also normally taken into consideration. Examples are oxidation of methionine and the formation of pyroglutamic acid from N-terminal glutamine in the peptides. However, missed cleavages and different modifications (points 1–4) greatly increase the number of theoretical peptides, thus also increasing the chances of random matches with the experimental data. If the number of cleavage sites in a sequence is n and the number of missed cleavages allowed in a peptide is k, then the number of theoretical peptides is given by

$$\frac{1}{2}(2n+2nk-k^2+k+2) \tag{3.1}$$

However, some of them may have identical masses (to the chosen accuracy).

The usual procedure for taking modifications into account is to specify a set of possible modifications, and the scanning procedure recognizes for each modification its sites on the sequence. Therefore, if n modification sites are observed for a theoretical peptide, then the number of calculated masses for the peptide is given by

$$\sum_{i=0}^{n}\binom{n}{i} = 2^n$$

If there is a limit h on the number of modifications per peptide, the number of masses will be

$$\sum_{i=0}^{h}\binom{n}{i}$$

For identification purposes we therefore only include in the *in silico* digestion the modifications that have a reasonable certainty of being present.

Points 5, 6, and 7 can only be considered after the candidate protein has been found, and would be a part of the further characterization of the protein.

Exercises

3.1 The introduced language for cleavage specificity is rather simple, and cannot describe the complete specification for several of the proteases. It can for example not describe that, for trypsin, `P` does not block the cleavage at `WKP`. Can you suggest how to extend the language to be able to deal with such situations?

3.2 How many peptides can be expected when using chymotrypsin for the digestion of a protein of length 400? Use the occurrence probabilities in Table 1.1, and assume that the positions of the amino acids are random.

3.3 Explain how the combined use of trypsin and chymotrypsin can (ideally) constrain the area in which there might be a modification, when performing characterization on a known protein. Illustrate by a sequence where a part is `...RCVLYKSTH WIVACRTSFPK...`, and there is a modification on the second `C` residue.

3.4 Prove the result in Equation 3.1.

Bibliographic notes

Protease/peptidase database
MEROPS http://merops.sanger.ac.uk/, Rawlings *et al.* (2004)

Cleavage specificities of proteases
Barrett *et al.* (1998); Earnshaw *et al.* (1999); Keil (1992);
Thornberry *et al.* (1997)
PeptideCutter http://us.expasy.org/tools/peptidecutter/

Prediction of Protease Specificity
PoPS Boyd *et al.* (2005), http://pops.csse.monash.edu.au/home.html

Random cleavage
Wu *et al.* (2003)

***In silico* digestion and missed and unexpected cleavages**
Cagney *et al.* (2003); Olsen *et al.* (2004); Thiede *et al.* (2000)

Programs for *in silico* digestion and modeling
MS-Digest http://prospector.ucsf.edu/ucsfhtml4.0/msdigest.htm
PeptideCutter http://us.expasy.org/tools/peptidecutter/
PeptideMass http://au.expasy.org/tools/peptide-mass.html

4 Peptide separation – HPLC

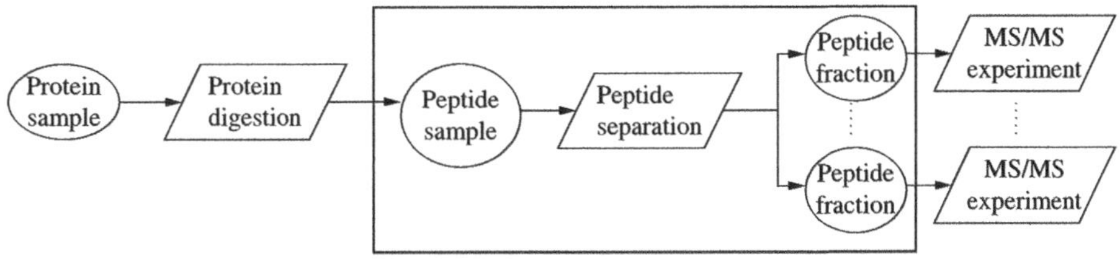

The term *chromatography* includes a family of techniques that are used to separate a mixture into its individual components. The name comes from early work on the separation of differently colored components. The separation is performed by passing the mixture through an immobilized porous substance with which the individual components will interact to different degrees. Ideally, it takes each component a characteristic time to pass through such a system, called the *retention time*. Chromatography is a large scientific field in itself, with several applications. In proteomics it is mainly used for separation, especially of peptides, but also occasionally of proteins.

Simply put, the separation is performed by injecting the sample (usually a mixture of many different components, for example peptides) to be separated into a *mobile phase*. The mobile phase (containing the sample) moves through a *stationary phase* (the immobilized porous substance), and at any time a component is interacting either with the stationary phase, or with the (moving) mobile phase. The more a component interacts with the stationary phase relative to the mobile phase, the more time it takes for migration. Individual components have different retention times because of differences in their interactions with these phases.

Column chromatography is the most commonly used type of chromatography in proteomics. Other types are paper and thin-layer chromatography. Column chromatography consists of a column made of glass, metal, or a synthetic material, containing the stationary phase, through which the mobile phase is moving.

Two methodologies of column chromatography are in use for organic components, *gas chromatography* (GC) and *liquid chromatography* (LC). Gas chromatography uses gas for the mobile phase, and almost always a liquid, gum, or elastomer as the stationary phase. The instruments for gas chromatography have a higher efficiency of separation, and are

Computational Methods for Mass Spectrometry Proteomics I. Eidhammer, K. Flikka, L. Martens and S.-O. Mikalsen
© 2007 John Wiley & Sons, Ltd

simpler to use than instruments for liquid chromatography. However, gas chromatography is unsuitable for non-volatile or thermally labile components such as peptides and proteins. Therefore, liquid chromatography dominates the field of proteomics.

4.1 High-pressure liquid chromatography, HPLC

Liquid chromatography uses a liquid as the mobile phase and a porous solid as the stationary phase. The surface of the solid is usually modified to achieve certain desired properties. The stationary phase is sometimes called the *packing*, and the mobile phase the *solvent* or the *eluent*. When the mobile phase has passed through the column, and carries the separated components out of the column, it is often named the *eluate*, and the components are said to *elute* (from the ancient Greek word meaning 'to wash out') from the column. The components to be separated are also called *analytes* or *solutes*.

Originally, the mobile phase flowed through the stationary phase under the force of gravity alone. However, to get a more reasonable flow rate, high-pressure pumps are now used, hence the name *HPLC*. HPLC also increases the separation efficiency. (Note that HPLC is also often interpreted as an abbreviation of High-Performance LC.) In the literature, HPLC and LC are often used interchangeably. Columns used in LC are typically 10–25 cm long. Columns with internal diameters of one to a few millimeter are called *micro LC columns*, and are used for so-called preparative purposes. For *capillary LC columns* the diameter is less than 300 μm, and these are used for the separation of small molecules (peptides). *Nano LC columns* have an inner diameter of 50–100 μm, and are ideal for separation coupled to MS/MS, where the eluate from the column is infused directly into an MS/MS instrument (as explained in Section 8.5).

Figure 4.1 shows the principle of HPLC. The components turn up as *bands* in the column. Each of the components to be separated consists of a large number of molecules. As these molecules are interacting with the stationary phase and the mobile phase on an individual basis, the molecules will not move through the column at an exactly identical rate. The individual rates will rather vary around an average value. Thus, the components will form a concentrated *peak* rather than one sharp, thin line as shown in Figure 4.2. The aim thus is to separate different components into distinct peaks.

A *detector* may be placed at the outlet of the column. The detector registers the components as they pass by. The detector is commonly based on the ability of the components to absorb light of certain wavelengths, usually in the ultraviolet region. The most common wavelengths used in proteomics are 280 nm (absorbed by aromatic amino acids) or 210–220 nm (absorbed by the peptide bond). A computer is typically connected to the detector, and a diagram is constructed for the presentation of the eluted components. This diagram is called a *chromatogram*, and has (retention) time along the horizontal axis and the intensities of the measured components along the vertical axis. If the HPLC infuses the eluate directly into a mass spectrometer, the ion current (amount of ions) can be measured instead of using an optical detector.

A good HPLC separation depends on finding a balance in the affinity of the components for the stationary phase, and the solubility of the components in the mobile phase such that the different components migrate at different rates. Also, the band in which each component elutes should be as narrow as possible, to prevent different bands (and therefore

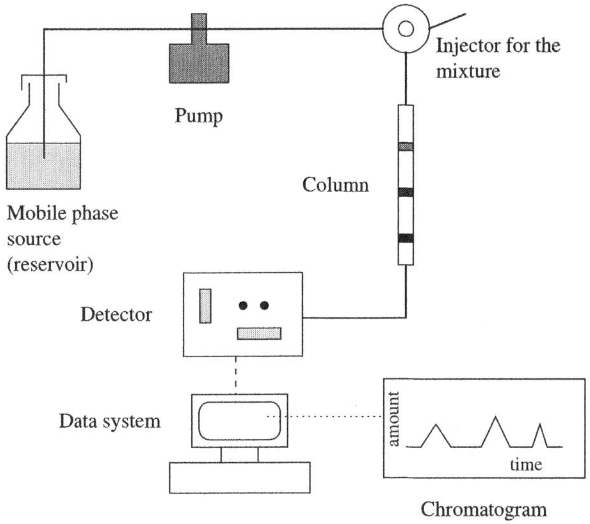

Figure 4.1 Illustration of the principle of LC. The mixture contains three components, and three peaks are correspondingly seen in the chromatogram when all the components have been detected

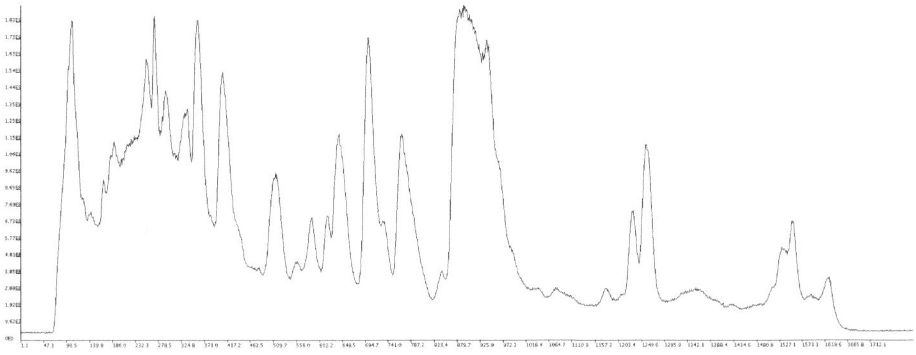

Figure 4.2 An example of a chromatogram. Retention time is on the horizontal axis, and total intensity on the vertical axis

peaks) overlapping. The challenge of chromatography is therefore to achieve different rates of migration for the different components, and narrow bands. The interpretation of a chromatogram can be difficult due to noise and *baseline drift*. The noise is often quantified in a *signal-to-noise* (S/N) ratio, which is a measure of how well a real signal (from a component) is differentiated from the background signals. The noise may be due to electronic noise in the system, or due to contaminating components in the sample or mobile phase. The *baseline* is the part of the chromatogram recorded by the detector when only the mobile phase emerges from the column. As seen in Figure 4.2, this baseline can vary (drift) over time.

4.2 Stationary phases and separation modes

The interaction of the stationary phase with the components, and their solubility in the mobile phase, determine the degree of migration and separation of the components in a sample. This is further dependent on the properties of the stationary phase, the mobile phase, and the components of the sample. Different types of stationary phases and separation modes are used for HPLC. HPLC is often classified according to the principle of separation applied in the columns: hydrophobicity, charge, affinity to specific functional groups, or size of the components. In the context of proteomics, nearly all work with HPLC is based on the two former principles, and especially if the HPLC is coupled directly to the mass spectrometer. The actual separation principles based on either hydrophobicity or charge are further subdivided into several methods.

The two most common HPLC methods in proteomics, *reverse phase chromatography* (separating on hydrophobicity) and *strong cation exchange chromatography* (separating on charge), will be described in some detail. The other separation principles will only be briefly mentioned. It should be noted that the same HPLC instrument can be used for all mentioned types of chromatography by exchanging the columns and mobile phases.

4.2.1 Reverse phase (RP) chromatography

RP-HPLC is a versatile method. It gives good resolution and separation, and has high reproducibility and good recoveries. The eluate may be infused directly into a mass spectrometer by electrospray ionization (see Section 5.2.2), or fractions of the eluate may be collected off-line for subsequent analysis. The stationary phase is usually made from surface-modified silica beads, and these modifications cause the adsorbance of the sample components. The modifications are therefore sometimes called *sorbents*. The most common modifications are based on carbon chains (similar to the carbon chains in fatty acids) of different lengths, for example 4 (C_4), 8 (C_8), or 18 (C_{18}) carbon atoms. Other modifications are sometimes also used, for example phenyl groups (carbon rings of 6 atoms). The longer the carbon chain, the stronger the hydrophobic interaction that can occur between the stationary phase and the peptides. C_{18} is most frequently used in RP chromatography of peptides.

Most of the separation methods in HPLC, RP chromatography or otherwise, use a mixture of two solutions as mobile phase. By convention, the solutions are called A and B. In RP chromatography solution A is usually water with a small amount of (organic) acid (often 0.1 % formic acid if the sample is directly infused into the mass spectrometer). The sample is injected into the column in solution A, and as the liquid is forced through the column, the peptides interact with and attach to the carbon chains, and are retained on the stationary phase. Solution B is mainly an organic solvent (for instance, 90 % acetonitrile, 0.1 % formic acid, and water). After the peptides have attached to the stationary phase, solution B is gradually mixed into solution A, thereby slowly increasing the percentage of organic solvent in the mobile phase. As the percentage of B is increasing, the less hydrophobic peptides detach first, and start moving together with the mobile phase. The more hydrophobic peptides detach at a higher percentage of B. This change in solvent strength over time is called a *gradient*. When the gradient reaches 60 % organic solvent, all peptides are normally eluted from the stationary phase.

Hydrophobic interaction (HI) and *normal phase (NP) chromatography* are two variants of LC that use the principle of hydrophobicity for separation. HI chromatography may use the same types of stationary phase as RP chromatography, but the sample is injected in a high salt solution (for example, $(NH_4)_2SO_4$), and the analytes are eluted by decreasing salt concentrations. This type of chromatography is mainly used for proteins and rarely for peptides. NP separates by using a polar stationary phase and a non-aqueous organic solvent as the mobile phase. Hydrophobic components elute most quickly in this approach.

4.2.2 Strong cation exchange (SCX) chromatography

SCX is a form of *ion exchange chromatography*. Ion exchange chromatography is based on the principle that opposite charges will attract each other. As explained in Section 1.3.3, the proteins/peptides in solution possess a net charge that depends on the pH of the solution and the isoelectric point of the protein/peptide. A protein/peptide with a pI above the pH of the solution will possess a net positive charge, while one with a pI below the pH will possess a net negative charge. When the pI is equal to the solution's pH the protein/peptide will be neutral.

For strong cation exchange the stationary phase is often a surface modified by sulfonic acid groups. Sulfonic acid becomes negatively charged at a pH above 2–3. The peptide sample is injected into the column at a low pH (often 3–3.5) solution. Thus, the peptides will be positively charged, and they will interact with the negative charges of the stationary phase. Positively charged ions are called *cations*. The more positive charges the peptide contains, the stronger it will interact with the stationary phase. Uncharged peptides or other components that do not bind to the stationary phase are washed out of the column.

Similar to what is described above for RP chromatography, another solution is gradually mixed into the mobile phase to perform the separation of the peptides. For strong cation exchange it is common to increase the *ion strength* of the mobile phase by using a salt in solution B, while keeping the pH constant. The salt (which contains both positively and negatively charged ions) will compete with the negative charges (on the stationary phase) and positive charges (on the peptides). The peptides with the weakest binding to the stationary phase will detach first and subsequently start moving with the mobile phase. The peptides with a stronger binding will only detach at higher salt concentrations. An alternative method of elution is to gradually change the pH of the mobile phase. This will cause the charge of the peptides to change, and at a certain pH the increasing number of negative charges and decreasing number of positive charges will release the peptide from the stationary phase.

In contrast to the eluate from RP chromatography, SCX is normally not infused directly into the mass spectrometer. This is because the mobile phase in SCX contains salts or other ingredients that will disturb the analysis of the peptides.

SCX is one of four main types of ion exchange chromatography. The others are *weak cation exchange* (characterized by a pH above 7–8 before the surface modifications of the stationary phase become negatively charged), *strong anion exchange* (the surface modifications become positively charged at a pH below 10–11), and *weak anion exchange* (surface modifications become positively charged at a pH below 7–8). Thus strong and weak in the context of ion exchange indicate whether the surface modifications are ionized in a wide (strong ion exchange) or narrow (weak ion exchange) pH range.

4.2.3 *Other types of chromatography for proteomics*

Two other types of chromatography should be mentioned.

Affinity chromatography is a type of chromatography that separates analytes by their interaction with specific functional groups that have specific affinity to them. The components of interest are retained while non-interacting components pass through the column. The retained analytes are afterwards eluted by changing the mobile phase conditions.

Size exclusion chromatography is a type of chromatography that separates analytes based on size. The stationary phase here consists of porous beads, which the smaller components are allowed to enter. This will lead to much larger path lengths for smaller components (they will criss-cross the porous beads rather than eluting straight past them) and will therefore elute larger analytes first. Size exclusion chromatography is not appropriate for peptide separation (due to the low resolution). In contrast to the other separation modes mentioned above where the mobile phase usually is a combination of two solutions, the mobile phase for size exclusion chromatography consists of one solution alone.

4.2.4 *Tandem HPLC*

To increase the separation power, two or more different separation techniques are sometimes used in series. This is explained in Section 14.3.

4.3 Component migration and retention time

The speed of a component's migration through the column depends on the chemistry of the stationary and mobile phase, the temperature, the dimension of the column, and the speed of the mobile phase, in addition to the properties of the component itself. This is reflected in the following definitions:

(Total) retention time, t_R, is the time between sample injection and the detection of the band (peak) maximum.

Dead time, t_0, is the time a component uses for migration from the injection until the appearance of the band maximum at the detector, assuming that the component did not have any interaction with the stationary phase. This is the time the component is in the mobile phase, and is equal for all components. In other words, dead time is the time it takes for a non-interacting component to elute through the column from the point of injection.

Capacity factor (retention factor), k', is defined as

$$k' = \frac{t_R - t_0}{t_0}$$

Thus, the capacity factor is the ratio of the time the component interacts with the stationary phase relative to the time it is in the mobile phase. It is a measure of

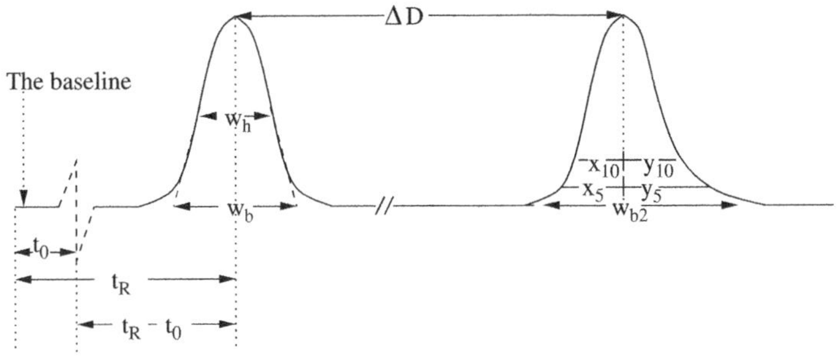

Figure 4.3 Illustration of different terms characterizing chromatography. t_i are different times, w_b is the baseline width of the peak, and w_h the half-height width of the peak. x_i and y_i are the distance from the middle line to the front and back of the peak at i % of the maximum peak height

the retention of a component in the column, and is generally different for different components.

Relative retention (separator factor) is the ratio between two capacity factors. It has the advantage that it only depends on the temperature and the chemistry of the two phases. It describes the discriminating ability of a system, and the denominator is often a component used as a reference (standard).

The retention is illustrated in Figure 4.3. The retention time t_R is used as the x-axis for the chromatograms, the other terms are used for characterizing and comparing columns.

4.4 The shape of the peaks

Ideal peaks are narrow and symmetrical, since such peaks imply better separation between different components, and it will be possible to separate more components in one run (one chromatogram).

4.4.1 *The width*

Generally the peak width increases with the retention time, and is commonly measured in one of two ways:

Baseline width, w_b, is the width at the *baseline*.

Half-height width, w_h, is the width at 50 % of the maximum peak height.

Both widths are illustrated in Figure 4.3.

Plate number A good separation requires narrow peaks in the chromatogram. However, since peak width increases with time (when all conditions remain equal), peak width is not a sensible term for describing how good a column is at producing narrow sample bands. Instead, a ratio of retention time and peak width is used, called the *plate* *number, N*. The plate number is almost constant for all peaks in a specific chromatogram, and is calculated from the baseline width as $N = 16(t_R/w_b)^2$, or from the half-height width as $N = 5.54(t_R/w_h)^2$. A column with a large plate number gives narrow bands. The constants used (16 and 5.54) are calculated from a fitted Gaussian distribution curve.

4.4.2 Asymmetry

Real peaks are seldom entirely symmetrical, and this creates several problems. Peak size calculation for example will be more difficult. The most general type of deviation from symmetry is *tailing*, where there is a 'tail' at the back, as shown for the second peak in Figure 4.3. Two measures are used for measuring the asymmetry of a peak:

Asymmetry factor, ***As***, is measured at 10 % of max peak height. Using the symbols shown in Figure 4.3, it is defined as $As = y_{10}/x_{10}$.

Tailing factor, ***TF***, is measured at 5 % of max peak height. Using the symbols shown in Figure 4.3, it is defined as $TF = (x_5 + y_5)/2x_5$.

The values calculated for the two measures are similar for the same peak, but not identical. Complete symmetry is reported by a value of 1.0, and the value should not exceed 1.5 for a reasonable chromatogram.

4.4.3 Resolution

Resolution, R, is a measure of how well separated two adjacent peaks are. In chromatography it is, by convention, defined as the ratio of the distance between the two peak maxima to the average baseline width of the two peaks. Again referring to Figure 4.3, $R = 2\Delta D/(w_b + w_{b2})$. A larger resolution value means a better separation. If the peaks are approximated by symmetric triangles, an R greater than or equal to one means that the components are completely separated. By definition, a resolution of more than 1.5 is considered *baseline* resolution. This is the value that will result in less than 1 % of mutual overlap between two peaks of equal size and ideal shape.

4.5 Chromatography used for protein identification

Chromatography in proteomics is mainly used for the separation of peptides, but it may also aid in the identification and characterization of the peptides. Reproducibility is tantamount for this purpose. HPLC has fairly good reproducibility, meaning that when all conditions are held constant, the retention times for a specific analyte remains fairly constant between different experiments. Therefore, with a theoretical model for calculating the retention time from a peptide sequence, a measured retention time can be compared to calculated retention times for possible theoretical peptides. Models have been developed for RP chromatography, while this is not the case for SCX chromatography.

4.5.1 *Theoretical calculation of the retention time for RP chromatography*

The retention time for RP chromatography depends mainly on the hydrophobicity of the peptides. The exact dependence is, however, not fully understood, and models have mainly been developed based on observing what effect the different amino acids have on the retention time. A *retention coefficient* is therefore determined for each amino acid. The retention time for a peptide is dependent on the sum of the retention coefficients of all the residues, plus coefficients for the end groups, plus a constant representing the time needed for the first part of the mobile phase to reach the detector (gradient delay time or dead time).

Retention coefficients

The retention coefficient for an amino acid should be a value for the relative impact the amino acid has on the total retention time for a peptide. Several sets of retention coefficients exist, and they have mainly been determined by one of two approaches:

- find the retention time for a set of peptides (training set), and calculate the contribution of each amino acid, either by regression analysis or by neural networks;
- construct synthetic peptides, where certain positions are substituted by each of the amino acids, and measure the retention times for each of the substituted amino acids.

Retention coefficients are primarily determined for a pH of 2.1 in the mobile phase.

SSRC – a model for calculating the retention time

The SSRC (Sequence Specific Retention Calculator) probably represents one of the best models for RP chromatography, and it is available on the Internet. We will therefore use this model as an example. The model is developed based on tryptic peptides, and may not be fully transferable to peptide samples generated by other proteases.

When developing a model, it will not be possible to take into account all variables that can affect the results. Since the aim is to develop a model that can be used in an experimental setting, it is important to choose one or a few conditions that are commonly used.

Assuming that the sample is constant in this context, the variables include the column size and choice of stationary phase, the acid used in solutions A and B (the acid is often called an *ion-pairing agent*, with different acids used depending on the experimental setup), the organic solvent used in solution B, the gradient slope, flow rate, temperature, on-line or off-line analysis of the eluate from the column, etc. Some of the variables mainly affect the time for the components to pass through the column, but not their relative speed and sequence. Such variables are the delay time, slope of the gradient, and flow rate (at least when they are varied within reasonable ranges). Other variables, like the choice of stationary phase, the ion-pairing agent, and the organic solvent, may change the relative speed of the components, and thereby the sequence of elution of the different components.

SSRC is based on the use of a C_{18} RP column with 3–4 μl/min flow rate, off-line sampling of fractions, use of trifluoroacetic acid (TFA, a strong organic acid) as ion-pairing agent, and acetonitrile as organic solvent.

In SSRC the user specifies the gradient delay time, the slope of the gradient, and 100 or 300 Å pore size for the stationary phase. SSRC assumes either that other experimental conditions are constant, or that they have no influence on the retention time.

SSRC then uses the properties of the peptide to calculate a hydrophobicity H of the peptide, an index specifically adapted to RP chromatography. The retention time is then calculated as $a+bH$, where a is the gradient delay time, and b is related to the gradient slope. These constants must be determined by calibration of the equipment used.

There are nine different peptide properties used in version 3 of SSRC. The first six are used to calculate a basal hydrophobicity index, and the last three are used for corrections to the index.

1. Retention coefficients. For each amino acid there are five different retention coefficients, depending on where on the peptide sequence they are located: position 1 (N-terminal), position 2, positions $3...n-2$, position $n-1$, and position n, where n is the number of amino acids in the peptide. The individual amino acid coefficients are added to get the total contribution from this property. Note that because all individual coefficients are summed, and most of the coefficients are positive, longer peptides will generally have a higher H than shorter peptides. This contrasts with a hydrophobicity calculation based on GRAVY, in which the coefficients can be negative as well. Furthermore, different retention coefficients are found for 300 and 100 Å pore sizes.
2. Nearest-neighbor effect. Amino acids adjacent to H, R, or K inside the peptide will have a different effect on the hydrophobicity.
3. Clusters of hydrophobic amino acids. The clusters will decrease the hydrophobicity of the peptide.
4. Specific role of proline. Proline is the most structurally rigid of the amino acids, and the hydrophobicity is decreased for successive occurrences of proline.
5. Influence of isoelectric point of the peptide. Lower pI favors retention of hydrophobic peptides and decreases retention for hydrophilic ones. For high pI the situation is reversed; hydrophilic peptides are preferentially retarded.
6. Separate set of retention coefficients for short peptides (< 9 residues).
7. Correction for peptide length. The hydrophobicity is decreased for peptides of less than 8 and increased for those larger than 20 residues.
8. Correction for overall peptide hydrophobicity. The hydrophobicity is decreased for peptides with a calculated H larger than 20, and is more decreased for even higher calculated H values.
9. Influence of peptide propensity to form helical structures. The H value is increased if the peptide contains small stretches of hydrophobic amino acids, separated by any other amino acid.

An example of the use of the SSRC is given in Exercise 4.3.

4.6 Chromatography used for quantification

The size of the peaks in a chromatogram is related to the injected amount of the components. This can be used for quantitative analysis. If standards of known amount have previously been run, the size of a component peak can be related to the peak size of a standard, and the amount of the component determined. Two measurements are used for the peak size:

Peak area is the area under the peak, and is the most fundamental measure. For well-designed instruments and reproducible conditions, the area is directly proportional to the mass of injected component.

Peak height is easier to measure, and when the width and asymmetry of a peak are constant from run to run, the height of a peak is proportional to its area.

Exercises

4.1 In ion exchange chromatography, either the salt concentration *or* the pH is varied in order to elute peptides from the stationary phase. Can you explain what would happen if both were altered at the same time?

4.2 Figure 4.4 contains a part of a chromatogram. Find approximate values for the necessary parameters, and calculate plate numbers (using both formulas), asymmetry factors, and tailing factors for the two highest peaks.

4.3 (a) Calculate the retention time for the peptide `GWFLLWFARLGK` using the SSRC procedure:

1 The retention coefficients (300 Å pore size) are shown in Table 4.1.

2 If `R` is followed by `L` the hydrophobicity is decreased by 0.3.

3 A stretch of six very hydrophobic residues (`WFLLWF`) decreases the hydrophobicity by 1.8.

4 Is not considered since the peptide does not contain any proline.

5 Ignore.

6 Is not considered since the peptide is longer than nine amino acids.

7 No correction.

8 Correct the hydrophobicity H according to the following table:

H	Corrected H
$H < 20$	H
$20 \leq H < 30$	$H - 0.27(H - 18)$
$30 \leq H < 40$	$H - 0.33(H - 18)$
$40 \leq H < 50$	$H - 0.38(H - 18)$
$50 \leq H$	$H - 0.45(H - 18)$

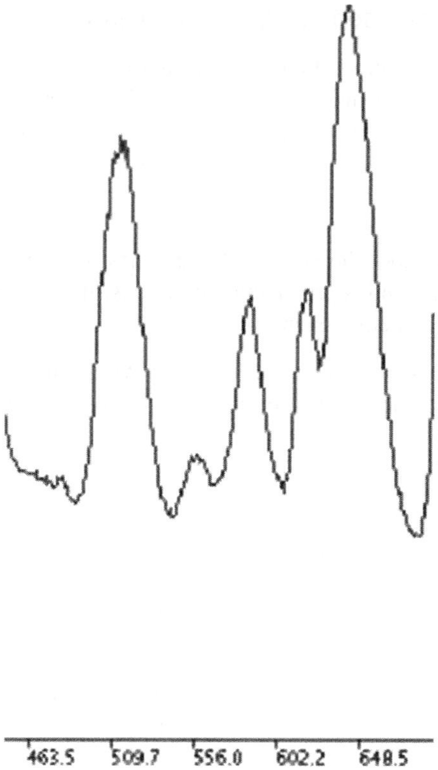

Figure 4.4 A part of a chromatogram

Table 4.1 Retention coefficients for some amino acids

Residue	Pos. 3 . . . $n-2$	Pos. 1	Pos. 2	Pos. n-1	Pos. n
A	1.1	0.35	0.5	−0.1	1.1
F	10.9	7.5	9.5	10.3	10.9
G	−0.35	0.2	0.15	−0.7	−0.35
K	−2.05	−0.6	−1.5	−1.45	−1.9
L	9.3	5.55	7.4	9.3	9.3
R	−1.4	0.5	−1.1	−1.3	−1.1
W	12.25	11.1	11.8	12.1	12.25

9 Is not considered for the present peptide

Use 5 (minutes) for a and 0.5 for b.

(b) Also run the on-line version of SSRC (see bibliographic notes) for the same peptide, using the same parameters.

Bibliographic notes

Retention coefficients

Guo *et al.* (1986b); Meek (1980); Sasagawa *et al.* (1982)
Krokhin *et al.* (2004); Petritis *et al.* (2003)

Retention time prediction

SSRCalculator	Krokhin (2006); Krokhin *et al.* (2004)
	http://hs2.proteome.ca/SSRCalc/SSRCalc.html
PeptideSort	http://www.hku.hk/bruhk/gcgdoc/peptidesort.html
ExPASy Tools	http://expasy.org/tools/pscale/HPLC2.1.html
Use of ANN	Petritis *et al.* (2003)

Retention time and use in identification

Guo *et al.* (1986a); Krokhin *et al.* (2004); Palmblad *et al.* (2002)
Chen *et al.* (2005); Wang *et al.* (2005)

5 Fundamentals of mass spectrometry

Mass spectrometry (MS) is used for measuring the mass, or more strictly speaking, the mass-to-charge ratio (m/z), of the components in a sample. The instruments are called *mass spectrometers*. This chapter focuses on two main topics of mass spectrometry:

- The common principles. There are many classes of instruments, some of which will be discussed in more detail in the subsequent chapters, yet all of these operate on the same underlying basis.
- The isotopic distribution of elements in peptides and proteins, and how this knowledge can be used to describe the performance of the instruments.

5.1 The principle of mass spectrometry

Very schematically, a mass spectrometer consists of three main parts: the *ionization source*, the *mass analyzer*, and the *detector*. All mass spectrometers use electric or electromagnetic fields to handle the components of the sample. The components under analysis therefore have to be ionized before their masses can be measured, and this ionization is performed in the ionization source. As mentioned in Chapter 1, when dealing with peptides and proteins, ionization is commonly achieved by the addition of protons to the molecules. This addition also increases the mass of the molecule by the nominal mass of 1 Da per charge (per proton).

The sample is then transferred to the mass analyzer that separates the components of the sample according to the mass-to-charge ratio (m/z) of the ions. After separation the components hit the detector, and a *mass spectrum* is constructed by a connected computer. A mass spectrum is a diagram, with m/z along the horizontal axis and the intensity of the signal for each component along the vertical axis. Since the analyzers operates on m/z rather than on the mass directly, the charge of a component must be known before the mass can be determined. m/z is usually reported as a dimensionless number, but units like the thomson (Th), u, or even Da are sometimes used. Note that the use of u or Da is essentially incorrect, whereas thomson is correct, but not recognized by standardization bodies.

Computational Methods for Mass Spectrometry Proteomics I. Eidhammer, K. Flikka, L. Martens and S.-O. Mikalsen
© 2007 John Wiley & Sons, Ltd

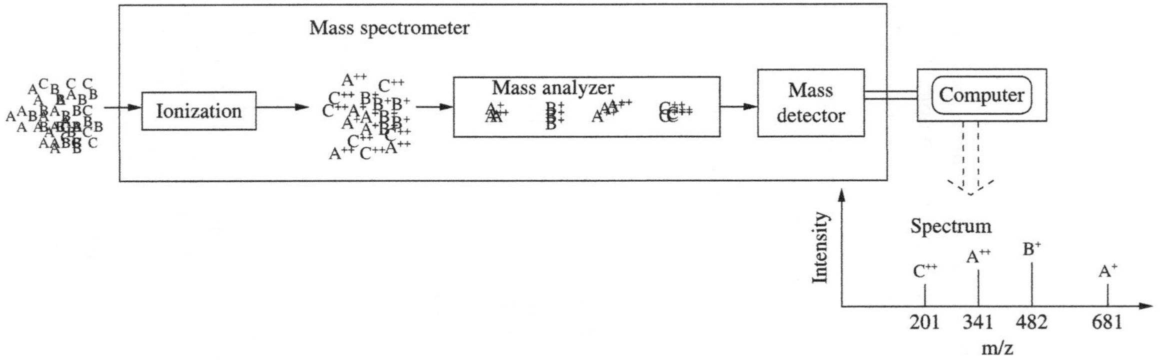

Figure 5.1 Illustration of the principle of mass spectrometry. In the ionization source, the sample components acquire their charges. In the mass analyzer the components are separated according to their m/z values before they hit the detector. A computer connected to the instrument constructs the mass spectrum. An example in the text describes in more detail the measurement of peptides A, B, and C

Example Suppose we have a peptide with mass 2000.0 Da, and that the ionization yields peptide ions of charge +1, +2, *and* +3, by the attraction of one, two, or three protons, respectively. The peptide ions will then be detected at

ions with charge +1: $m/z = (2000 + 1)/1 = 2001$

ions with charge +2: $m/z = (2000 + 2)/2 = 1001$

ions with charge +3: $m/z = (2000 + 3)/3 = 666.7$

△

Example Figure 5.1 illustrates the main principle of mass spectrometry. Suppose we have a sample of three peptides (A, B, C), with nominal masses A: 680, B: 481, C: 400. Assume further that some of the A peptides become singly charged, and some doubly charged, while all B peptides are singly charged, and all C peptides doubly charged. The m/z values for the ions thus become A: (680+1)/1=681 and (680+2)/2=341, B: 482, C: 201, as shown in the spectrum in Figure 5.1.

△

5.2 Ionization sources

Schematically, as described in Chapter 1, there are two main classes of mass spectrometers for proteomics: those that perform single mass spectrometry and those that perform tandem mass spectrometry (MS/MS). The former measure the m/z ratio of intact peptides. It is an advantage if the peptides mainly have a single (positive) charge. In tandem mass spectrometry the peptides are intentionally fragmented (mainly due to peptide backbone breaking, resulting in two fragments per fragmentation event) to measure the masses of fragments. Ideally, one wants to detect the mass of both fragments, which can only be achieved if both are ionized. Thus, in tandem MS it is an advantage if the peptides carry several charges (often +2) as this increases the chance that each of the fragments will be ionized. The two classes of instruments therefore generally use different ionization sources, but exceptions exist.

In the ionization source, the sample components are brought into the gas phase while they acquire their charge. There are many methods to achieve ionization, with different advantages and drawbacks. Some desired properties for the ionization process in proteomics are as follows:

- All the components in the sample should be ionized in a detectable amount.
- The ionized amount should be proportional to the sample component amounts.
- There should be no *fragmentation* of the components unless we want to analyze the fragments. Fragmentation here means breaking components into smaller parts which may or may not be ionized.
- There should be no unwanted *adduct ions*. An adduct ion is an ion comprised of a component and one or more additional atoms. Typical unwanted adduct ions in

proteomic experiments would be the potassium or sodium adducts of peptides. Such ions can easily arise from samples that contain even low amounts of salts.

- There should be no ions from other molecules (contaminants).

No ionization source completely satisfies all these desires, however. The actual source to use is therefore usually chosen based on the type of analysis that is to be performed (obviously often limited by the available instrument(s)). The *protonated* molecular ions are often called *quasi-molecular* ions, and are commonly denoted by $[M+nH]^{n+}$, where n is the number of protons added. Sources which cause only limited fragmentation are called *soft ionization* sources, as opposed to *hard ionization* sources, in which components typically fragment upon ionization. Soft ionization sources are used for peptides and proteins, and if fragmentation is desired afterwards (as in MS/MS) other (post-source) methods are used to achieve fragmentation. Below, we will briefly describe the two most commonly used ionization sources for proteomics, MALDI and ESI, which both perform soft ionization, and also mention some others.

5.2.1 MALDI – Matrix-Assisted Laser Desorption Ionization

MALDI is the dominating ionization source for (single) mass spectrometry, but is also sometimes used in instruments that can perform MS/MS. The name provides a description of the ionization process. The matrix consists of small organic molecules that absorb light at specific wavelengths, usually in the UV area. There are hundreds of organic compounds that have been used as the matrix for MALDI, but in proteomics only a handful or so are commonly used. The matrix is dissolved in an organic solvent under acidic conditions, and mixed with the sample. A small drop (microliter volume) is spotted on a sample plate, and the organic solvent is allowed to evaporate. During the evaporation, the matrix forms small crystals, and the sample components are incorporated into these crystals. This is called *crystallization*. A laser then fires light (energy) onto the plate in very short pulses (a few nanoseconds), which is absorbed by the matrix. It is clear that optimal results can be achieved if the matrix has a strong absorption peak at the wavelength of the laser light. The matrix essentially fulfills two roles at this stage: it captures the laser light and becomes ionized while at the same time protecting the analyte molecules (peptides or proteins) from the disruptive energy transferred by the laser light.

The absorbed energy causes the matrix molecules and sample molecules to eject from the plate. The peptides are able to receive protons from the ionized matrix molecules, and they become ionized in the gas phase. Most of the ionized peptides carry only one adduct proton. Thus, the peptide ions generally have the charge +1. Under the influence of an electric field, the ions are transported to the mass analyzer. In most MALDI instruments, the generation of ions occurs under vacuum (or rather, very low pressure). The principle is illustrated in Figure 5.3 below.

5.2.2 ESI – Electrospray Ionization

ESI is primarily used for MS/MS analysis. The peptides are brought into the ionization source by a liquid flow. Often the liquid is the eluate from an HPLC instrument. The liquid is sprayed from a heated needle or capillary into a strong electromagnetic field, resulting in a mist of small droplets with a charged surface. As the solvent in the droplets evaporates, the droplets get smaller and smaller, increasing the electric field on the surfaces. When the electric field becomes strong enough, charged peptides desorb from the surfaces. Under these conditions, most of the ionized peptides will carry two or more protons. Under the influence of the electric field, the generated ions are transported to the mass analyzer. In most ESI instruments, the generation of the ions occurs under atmospheric pressure, while the mass analyzer operates under low pressure. Therefore, the ions must be transported from atmospheric pressure to vacuum. The principle for ESI is shown in Figure 5.2.

5.2.3 Other ionization sources

Several other types of ionization sources are used in mass spectrometry, although they are more rarely employed for proteomics experiments. Photoionization uses UV light from a krypton lamp. In chemical ionization, the solvent is able to ionize the sample. Fast atom bombardment (FAB) ionization uses argon or xenon atoms that are shot at the liquid surface of the sample.

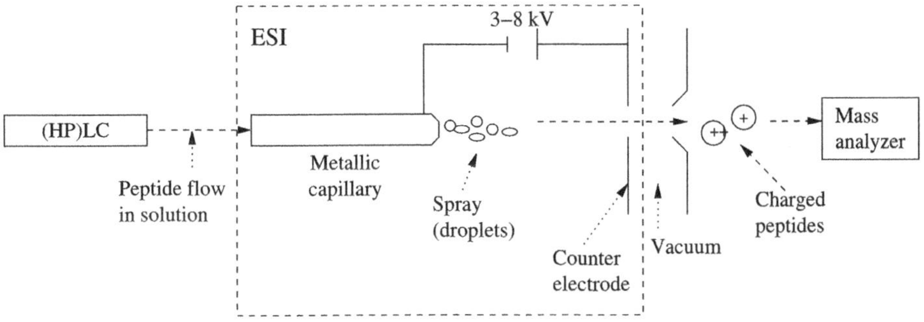

Figure 5.2 The principle of ESI

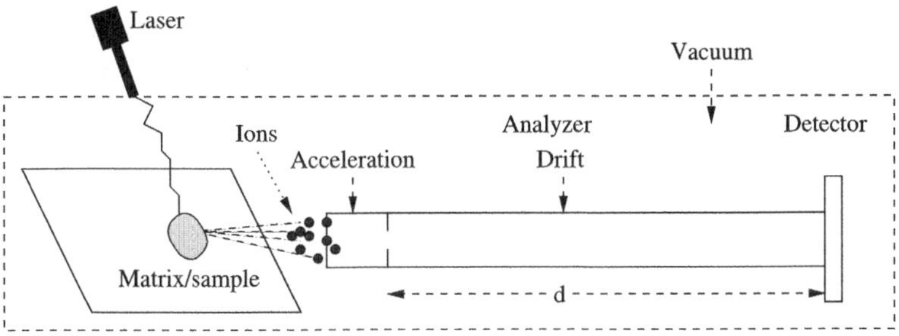

Figure 5.3 Illustration of a linear MALDI-TOF analyzer

5.3 Mass analyzers

There are several principles of mass analyzers used in proteomics, but here we will briefly describe the simplest one to understand. A more detailed description of the different mass analyzers is given in the following chapters.

In the MALDI *time-of-flight (TOF)* mass analyzer, ions are sent to the analyzer in short pulses due to the short laser pulses described for MALDI. The ions are accelerated by an electric field, and then they enter a field-free drift tube. The velocity that the ions have achieved during the acceleration is dependent on the mass and the charge of the ion, and this velocity is kept during travel through the drift tube. Naturally, the time needed to pass through the drift tube is dependent on the velocity. When the ions hit the detector at the end of the drift tube, the flight time is registered, and the m/z value can be calculated. The principle of such a linear TOF analyzer is illustrated in Figure 5.3.

5.4 Isotopic composition of peptides

As briefly mentioned in Section 1.3.2, many elements exist naturally in several isotopes. Where these elements are incorporated into biological molecules, such as amino acids, the isotopes are represented in ratios that correspond to their abundance in nature. Since the chemical properties of the different isotopes of an element are identical, the chemistry of the biomolecules (and therefore their function) is not affected by isotopic differences. Six elements (hydrogen, carbon, nitrogen, oxygen, phosphorus, and sulfur) constitute the overwhelming part of the elements incorporated into proteins and their post-translational modifications. The abundances and masses of the stable isotopes of these six elements are shown in Table 5.1. Note that several elements also have radioactive isotopes (for instance, ^{3}H and ^{32}P), which are excluded from this discussion as they are subject to decay over time.

Table 5.1 Abundances and masses of the stable isotopes of elements naturally occurring in proteins and their posttranslational modifications

Element		Abundance (%)	Mass
Hydrogen	^{1}H	99.99	1.007 83
	^{2}H	0.01	2.014 10
Carbon	^{12}C	98.91	12.000 0
	^{13}C	1.09	13.003 4
Nitrogen	^{14}N	99.6	14.003 1
	^{15}N	0.4	15.000 1
Oxygen	^{16}O	99.76	15.994 9
	^{17}O	0.04	16.999 1
	^{18}O	0.20	17.999 2
Phosphorus	^{31}P	100	30.973 8
Sulfur	^{32}S	95.02	31.972 1
	^{33}S	0.76	32.971 5
	^{34}S	4.22	33.967 6

Numerically, carbon and hydrogen are the most common elements in proteins, but as the heavy isotope of carbon, ^{13}C, has a 100-fold higher abundance than deuterium, ^{2}H, the former has far more influence on the isotopic pattern of peptides than any of the other elements. We will therefore focus on isotopes of carbon, and for the present purpose suppose that all the other elements are always represented by their lightest isotope.

Assume that we are working with a small peptide with a mass of approximately 600 Da (below called mass M). Such a peptide will contain around 30 carbon atoms. Approximately one-third of the peptide molecules will therefore contain one ^{13}C atom with the remainder being ^{12}C. Since only a very small amount of the peptides will contain two (or more) ^{13}C atoms, approximately two-thirds of the peptide molecules will only contain ^{12}C. As there is a 1.0034 Da mass difference between the peptide molecules that only contain ^{12}C and the peptide molecules that contain one ^{13}C atom, the mass spectrum will show one high peak at mass M, which is called the monoisotopic peak. At mass M+1 there will be a peak of approximately one-third the intensity of the monoisotopic peak. A very small peak may be seen at mass M+2, corresponding to the peptide with two ^{13}C atoms.

If we work with a peptide with a mass of approximately 3000 Da, there will be a low percentage of the peptide molecules that only contain ^{12}C atoms. A higher percentage of peptide molecules will contain one or two ^{13}C atoms. Peptide molecules containing three, four, or five ^{13}C atoms can be seen in decreasing amounts. Figure 5.4 illustrates the different behavior of the isotopic peaks of a relatively small and a relatively large peptide. Thus, with increasing mass of the peptide, fewer and fewer molecules will contain only ^{12}C atoms. For peptide masses above 5000 to 7000 Da, most instruments will no longer be able to distinguish the monoisotopic peak, and the other peaks will be poorly resolved. In experimental settings, we therefore rarely use the monoisotopic mass of large peptides or proteins (although it can easily be calculated), but rather the average mass. A collection of isotopic peaks from the same peptide is called an *isotopic envelope*. Knowledge of the isotopic peak pattern helps to interpret mass spectra, for example in estimating the charge.

5.4.1 *Estimating the charge*

A singly charged peptide will have Δ[1] $(m/z) = 1$ between the peaks in the isotopic envelope as illustrated in Figure 5.4. Doubly charged peptides will have $\Delta\ (m/z) = 0.5$ between the peaks, and triply charged peptides will have $\Delta\ (m/z) = 0.33$. At the same time, the m/z position of the peak in the mass spectrum is moved according to the formula $(M + nH^+)/nH^+$, where M is the mass of the peptide and n is the charge. This can be used to estimate the charge of the peptides.

5.5 Fractional masses

It is well known that peptide masses occur in clusters with distances a little more than 1 Da apart, as illustrated in Figure 5.5.

[1] Note that delta means difference.

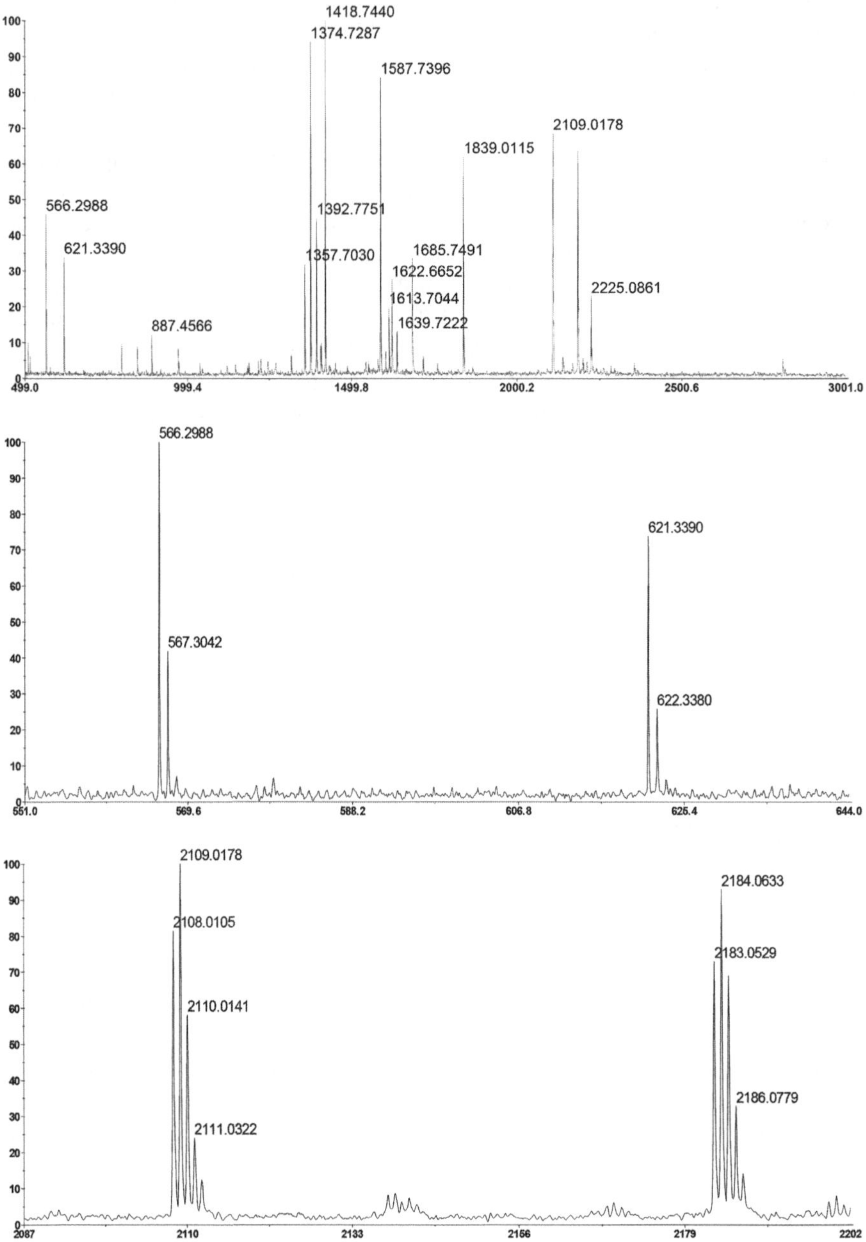

Figure 5.4 Example of a peptide mass fingerprint of a tryptic digest of the matrix metalloproteinase 2 protein. For ease of viewing, not all peaks have been labeled in the upper spectrum. Two peaks from the lower and two peaks from the higher mass range of the spectrum are shown in the two zoomed views below. Note the differences in the isotopic peak patterns between the low-mass and the high-mass peptides. Also note that the isotopic peaks have a spacing of approximately 1, indicating that all these peptides carry a single charge

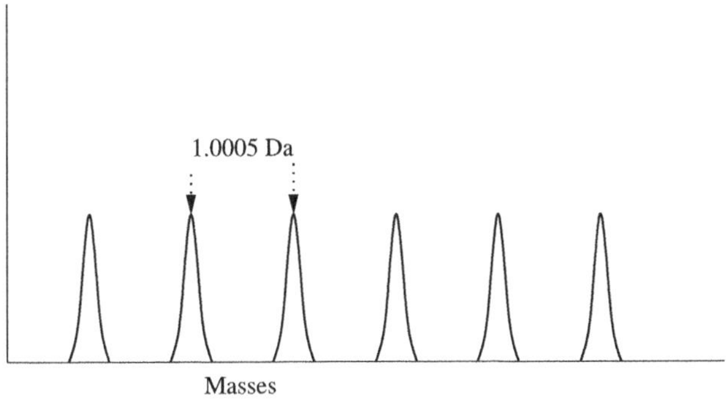

Figure 5.5 Distribution of the peptide masses. The distance between the apex of two neighboring clusters is approximately 1.0005 Da

The reason for this lies in the distribution of the amino acid masses. The amino acids are built from only five different atom types, each with a mass nearly equal to an integer. Consequently, the amino acids also have masses that are nearly equal to an integer. The integer is called the nominal mass of the amino acid, and the numbers after the decimal point are called the fractional mass. From Table 1.1 one can see that the fractional masses of the individual amino acids are in the interval [0.01, 0.1] Da. One can also see that the *normalized* fractional mass (the fractional mass divided by mass) varies over $(0.09\text{–}0.74)\cdot 10^{-3}$. The lowest normalized fractional mass is found for cysteine, and the highest for leucine, isoleucine, and lysine. The weighted average of the normalized fractional mass is approximately $0.47\cdot 10^{-3}$, using the amino acid abundances for the weighting. (Because of the weighting, the exact number will vary slightly according to the species and protein database used for the calculations as the amino acid abundances will vary slightly.) When the fractional mass is plotted as a function of the peptide mass, a plot like Figure 5.6 is obtained. The average fractional mass regression line (middle line of Figure 5.6) has a slope that corresponds to the weighted average of the fractional mass. The average fractional mass regression line is broken at a mass of approximately 2100. This corresponds to the point where the sum of the individual fractional masses of the amino acids in the peptides exceeds one. This is also seen by the calculation $(0.47\cdot 10^{-3}\cdot 2100 \approx 1)$.

It is found that the masses of most of the modified peptides also follow this clustering. The clustering effect is interesting because it can be taken into account when considering the correctness of identifications, and can also be used to perform calibrations. Conversely, it can also be used to identify and remove non-peptide peaks, like those arising from matrix–alkali clusters, non-biological polymers, Coomassie Blue stain, etc. The fractional mass of these contaminating peaks will be located outside the distribution of the peptide peaks. Without going into further details, note that different proteases will give slightly different fractional masses of the resulting peptides.

Example One research group (Wool and Smilansky (2002)) has found that the centroid cluster peptide mass M_c can be calculated from the nominal mass M_n as $M_c = 1.000\,495 M_n$. The nominal mass for a peptide is calculated by adding the nominal masses

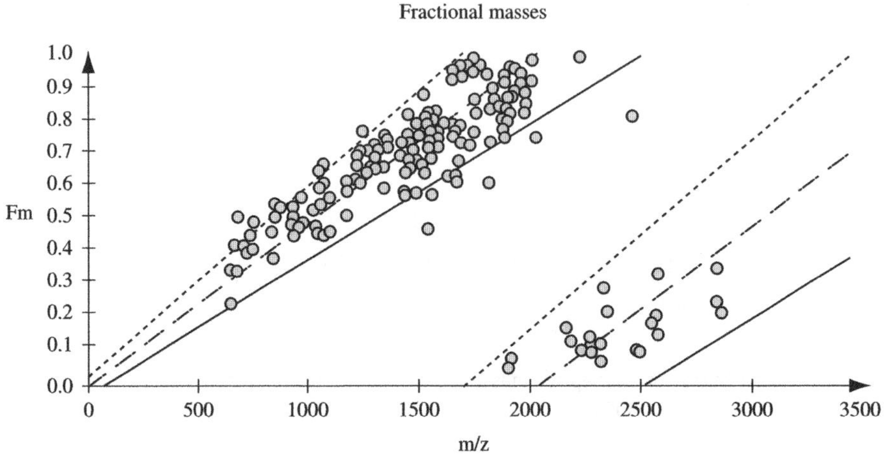

Figure 5.6 Illustration of (experimental) fractional masses of tryptic peptides. The middle line corresponds to the average regression line for the fractional masses of tryptic peptides, $Fm = 0.000\,495M_n$, where M_n is the nominal mass. The upper and lower lines correspond to plus or minus one standard deviation. The relationship shown is valid only for singly charged peptides

of all the residues. The standard deviation D_m of the fractional mass distribution was found to be $D_m = 0.03 + 0.02M_n/1000$. This means that for peptides with nominal mass $M_n = 2000\,\text{Da}$ we find $M_c = 2000.99\,\text{Da}$ and $D_m = 0.07\,\text{Da}$. This means that 99.7 % (corresponds to $3D_m$, assuming a normal distribution) of the peptides with nominal mass 2000 Da have masses in the interval $2000.99 \pm 0.21\,\text{Da}$.

△

5.5.1 *Estimating one or two peptides in a peak complex*

The existence of isotopes and clusters of the fractional masses has the consequence that isotopic envelopes from one peptide may overlap with the envelope from another peptide. Such cases can be detected during the interpretation of the spectrum because the isotopic intensity distribution of overlapping envelopes will be different from the isotopic intensity distribution of a single peptide.

Assuming that two peptides are spaced approximately 1 Da apart, the second (overlapping) peptide will add intensity to the isotopic peak pattern of the first peptide. The first peak in the isotopic peak complex will be the monoisotopic peak of the lighter peptide. The second peak will be the sum of the monoisotopic peak intensity of the second peptide plus the peak intensity of the first peptide with one ^{13}C atom, and so on. One such example that may be encountered relatively often in practice is the deamidation of asparagine in NG sequences. This is a spontaneous reaction that may occur under slightly alkaline conditions. The deamidation of asparagine generates aspartic acid with a concurrent mass increase of 1 Da. The result consists of two peptides, spaced 1 Da apart: one peptide contains the original sequence with an N, while the other carries the spontaneously generated D at that position.

5.6 The raw data

As explained previously, a detected compound is presented as a peak stretching over a (short) range of m/z values, rather than a single line at one exact m/z value (that is, a peak without spreading). Thus for each peak there are numerous intensity measurements at defined small increments of the m/z value (defined by the time resolution of the detector), and these should be combined into one intensity value as a peak without spreading. Hence we can roughly say that a mass spectrum can be presented in one of two forms: as raw data or as a *peak list*. The peak list is a processed form of the raw data. In its simplest form, it contains a list of m/z values of the detected peptides. Most peak list formats, however, also include the intensity value and possibly other data (for instance, charge state), for each of the detected peptides. The intensity (the number of detected ions) is proportional to the area under the curve, and the physical value at the centroid of the peak is usually used to calculate the ion's actual m/z value. A typical raw data spectrum can for example contain about 3000 pairs (m/z, intensity) inside a 100 unit interval, while the derived peak list contains only four pairs (corresponding to four peptides). The ratio between these numbers depends on the instrument used. Figure 5.4 shows an example of a raw data spectrum with little noise.

The processing steps involved in transforming raw data into a peak list are further described in Section 6.2.

5.7 Mass resolution and resolving power

Resolution and resolving power in mass spectra are terms for describing the possibility to discriminate between different components with small differences in the masses. A commonly agreed definition of how to measure them does not exist, and in some milieus the two terms mean the same, while they have slightly different meanings in others. We will here present two definitions for resolution that are used to some extent (taken from an earlier version of IUPAC definitions).

Resolution (in mass spectroscopy) – *p* percent valley definition Let two peaks of equal height in a mass spectrum at masses m and $m - \Delta m$ be separated by a valley that at its lowest point is just p percent of the height of either peak. For similar peaks at a mass exceeding m, let the height of the valley at its lowest point be more (by any amount) than p percent of either peak height. Then the resolution (p percent valley definition) is $m/\Delta m$. It is usually a function of m. The ratio $m/\Delta m$ should be given for a number of values of m. The commonly used value for p is 10.

Resolution (in mass spectroscopy) – peak width definition For a single peak made up of singly charged ions at mass m in a mass spectrum, the resolution may be expressed as $m/\Delta m$, where Δm is the width of the peak at a height which is a specified fraction of the maximum peak height. It is recommended that one of three values, namely 50 %, 5 %, or 0.5 %, should always be used. A common standard is the definition of resolution based upon Δm being full width of the peak at half its maximum height, sometimes abbreviated 'FWHM' (50 %).

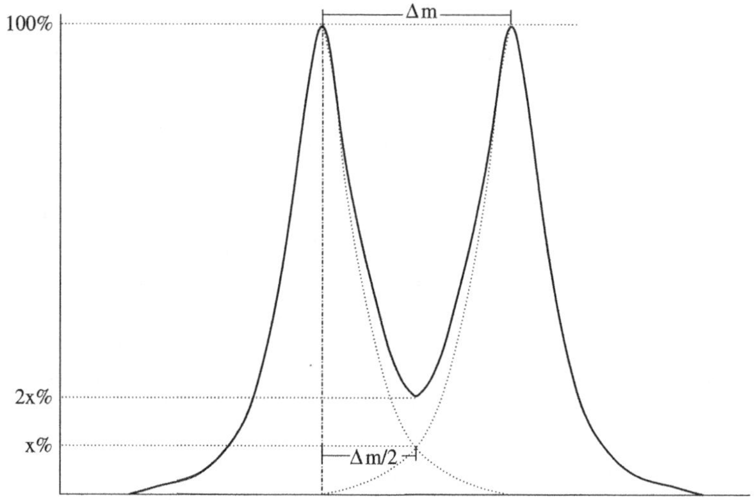

Figure 5.7 Figure showing the similarity between the two definitions of resolution, peak width definition ($\Delta m/2$ at x %) and valley definition (Δm at $2x$ %). The full line curve corresponds to the added intensities from two similar and symmetrical overlapping peaks, and the dotted subcurves to the intensity from each peak (drawn from a figure in de Hoffmann and Stroobant (2002) with permission)

Note that the percent valley definition depends upon two adjacent mass spectral peaks of equal size and shape. This can sometimes be found within the isotopic envelope of a single peptide, but rarely for two different peptides. The resolution is therefore commonly calculated from a single peak. However, for the most common peak shapes occurring in mass spectrometry, it is shown that for an isolated symmetrical peak recorded with a system which is linear in the range between x % and $2x$ % levels of the peak, the x % peak width definition is technically equivalent to the $2x$ % valley definition. This is illustrated in Figure 5.7.

The spectrometers are often classified as being of low, middle, or high resolution, depending on which mass spectra they are able to produce. Commonly, a resolution of up to 2000 is denoted as low, and those over 20 000 as high, but these limits are not fixed and they evolve as instruments evolve (typically shifting towards ever higher absolute resolutions).

5.7.1 *Isotopic resolution*

The determination of isotopes is important in the interpretation of mass spectrograms, as explained above. The resolution should therefore be high enough to recognize the isotopes. For a peak of m/z, the resolution R_m must satisfy $R_m \geq m/z \cdot z$.

Example Suppose that we have an ion of $m/z = 600$, and that z can be up to three. Then for $z = 3$ the distance between isotopic peaks is 1/3. For the peaks to be resolved at

FWHM, the width Δm of the peaks must satisfy $\Delta m < 1/3$. This means that the resolution $(m/z)/\Delta m$ measured by FWHM at 600 must be greater than $600/0.333\,33\ldots = 1800$.

△

Exercises

5.1 In a spectrogram there are two peaks at m/z values 239 and 358. Examine if these can come from the same peptide, but with different charges, and determine in the charges in that case.

5.2 The average residue mass is approximately 110, and the average number of C atoms per residue is 5.4. Use this to calculate the relative abundances of the isotopes for a peptide of mass 3000 Da and one of mass 500 Da.

5.3 In an MS spectrum generated by MALDI there are peaks at m/z values 600.2 and 600.7. Do you think any of them is noise?

5.4 Suppose that a resolution (FWHM) of 7000 is obtained for peaks at $m/z = 3000$. What is the maximum charge at which isotopic resolution can still be obtained?

Bibliographic notes

Fractional masses

Gay *et al.* (1999); Gras *et al.* (1999); Lehmann *et al.* (2000);
Wool and Smilansky (2002)

6 Mass spectrometry – MALDI-TOF

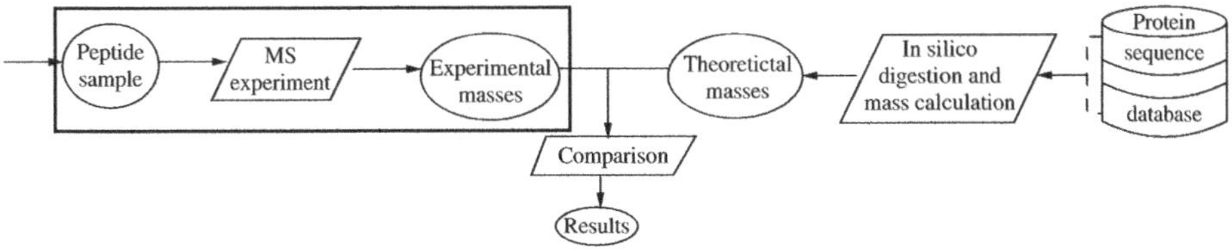

The MALDI-TOF instruments are the simplest MS instruments suitable for protein and peptide analysis. The instruments are easy to handle, and results are rapidly obtained. The sample for MALDI-TOF instruments should preferably contain the peptides from a single protein or a very low number of proteins (two or three at maximum). Thus, extensive sample preparation must be performed before the analysis can take place in the instrument. Analyses by MALDI instruments are therefore often combined with 2D gel electrophoresis.

Some important properties of MALDI and MALDI instruments are as follows:

- MALDI instruments are robust, fast, and easy to use.
- Not all of the peptides become ionized. There is a 'competition' between the peptides for the available charges (*competitive ionization*), and which peptides are preferentially ionized depends partly on the matrix used. The ionization process by MALDI is not fully understood, but several studies have empirically investigated how the ionization depends on peptide properties. This is described in more detail in Section 6.1.3.
- MALDI is able to ionize molecules with masses from a few hundred daltons up to several hundred thousand daltons, although the mass range studied during a single analysis is much more restricted. It is usually known whether we study peptides or large proteins, and the instrument settings can therefore be optimized for each purpose. The lower mass range is limited by noise generated from matrix ions and other unwanted ions.

Computational Methods for Mass Spectrometry Proteomics I. Eidhammer, K. Flikka, L. Martens and S.-O. Mikalsen
© 2007 John Wiley & Sons, Ltd

- Almost all of the ionized peptides have a single positive charge due to the acquisition of a proton. As a general rule, only very large peptides or proteins may pick up two or three protons. The spectra are therefore relatively easy to interpret.
- The accuracy in the peptide mass area (500–6000 Da) is good. The exact accuracy obtainable is dependent on the instrument and the calibration procedures. When internal calibration is used, the accepted accuracy during database searches is often set to 50 ppm, meaning that the difference between the experimental m/z value and the theoretical m/z value must be less than 50 ppm.
- The resolution is good, meaning that all newer instruments can distinguish the isotopic peaks in the peptide mass range and correctly assign the monoisotopic peak.
- The peptides ionized by MALDI are generally quite stable, and there is little fragmentation.
- There can be adduct ions from contaminating salts. However, MALDI tolerates salt contamination better than ESI instruments.
- There can be other contaminants. We may distinguish between three types of contaminants: the matrix (which is unavoidable), contaminants introduced by sample handling, and contaminants due to insufficient sample separation.
- It is possible to connect to other 'intelligent systems' (robots) for automatic proteomics. Thereby it is possible to automate sample handling, strongly increasing the sample throughput.
- MALDI instruments are generally quite sensitive. Identification is often obtained if the sample initially contained as little as 10 to 100 femtomoles of protein, and with newer instruments the sensitivity may enter the attomole area. Sensitivity here is a measure of the amount of a component that is needed for that component to produce a signal in the spectrum.
- Contrary to ESI sources, in which the sample is spent during the analysis, MALDI targets that are appropriately stored after analysis can be reanalyzed at a later time. The only caveat is that each laser shot on the target consumes some of the sample at that location. However, the spots are typically large enough to hold much more sample than is consumed by a single analysis. This is sometimes called the *sample archiving* property of MALDI sources. An important fact to consider with regard to sample archiving is that the peptide bond absorbs UV light, and can be broken as a result of the energy absorbed. The target should therefore always be archived in the dark.

6.1 TOF analyzers and their resolution

TOF analyzers are the dominating analyzers for single MS. When connected to a MALDI ionization source, the instrument is called a MALDI-TOF mass spectrometer. The simplest form of TOF analyzer is the linear TOF, schematically described and illustrated in Section 5.3. Ions are sent to the analyzer in short pulses due to the pulses of the laser and the synchronized changes in the electric field. The ions are accelerated in the electric field. Let the potential of the acceleration field be P, the velocity at the end of the acceleration v, the distance to the detector d, and e the charge of an electron. The kinetic energy is then

$$\frac{mv^2}{2} = zeP \tag{6.1}$$

After the acceleration, the ion enters a field-free drift tube and continues with the obtained velocity until it hits the detector. The time for the ion to fly to the detector is $t = d/v$. From this we get

$$t^2 = \frac{d^2}{v^2} = \frac{m}{z}\frac{d^2}{2eP} \quad \Rightarrow \quad t = \frac{d}{\sqrt{2eP}}\sqrt{\frac{m}{z}} = C\sqrt{\frac{m}{z}} \tag{6.2}$$

for a constant C. We see that m/z can be calculated from the flight time.

As described in the previous chapter, the resolution of an MS instrument is its ability to discriminate between different components with small differences in their m/z values. We will illustrate how this is related to time and masses for TOF analyzers.

Example Consider the difference in flight time at one mass unit difference:

$$\Delta_t = \frac{d}{\sqrt{2eP}}\left(\sqrt{\frac{m+1+z}{z}} - \sqrt{\frac{m+z}{z}}\right) = C\left(\sqrt{\frac{m+1+z}{z}} - \sqrt{\frac{m+z}{z}}\right) \tag{6.3}$$

We see that the time difference (and hence the discrimination ability) decreases with increasing mass. For z equal to one, we get $\Delta_t = 0.02C$ for $m = 500$ Da, and $\Delta_t = 0.007C$ for $m = 5000$ Da. This means that as the mass increases it becomes more difficult to differentiate between peptides with mass differences of 1 unit.

△

Another important factor in relation to resolution is that not all ions of a peptide (or several peptides of the same m/z) will reach the detector at exactly the same time. Therefore, there will be a spreading in the measured time (and therefore the calculated m/z) of a peptide. The reasons for these different flight times are mainly derived from the fact that not all ions come from exactly the same point in the matrix/sample, and they are therefore spread out in a 3D space when they enter the electric field. The velocities after acceleration may thus vary slightly. As a result, overlaps can occur between different peptides, as shown in Figure 6.1. This spreading is decisive for the resolution, and the most important factors that determine the spreading are:

- the length of the laser pulse for ion creation;
- the size of the matrix/sample volume;
- the variation in kinetic energy.

We also see from Equation 6.3 that we can increase the resolution by increasing d, or decreasing the acceleration voltage. However, increase of d is limited by practical considerations, while limiting the voltage has a negative impact on the *sensitivity*.

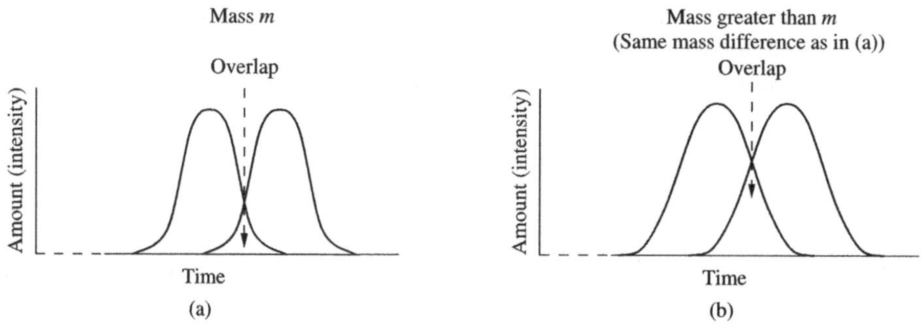

Figure 6.1 Illustration of the overlap in signal for two components with small mass differences. It also shows that the overlap increases for higher masses, if the mass differences are the same

Two strategies are commonly used to reduce the spreading (thus increasing the resolution).

Delayed (pulsed) extraction, DE When the ions of a single peptide are ejected from the plate after the laser pulse, they exhibit a range of velocities, and would therefore each reach the detector at different times. To remedy this, a delayed extraction voltage can be used. Delayed extraction, as its name suggests, allows for a period of field-free drift before switching the extraction voltage on. During this field-free time interval, faster ions move farther than the slower ions. When the extraction voltage is subsequently applied, the faster ions find themselves further into the potential field, and thus at a lower potential, than the slower ions. The faster ions thus acquire less additional velocity than the slower ions, which helps to reduce the spreading on the ion velocities. This extraction voltage pulse is applied from a few tens to a few hundreds of nanoseconds after the laser pulse, trying to optimize the mass resolution.

Reflectron A reflectron is an electrostatic mirror placed at the end of the linear flight tube. When the ions enter the reflectron, they are exposed to an electric field that forces them to turn around and fly back, with their return path slightly angled relative to the direction they came from. The reflectron has two effects. First, it increases the flying distance d by sending the ions back through another TOF analyzer that has a detector placed at the far end. Second, the reflectron is able to 'collect' the ions of the same m/z that had slightly different velocities in the first drift tube. This is because ions with a higher velocity are able to penetrate farther into the reflectron than ions with slightly lower velocity. The fastest ion thus follows a longer overall flight path through the reflectron, while the slower ions are turned around more easily. This effectively focuses the ions that have the same m/z. In more formal terms, the reflectron compensates for the spread in kinetic energies that the ions may have acquired during their initial acceleration. The reflectron principle is schematically illustrated in Figure 6.2.

All advanced MALDI-TOF instruments have the possibility to choose between linear TOF and reflectron TOF in addition to delayed extraction. The linear TOF is mainly used for the analysis of proteins, while the reflectron is used for the analysis of peptides.

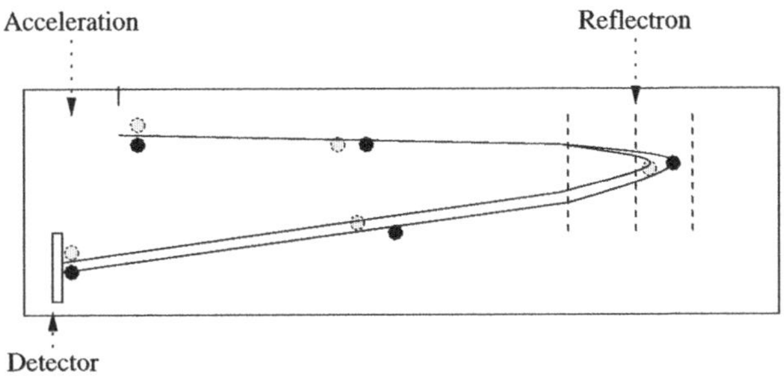

Figure 6.2 Illustration of a TOF analyzer fitted with a reflectron. The reflectron consists of a collection of grids that act as an ion mirror. For ions of the same m/z, a higher velocity will result in a longer path through the reflectron, which thus compensates for the spread in kinetic energy. As a result, ions of the same m/z should reach the detector almost simultaneously

6.1.1 Time-to-mass converter

From Equation 6.2 we have $m/z = at^2$ for a constant a. However, it is more realistic to account for a non-zero initial velocity and other factors in the instrument. The equation used for determining the m/z is therefore

$$\frac{m}{z} = at^2 + bt + c \tag{6.4}$$

The constants a, b, c must be determined for each instrument.

Using the simple form in Equation 6.2 for the conversion we have

$$t^2 = \frac{m}{z}\frac{d^2}{2eP}$$

Here $e = 1.6 \cdot 10^{-19}$ C (coulomb) and 1 Da $= 1.665402 \cdot 10^{-27}$ kg. Assume the potential $P = 1000$ V, $d = 1$ m, $z = 1$. A peptide with a mass of 500 Da will then use 51.01 μs, and a peptide with a mass of 501 Da uses 51.06 μs. For the detector to be able to differentiate between these two masses, the accuracy of the timer must be at least 0.05 μs. This requires the use of electronic devices. In addition to the TOF, the amount of ions reaching the detector must also be measured for each time interval.

6.1.2 Producing spectra

We have seen that ions are created by laser pulses, with each pulse sending ions to the detector, generating a signal from which a spectrum can be drawn. In practice, however, a spectrum for a sample of components (peptides) is obtained by averaging over the spectra generated from a number of pulses. The number of averaged spectra may vary according to the amount of peptides in the sample or the signal-to-noise ratio. Random noise is more likely to be partially cancelled out when many pulses are accumulated, while a weak but constant signal from a peptide will consistently be added up and can thus more readily be distinguished from the background noise. A spectrum may therefore be produced from a few pulses (10–100) or many pulses (more than 1000), depending on the circumstances.

6.1.3 *Ionization statistics*

It is commonly observed that some peptides from a digested protein are absent in MALDI-TOF spectra. Rather, a typical sequence coverage of 20 to 50 % is obtained. The detection or lack of detection of a peptide may have different reasons. Some peptides might be too small, while others are too large for detection. However, a major reason is that not all peptides are ionized. As explained earlier, the ionization involves a competition for the available charges, and which peptides are ionized depends partly on different peptide properties and partly on matrix properties. Most studies have used tryptic peptides in concert with α-cyano-4-hydroxycinnamic acid, a very common matrix for peptides. Tryptic peptides have `R` or `K` as their C-terminal amino acid (except the C-terminal peptide of the protein). Peptides with a C-terminal `R` generally have above-average intensities, so consequently those containing `K` have lower intensities. The intensity of `R`-containing peptides depends on the neighboring residue to `R`, with `G` having a positive effect, and `D` and `E` having a negative effect. It is generally accepted that `F` has a positive effect on peak intensity. `Y`, `W`, `P`, and `L` also appear to improve intensity, and thereby the chance of detection. It therefore seems that the signal intensity of a peptide depends mostly on the presence of certain amino acids rather than on the overall physicochemical properties of the peptide. There are examples indicating that the matrix may influence the ionization of peptides. Phosphopeptides are generally considered as 'bad flyers' using α-cyano-4-hydroxycinnamic acid (that is, these peptides are often difficult to detect). Certain matrices like 2,4,6-trihydroxyacetophenone or 2,5-dihydroxybenzoic acid, or supplements in the matrix, like phosphoric acid or ammonium salts, have been reported to improve the ionization capabilities of phosphopeptides.

6.2 Constructing the peak list

The raw data spectrum contains signals from the peptides, as well as signals derived from different forms of noise. Specifically revealing the signals (peaks) corresponding to the sample peptides is a multistep task. The detailed procedures for peak list construction vary with the instrument and software used, as well as with the type of spectra. However, some basic (sub)tasks have to be performed, either as separate tasks, or performed in an integrated procedure. The main challenges are the removal of noise peaks without removing any of the peptide peaks, and to determine the m/z and intensity values with the best possible accuracy.

6.2.1 *Noise*

The quality of the obtained spectra may range from excellent to poor. The amount of noise in the spectrum relative to the amount of sample compounds strongly influences the quality of the spectra.

Figure 6.3 shows a spectrum with much noise, and its processing by techniques is explained later in this section. It can be understood that when the peptide peaks have low intensity and are surrounded by much noise, this can influence both the exact m/z value and the peak pattern. Noise may be defined in a narrow manner as random fluctuations

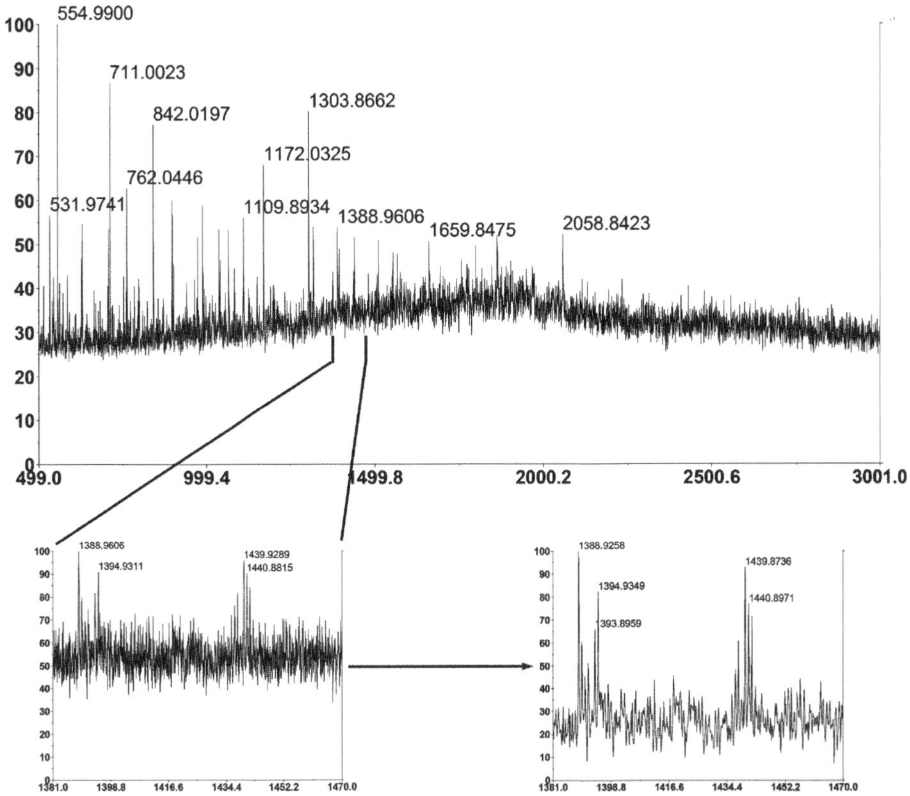

Figure 6.3 A spectrum with much noise, a high baseline, and low-intensity peptide peaks. A detail of the spectrum is zoomed into, and the effect of baseline correction and noise filtering is illustrated for this detail. The spectrum has not been calibrated. The high noise level may influence both the exact m/z value of the peaks and the detection of the isotopic peak distribution. In the zoomed part of the spectrum, peaks 1388.9 and 1439.9 have isotopic distributions that could be commensurate with peptides, while 1393.9 and 1437.8 (not labeled, approximate m/z value) do not have such envelopes

between two spectra of the same sample, or between two laser shots on the sample. Noise may also be defined in a wide manner as all unwanted peaks not belonging to our sample. This latter definition can for example include unwanted contaminants.

Noise may be divided into two main types.

Chemical noise may have several sources, as follows:

- It may come from chemical *contaminants* introduced during the sample handling, for example detergents that have not been removed from the samples and polymers that have leached out from low-quality plastic tubes.
- It may be derived from proteins or peptides that have been unintentionally introduced during sample handling, for example human keratins from the skin or hair.

- In MALDI-TOF analyses it may also come from the instability of peptides after receiving energy from the laser pulse. This may result in fragmentation along the peptide backbone, loss of parts of the amino acid side chains (loss of water and ammonia are the most prevalent), or loss of some of the posttranslational modifications (especially phosphorylations or glycosylations attached to serine or threonine residues).

In all of the above cases unexpected peaks are produced.

Electronic noise is the result of electronic disturbances, and occurs with random fluctuations between the chemical noise.

The noise level of a spectrum (or part of a spectrum) is typically calculated as the standard deviation of selected points assumed to be noise. The signal-to-noise ratio (S/N ratio) is also often used.

Noise and contaminations may cause so much disturbance that the MS analysis is unusable. Both sample handling and instrument settings may therefore be critical during an experiment.

6.2.2 Baseline correction

A specific form of noise occurs as the *baseline* of a spectrum, derived predominantly from chemical noise. The baseline is an offset of the intensities of masses, and should be subtracted from the measured intensities. It often shows a dependency on the m/z value such that it is highest at low m/z values, and shows an exponential decay towards higher masses. The baseline varies from spectrum to spectrum. Each spectrum must therefore be treated separately.

Many algorithms for baseline correction have been developed. The simplest baseline correction finds the lowest point in the spectrum and drags this point down to zero. At the same time, the highest point in the spectrum (the *base peak*) is kept at 100 % intensity. The result is that the intensities in the spectrum are 'extended' along the vertical axis. More advanced methods take into account that the baseline varies across the spectrum, and try to locally (in windows along the m/z-axis) fit the baseline points to polynomial or exponential functions. The windows should preferably be intervals that do not contain real peaks, such that only baseline points contribute.

6.2.3 Smoothing and noise reduction

As illustrated in Figure 6.3, a spectrum can be jagged, making it difficult to detect peaks amongst the noise. A smoothing is therefore often performed on the spectrum. Again this can be done in different ways. The straightforward way is to use a sliding window, where a new value is calculated for the point in the middle of the window, based on the values for the points in the window. The calculation is very often a weighted average.

Example Suppose we have the intensity values 20, 16, 19, 18, 15 for five succeeding points of measurements, and use filter length five with weights 1:2:4:2:1. Then the new value for the middle point is 18 (17.9).

△

The length of the window and the weights are determined from known forms of the noise, often taking the resolution into account. A Gaussian filter is also often used, based on the Gaussian function

$$G(x) = \frac{1}{\sqrt{2\pi}\rho} e^{-\frac{x^2}{2\rho^2}}$$

which is symmetric. ρ determines the length of the window, and the weights are calculated by a discretization of the function values. Due to variation of the noise level along the mass range, they are calculated for small intervals (down to under 1 unit). More advanced smoothing functions include Fourier transformations and wavelet transformations.

6.2.4 Peak detection

Peak detection is the process of distinguishing interesting peaks from noise. The aim is that every isotopic variant of each peptide should be represented by exactly one data point in the spectrum, and several techniques are used. One way is to first identify the apex of a peak. This is the point where the intensity stops increasing and starts decreasing. The intensity at this apex is then compared to the surrounding noise level. Alternatively, the valleys on each side of the apex can be used to determine the start and end of the peak. The area of the peak is then calculated, and is compared to a minimal 'true peak area threshold'.

Other methods consider a peak as a continuous range of points with intensity above the noise level. Some also consider the *shape* of the peak to distinguish peptides from contaminants and noise, as peptide peaks tend to have different shapes than noise peaks. The number of expected raw data points per peak (which can vary over the spectrum) is also used in some peak detection algorithms. One such method is described in Gentzel *et al.* (2003), which takes the resolution of the mass spectrometer into account. The resolution is mainly determined by the geometry and the settings of the instrument. At fixed settings, the function describing the peak width, pw, as a function of the mass is also fixed. This function is set to $pw(x) = 0.08 + 0.0004x$, where x is the m/z value. (We see that the resolution varies with x.)

By denoting the intensity at mass x_i by y_i, a (raw) spectrum is represented by a set of (x_i, y_i) values, where the difference between succeeding x_i can vary. A centroid is started at x_i if

$$\sum_j y_j \geq t$$

where the sum is over j such that $x_i \leq x_j \leq x_i + pw(x_i)$ and t is a predefined threshold. The maximum of the rising peak is determined to be x_m, and the centroid peak (x_c, y_c) is calculated as

$$x_c = \frac{\sum_j x_j y_j}{\sum_j y_j}; \quad y_c = \sum_j y_j$$

where the sums are over j such that $(x_m - pw(x_m)/2) \leq x_j \leq (x_m + pw(x_m)/2)$. x_c and y_c are then the m/z value and the intensity of the centroided peak. By comparing the borders of two succeeding centroid peaks, overlapping can be discovered, and the overlapping intensities are split between the two (centroid) peaks.

Whatever peak detection algorithm is used, high-intensity peaks of single peptides are usually detected without failure. However, when the peaks have low intensities and there are complex patterns of overlapping peptides, errors are much more likely to occur.

6.2.5 *Example*

We have the following part from a raw data spectrum (m/z, intensity; with the values limited to two decimals for presentation reasons):

1939.00	64.72	1939.62	107.91	1940.25	45.24	1940.87	227.58
1939.04	59.56	1939.67	129.19	1940.29	46.59	1940.91	210.30
1939.09	54.14	1939.71	150.26	1940.33	49.28	1940.96	190.47
1939.13	48.67	1939.75	170.48	1940.38	53.26	1941.00	169.51
1939.17	43.54	1939.80	176.96	1940.42	58.55	1941.05	148.09
1939.22	39.25	1939.84	157.23	1940.47	65.16	1941.09	126.90
1939.26	39.51	1939.89	136.76	1940.51	73.71	1941.14	106.51
1939.31	41.84	1939.93	116.24	1940.56	94.22	1941.18	88.00
1939.35	45.41	1939.98	96.38	1940.60	124.24	1941.23	73.99
1939.40	50.21	1940.02	82.06	1940.65	154.87	1941.27	64.92
1939.44	56.16	1940.07	71.62	1940.69	185.41	1941.32	59.72
1939.49	63.37	1940.11	62.74	1940.74	215.17	1941.36	57.46
1939.53	71.99	1940.16	55.36	1940.78	234.24	⋮	
1939.58	87.17	1940.20	49.50	1940.83	238.55		

A part from the derived peak list is shown below:

Centroid mass	Peak height (to zero)	Relative intensity	Peak area (rel. local baseline)
1939.790 61	844.0	2.85	843.67
1940.844 95	1897.0	6.41	1897.29
1941.802 98	7717.0	26.05	7717.00
⋮			

6.2.6 *Intensity normalization*

Intensities are used to distinguish noise from real signals in the score calculations of comparisons, in the classification of spectra as good or bad, and in quantitative analyses. The intensities of peaks are highly variable from spectrum to spectrum, and in order to use them in a uniform way they should be normalized somehow. The normalization should be such that the normalized values reflect the probabilities that the ions are real ions (derived from peptides) and not noise. The traditional approach has been to transform to relative intensities, usually relative to the total intensity (*total ion count* (TIC)) or to the maximum peak intensity (which results in the same relations between the normalized peaks). This can then be used to remove noise, by considering peaks with a normalized intensity below a threshold as noise.

To obtain higher robustness across different spectra, rank-based intensity normalization has been proposed. This means that the most intense peak gets rank 1, the next highest rank 2, and so on.

The rank-based normalization does not take the magnitude of the intensities into account, therefore Na and Paek (2006) proposed a *cumulative intensity normalization*. The peaks are sorted after decreasing intensities, and the normalized intensity of peak with rank n is defined as

$$\frac{\sum_{rank(x)\geqslant n} I(x)}{TIC}$$

where $I(x)$ is the original intensity of peak x, and TIC is the total intensity of the spectrum. Normalization is also treated in Chapter 15.

6.2.7 Calibration

The spectra obtained from MS instruments must be calibrated to achieve the accuracy needed for database searches, as described in Section 1.8.1. Internal calibration is most commonly used, and is achieved by knowing the exact m/z values of certain peaks in the experimental spectrum. Such peaks are obtained in one of two ways:

- known standards may be added to the sample; or
- the protease used for digesting the sample will produce known autolytic peaks by self-digestion.

It is also possible to use external calibration by spotting known standards close to the sample on the sample plate of the instrument, and by acquiring the spectra from the sample spot and the standard spot consecutively. Measurement deviations as observed in the standards spectrum are then extrapolated to hold for the sample spectrum as well.

Different forms of regression curves for calculating the calibrated values are used, from a linear $m^* = am + b$ function to higher order functions, for example

$$m^* = (am + b\sqrt{m} + c)^2$$

where m^* is the calibrated m/z value. Which form of calibration to use depends on the instrument. In the case of internal calibration, the new constants are applied to the same spectrum, correcting the m/z values of all peaks in the spectrum. In the case of external calibration, the constants from the standard spectrum are applied to the sample spectrum. External calibration will usually be less exact than internal calibration.

The fractional mass distribution can also be used as a component of the calibration, relying on the observation that the distribution of the peptide masses is not uniform, as they

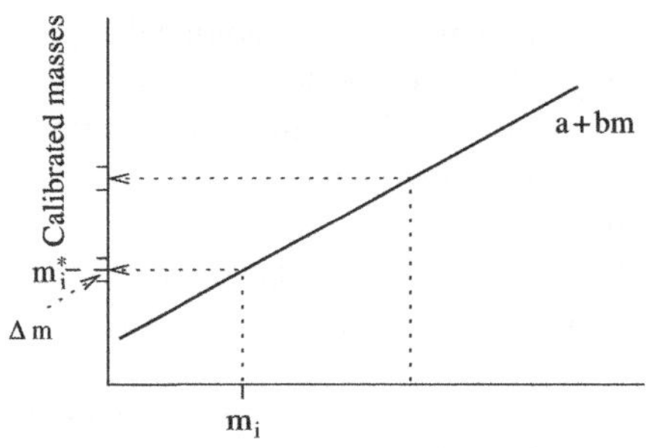

Figure 6.4 Calibrated masses are determined by maximizing $\sum_i P(a+bm_i)\Delta m$

occur in clusters. In addition, the number of peptides in such a cluster decreases at higher masses. The peptide mass range can therefore be divided into small intervals of length Δm. If P is the mass probability distribution, then the probability for a random peptide mass to occur in interval $[m, m+\Delta m]$ is $P(m)\Delta m$. A set of peaks with experimental masses $\{m_i\}$ can then be calibrated anew by determining the constants a, b such that the total probability $\sum_i P(a+bm_i)\Delta m$ is maximized, as shown in Figure 6.4. It has been shown that for this method to work the initial errors must be less than 0.5 (Gras *et al.* (1999)).

6.3 Peak list preprocessing

After construction of the initial peak list, it should be preprocessed to make it more appropriate for subsequent database searching.

6.3.1 Monoisotoping and deisotoping

This process reduces a cluster of isotopic peaks (an isotopic envelope) to a single peak, with intensity equal to the sum of the isotope intensities. *Monoisotoping* reduces the isotopic envelope to the peak with the lowest m/z in the cluster. This is usually the procedure for MALDI experiments where delayed extraction and reflectron mode have been used. *Deisotoping* reduces the isotopic envelope to a centroid peak, with the m/z value determined from the intensities of the individual isotopes. The centroid m/z value corresponds to the value that would be obtained if average masses of the atoms in the peptide were used for a calculation of its mass.

6.3.2 Removing spurious peaks

The fractional masses (Section 5.5) can be used to try to remove non-peptide masses, by removing those peaks whose fractional masses do not follow the pattern for peptides. An illustration of this can be found in the masses that lie outside the marked area in Figure 5.6.

The background noise increases at low m/z ratios, and masses below a specified limit (sometimes 500, but more typically 700 to 800) are therefore removed. High masses (typically above 3000 to 4000) are also removed, since the sensitivity and the accuracy are usually lower at such high masses. Additionally, there are relatively few peptides in this high-mass range as explained earlier.

It is also common to filter the masses for known contaminants, such as peptides from keratin or autolytic peptides from the protease used. For high-throughput protein identification, such contaminants can be found in an automatic way, for example by comparing many spectra at the same time. If many spectra contain the same peak (within a specified accuracy), that peak is likely to be a contaminant. More formally, the mass range is divided into bins of small size (for example, 0.6), and for a batch of spectra each bin gets a value equal to the number of spectra in which this mass occurs. If this number is above the expected number, the mass is considered a contaminant. The expected number can be estimated from the spectra in the batch, or calculated from a set of random spectra, generated from a non-redundant database.

6.4 Peak list format

There are several formats used for presenting the peak list data (pkt, pkm, dta, mgf, bdtx, mzXML, etc.). All include the m/z value and an intensity for each peak. Some also have additional data from the raw data peak, such as peak start, peak end, area, etc.

Note that there is an ongoing work for standardization of proteomics data: 'The HUPO Proteomic Standards Initiative' (PSI). This is explained in detail in Chapter 18. The MS data standard drafted by the PSI also aims to capture the experimental setup in addition to the more traditional m/z and intensity.

6.5 Automation of MALDI-TOF-MS

The MALDI-TOF instruments are robust and rapid, and they can be used in high-throughput proteomics. To this end, a considerable degree of automation is required at the level of sample handling, acquisition of the spectra, and the analysis of the spectra. As 2D gels often are the starting point of MALDI-TOF analyses, we will briefly describe a possible workflow.

Images of the stained 2D gels are run through an image analysis software package. The software is able to assign coordinates to the spots, and potentially compare the image with previous images of other gels for differential analysis. The user may manually determine the spots that are interesting, or the software can do so according to specified criteria. The coordinates of the interesting spots are sent to a spot-picking robot that excises the spots from the gel, and transfers the gel pieces to 96-well plates. Each spot is transferred to one well, such that one plate can contain the protein(s) from 96 spots. The plates are then transferred to a pipetting robot that can add and remove liquid from the wells with the gel pieces. Here the gel pieces are treated according to procedures described earlier (destaining, dehydration, protease treatment, extraction of peptides). The robot may then clean the peptide samples by reverse phase microcolumns (either in other 96-well plates or in pipette tips), and thereafter spot the samples on the sample plate (chip) of the MS instrument. Such a plate can contain several hundred spots (corresponding to several hundred 2D gel spots). The sample plate is transferred to the instrument, and each spot is exposed to laser shots. In this way a peptide spectrum is produced for each of the selected spots from the 2D gel. The spectra are then processed as described previously in this chapter, and the peak lists are submitted to the identification programs as will be described in the next chapter. In this way, hundreds of, or even a thousand, spots may be analyzed per day in an unattended manner. Note that the rate-limiting steps in such an approach are often the 2D gel separation of the proteins, and subsequent staining of the gel.

As can be appreciated, such full automation is well suited for the high-throughput identification of proteins, while detailed characterization of posttranslational modifications presently requires more interaction from the user.

Exercises

6.1 Use Equation 6.3 to calculate Δ_t for $m = 500$ and $m = 5000$ for $z = 2$. Use the values $e = 1.6 \cdot 10^{-19}$ C and 1 Da $= 1.665402 \cdot 10^{-27}$ kg. Furthermore, let $P = 1000$ V and $d = 1$ m.

6.2 Consider the following part of a raw data spectrum:

842.009 577	41.7505	842.303 592	127.796	842.597 659	874.646
842.038 976	39.4705	842.332 997	175.743	842.627 068	646.352
842.068 376	38.1118	842.362 401	390.189	842.656 478	431.301
842.097 776	37.9343	842.391 807	712.181	842.685 889	319.232
842.127 177	39.3472	842.421 213	1078.96	842.715 300	242.616
842.156 578	42.6215	842.450 619	1432.5	842.744 711	181.639
842.185 980	48.3637	842.480 026	1627.03	842.774 123	137.138
842.215 382	58.8778	842.509 433	1542.52	842.803 536	103.744
842.244 785	75.0604	842.538 841	1331.61	842.832 949	78.816
842.274 188	98.379	842.568 250	1103.36		

(a) Use the peak detection method described in Section 6.2.4 and $t = 4000$ to show that a centroid is started at $m/z = 842.038\,976$.

(b) Find (x_m, y_m) for this peak.

(c) Calculate (x_c, y_c) for the peak.

6.3 Suppose that the m/z values 399.98, 900.15, 1100.22 are found in a spectrum. Describe how you can use the fractional masses to calibrate the spectrum, illustrated by these values.

Bibliographic notes

Ionization statistics

Baumgart *et al.* (2004); Cohen and Chait (1996); Parker (2002); Weiller *et al.* (2001)

Peak list construction

Berndt *et al.* (1999); Gras *et al.* (1999); Sauve and Speed (2004)
Du *et al.* (2006); Samuelsson *et al.* (2004)

Data Explorer	http://cabm-ms.cabm.rutgers.edu/Exp_guide40.pdf
MassLynx	http://www.matrixscience.com/help/instruments_masslynx.html

Intensity normalization

Bern *et al.* (2004); Matthiesen *et al.* (2004); Na and Paek (2006)

Preprocessing

Gentzel *et al.* (2003); Gobom *et al.* (2002); Samuelsson *et al.* (2004)
Levander *et al.* (2004); Zhang *et al.* (2005)

Calibration	Chamrad *et al.* (2003); Gras *et al.* (1999); Wool and Smilansky (2002) Samuelsson *et al.* (2004); Wolski *et al.* (2005)
Contaminants	Levander *et al.* (2004); Schmidt *et al.* (2003)
Removing spurious peaks	Schmidt *et al.* (2003)

Standards for MS

Orchard *et al.* (2005), http://psidev.sourceforge.net/ms

7 Protein identification and characterization by MS

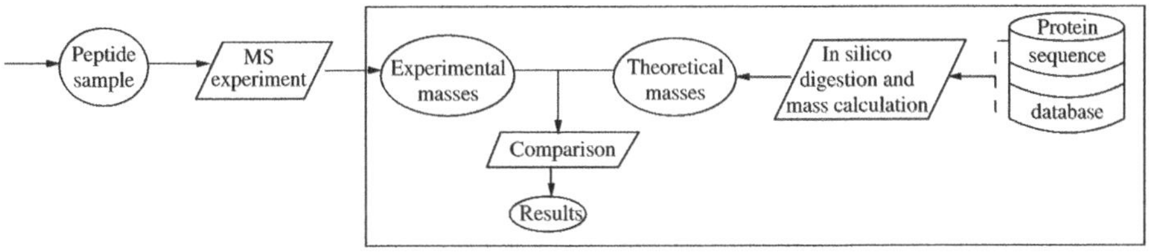

In this context, protein identification is a multistep process. Peptides originating from an unknown protein have been analyzed by mass spectrometry, and a spectrum has been acquired. The experimental data are compared with theoretical sequences from a sequence database. The aim is to find a protein in the sequence database that can explain the experimental data as far as possible, and in such a way that it is highly unlikely that the fit between the experimental and theoretical data is artifactual.

In practical terms, this is done by presenting the acquired peak list to a specialized software program that performs the comparison. Such software may be provided with the instrument, but there are also several freely available Internet tools as well as some stand-alone commercial programs that are designed for this task. Only some of the programs will be directly mentioned (see list at the end of the chapter), as we focus here on the principles and approaches, rather than on the details of specific programs.

Reliable identifications require high-quality MS experiments. This means that the spectra should contain clear peaks from the peptides of the protein sample, and ideally none from other molecules. As explained earlier, the quality of MS experiments depends on different factors such as the sample preparation methods, the matrix composition, the characteristics of the analyzed sample (such as sugar content or other posttranslational modifications, hydrophobicity, amino acid sequence, 3D structure), the amount of protein in the sample, the purity of the sample, the characteristics of the MS instruments (accuracy, precision, resolution), and the availability of internal or external calibration standards.

Computational Methods for Mass Spectrometry Proteomics I. Eidhammer, K. Flikka, L. Martens and S.-O. Mikalsen
© 2007 John Wiley & Sons, Ltd

7.1 The main search procedure

A generic identification procedure using MS data is illustrated in Figure 7.1. The different parts of the procedure will be explained in more detail next.

7.1.1 *The experimental data*

The minimum experimental data are a list of peptide masses. This list may be supplemented by the peak intensities. We then have a peak list $\{(m_1, I_1), \cdots, (m_s, I_s)\}$ of peptide masses and intensities. In addition we may know (or have estimated) one or several other types of data, for example:

- the origin species (or group of species) of the proteins;
- the apparent mass of the intact protein (for example, found by 2D gel separation);

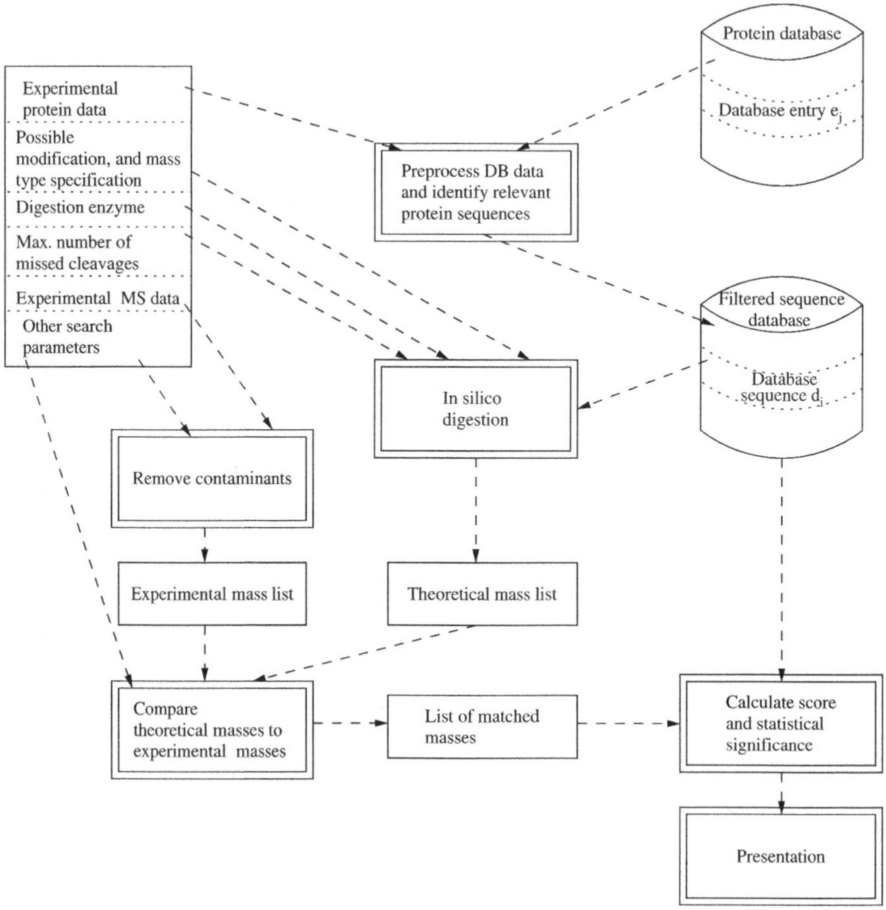

Figure 7.1 Illustration of protein identification by MS data

- the isoelectric point of the intact protein;
- the protein amino acid composition (though this is uncommon as a laborious chemical procedure is required to determine it).

Usually the peak list is processed as described in Chapter 6 before comparison with the theoretical data.

7.1.2 The database – the theoretical data

The theoretical data are a set of theoretical peptide masses derived from sequences in the database. The database is usually preprocessed and filtered before the *in silico* digestion.

Preprocessing the database

The sequences in the database may not correspond completely to the sequences of the sample proteins. For example, they may contain signal sequences, propeptides, or transit peptides. Such sequences are normally cleaved off from the protein during its life cycle and/or its transport in the cell. They are therefore often removed from the database sequences before *in silico* digestion (which can lead to unidentified peaks if the sample proteins still contain these subsequences). In order to obtain more reliable comparisons, one can also take into account other information about the sequences, for example reported sequence conflicts or variants, see the feature table of Swiss-Prot, Section 1.5. Thus there can be several alternative masses for some of the theoretical peptides.

Filtering the database

If some properties of the sample protein(s) are known (like mass, pI, amino acid composition, species of origin), the database sequences that do not satisfy these properties (within certain tolerances) can be filtered out. Filtering can, however, be risky if it is not taken into account that the database entry can contain a slightly different form for certain proteins, such as a fragment (or subsequence) of the experimentally observed protein. Search programs can take such problems into consideration, for example in the case of Mascot where applying a protein mass constraint results in a search against database sequences that have a theoretical mass that is less than or equal to the specified mass. Additionally, keywords in the specific search database can be used (such as biological process, cellular component, disease, etc.) if they are available (not all sequence databases carry such annotation). The amount of filtering applied will be a trade-off between ensuring that the correct protein sequences are included in the searched sequences on the one hand, and attempts to minimize the number of candidate sequences in order to reduce the search time, and avoid potential random matches on the other hand.

Peptide mass calculation

The mass of a theoretical peptide is calculated as the sum of the residue masses with the addition of the mass of H (nominal mass 1 Da) at the N-terminus, and OH (nominal mass 17 Da) at the C-terminus.

In silico digestion and construction of the theoretical peptide mass list

When the relevant sequences are extracted from the database, they are *in silico* digested and the peptide masses calculated, in accordance with several specified parameters, as follows:

- The enzyme used for digestion.
- The modifications to consider. Usually the search programs are linked to a database (or list) of different modifications, containing the mass of the modifications and the affected amino acids. Some search programs also allow for user-defined modifications.

  The modifications can be of two types:

  Fixed modifications These are modifications that will be present on any occurrence of the affected amino acids. Fixed modifications do not affect the search time or the size of the search space, since the only consequence is that the mass of the affected amino acid is changed (thus not extending the number of theoretical masses).

  Variable modifications These are modifications that may be present at some or all positions of the affected amino acid. Variable modifications can (dramatically) increase the search time and the size of the search space. If an amino acid with a possible modification occurs *n* times in a peptide, then the experimental mass is one of 2^n different possible theoretical masses (as shown in Section 3.2).

- The mass can be specified in different ways, with and without (positive or negative) charge:

  - $[M+H]^+, [M], [M-H]^-$
  - monoisotopic or average masses

- Maximum number of missed cleavage sites for a peptide. If the digestion is assumed to be complete, the number of missed cleavages is set to zero. Increasing this parameter will increase the number of theoretical masses constructed, and will therefore also increase the probability for random matchings. As shown in Section 3.2, if a complete digestion produces *n* (unmodified) peptides, than allowing for *k* missed cleavages will produce $\frac{1}{2}(2n+2nk-k^2+k+2)$ peptides.
- There might also be cleavages at unexpected sites. This can be implemented by searching for any subsequences that have a mass equal to an (unmatched) experimental mass. The subsequence may occur anywhere on the whole sequence, or must satisfy the cleavage rule of the protease at one terminus (the latter is sometimes called *semi-specific cleavage*). For trypsin this last alternative means that either the subsequence must end with one of {`R,K`}, or the residue before the subsequence must be one of these two.

7.1.3 Other search parameters

The search and presentation of results are controlled by specifying several *search parameters*, in addition to those described in the preceding subsections. Some prominent examples are:

- The expected number of proteins in the sample. This has consequences for the score calculation and statistical significance of the sequences found.
- The tolerance for comparing protein masses.
- The tolerance for comparing peptide masses. This value depends on the expected accuracy of the MS instrument; since no mass spectrometer has perfect accuracy, this parameter is always specified.
- Several parameters regarding the presentation of the results (for example, the minimum number of peptide mass matches required for a sequence to be presented). These parameters can be used further to automatically filter the results.

7.1.4 Organization of the database

In order to quickly obtain the protein sequences of interest (filtering), the database can be organized as a relational database, or with an explicit set of index tables. Each protein mass in such an index table then has pointers to the protein sequences with this mass.

Another means to increase the speed relies on saving the result of *in silico* digestions. The theoretical masses obtained are then sorted, and each mass is provided with indices that point to the sequences in which they occur, together with some peptide information (modifications etc.) as shown in Figure 7.2. Note that this approach requires separate index tables for each protease. Also note that the index tables have to be changed whenever the protein database is updated. Thus, the decision to use index tables relies on a balance between saving computer time during the search on the one hand, and the extra database administration and preprocessing time required on the other hand.

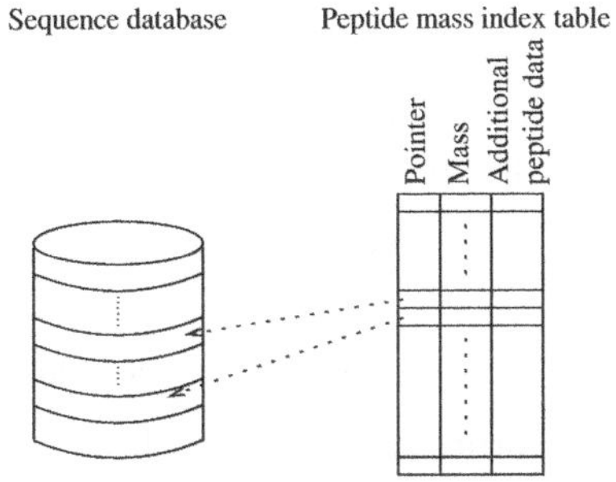

Figure 7.2 A peptide mass index table. Each peptide mass has pointers to database sequences as well as some additional information

7.2 The peptide mass comparison

The main task is to compare two lists of masses, and the straightforward approach is to sort the two lists on masses and perform a parallel comparison. Some aspects that have to be taken into account are as follows:

- An experimental mass may match more than one theoretical peptide mass within the given threshold.
- A theoretical mass may match more than one experimental peptide, i.e., several experimental peptides may have the same mass.
- A theoretical mass may match both an unmodified peptide and a second modified peptide.
- Both a concatenated theoretical peptide (missed cleavages) and one of its parts may find matches.
- Some of the experimental masses may come from noise or contaminants.
- Different peptides can have similar masses, due to permutations of the amino acids, or because different amino acids, doublets, or triplets of amino acids can have similar masses, such as the mass equality of `I` and `L`, and the mass similarity of `W` and `EG`, and of `HP` and `FS`.

Thus for each experimental mass there can be a number of false matches (matches to other peptides than the correct one), and this number depends on the accuracy of the measurements. To get a feeling for how many false matches one can expect, try Exercise 7.4.

7.2.1 Reasons why experimental masses may not match

Even if the correct protein sequence is in the database, usually not all of the experimental masses find a match in the database sequence, as also described in Chapter 3.

There are several reasons for this, the most common being:

- the unmatched masses might come from other proteins, noise, or contaminants in the protein mixture;
- unexpected posttranslational modifications (modifications that are not considered) may occur on some of the peptides;
- (unexpected) differences in the database sequence and the experimental sequence (variants or conflicts);
- an unexpected number of missed cleavages, or unexpected cleavages;
- a mass measurement that carries an error larger than the specified tolerance.

The number of unmatched masses therefore depends on how these points are considered in the searching.

7.3 Database search and recalibration

The masses from a spectrum may be incorrect, and a reasonable *recalibration* 'weaved' into the database search can often be useful. Such a recalibration utilizes the correlation between experimental and theoretical masses of recognized mass matches $\{(m_i, t_i)\}$. These correlations are used to calculate a regression line, and the experimental masses are changed, according to this, to $\{m_i^*\}$. A new search with the corrected masses is then performed. The idea is that more matches should be found after a reasonable recalibration, but there are also some of the original matches that can now be missed. This recalibration can be iteratively repeated. For the recalibration to be effective it requires that a sufficient number of the matches found in the first search are true matches, otherwise there is a substantial risk that the corrected masses will be farther away from the correct ones than the original masses. We now show how the idea of recalibration is used in two search programs.

7.3.1 The search program MSA (Mass Spectra Analyzer)

MSA is a program that uses recalibration implicitly, and the developers claim that this makes internal or external calibration unnecessary. It does not require manipulation of the experimental data, and makes use of the observation that the relative mass errors correlate linearly with m/z. MSA performs the search in two steps (before a score is calculated in a third step).

Step 1 *Search with large mass tolerance.* The database is searched for all sequences that contain at least five theoretical (tryptic cleavage) peptides that all have a mass similar to an experimental mass within a fairly large deviation (for example, ± 500 ppm). This results in a subset of the database sequences.

Step 2 *Implicit recalibration and removal of matches.* Each sequence extracted in Step 1 is individually examined to remove matches that obviously are incorrect, and also to calculate how good the matches are. This is (iteratively) performed through four substeps. The mass errors are calculated relative to the experimental mass as $e_i = (m_i - t_i)/m_i$.

1. The mean μ and standard deviation σ of the errors are calculated. In Figure 7.3(a) the dashed lines indicate the limits $\mu - 2\sigma$ and $\mu + 2\sigma$. Matches outside these limits are removed (three in the figure).
2. A linear regression determines a line $y = a + bm$, correlating the relative mass errors with the peptide m/z for the rest of the matches, as shown in Figure 7.3(b).
3. The distances from the relative errors of the matches to the line y are calculated and plotted, Figure 7.3(c). Outliers are again removed.
4. Substeps 2 and 3 can be repeated, to determine the final matches.

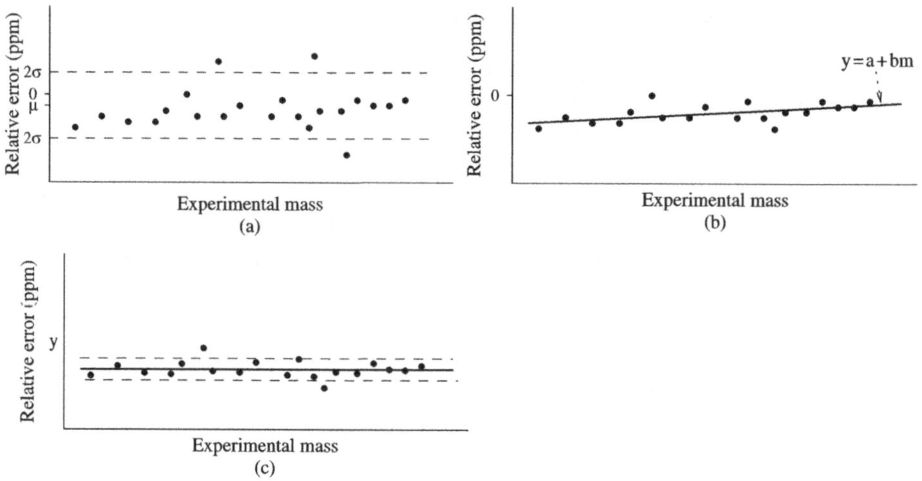

Figure 7.3 Illustration of Step 2 in MSA. (a) The relative error as a function of the experimental mass (m/z). (b) The linear relation between the relative error and the experimental mass, $y = a + bm$. (c) The relative error distances from the line $y = a + bm$. The dashed lines are at two standard deviations

7.3.2 Aldente

The program Aldente solves the multiple matching problem by calculating a line through an experimental/theoretical mass diagram. The line should maximize the number of matches, and this matching is combined with taking the calibration error into account. A diagram is shown in Figure 7.4(a). It shows that an experimental mass can match more than one theoretical mass, and the correct matches are to be determined.

It is assumed that, taking the calibration errors of the mass spectrometer into account, the experimental mass m can be described as a function of the correct mass r, as $m = h + (1 + s)r$, where h is the shift, and s the slope. h is an absolute value (given in daltons), and s a relative value, given in ppm. The user can specify maximum and minimum values for h and s (by using knowledge about the instrument). The maximum value of h is the maximum difference allowed between a theoretical and experimental mass in daltons. The maximum value of s is the maximum difference allowed between a theoretical and experimental mass in ppm. If both are given (not equal to zero), the largest value of these two is used as the threshold (see Figure 7.4(b)), where the most external lines define the area used for matching (h for lower masses, s for higher masses). If the theoretical masses are considered the correct masses, the relation between the experimental and theoretical masses (t) is $m = h + (1 + s)t$.

A straight line through the defined area is to be determined which should maximize the number of matches. The threshold for a match between the line and an (experimental, theoretical) point is called the *internal error* (of the instrument), and is specified in ppm. Figure 7.4(c) illustrates such a line. Hough transformation is used for determining the line. An example of the values for the parameters are $h_{max} = 0.2$ Da, $s_{max} = 200$ ppm, internal $\text{error}_{max} = 25$ ppm.

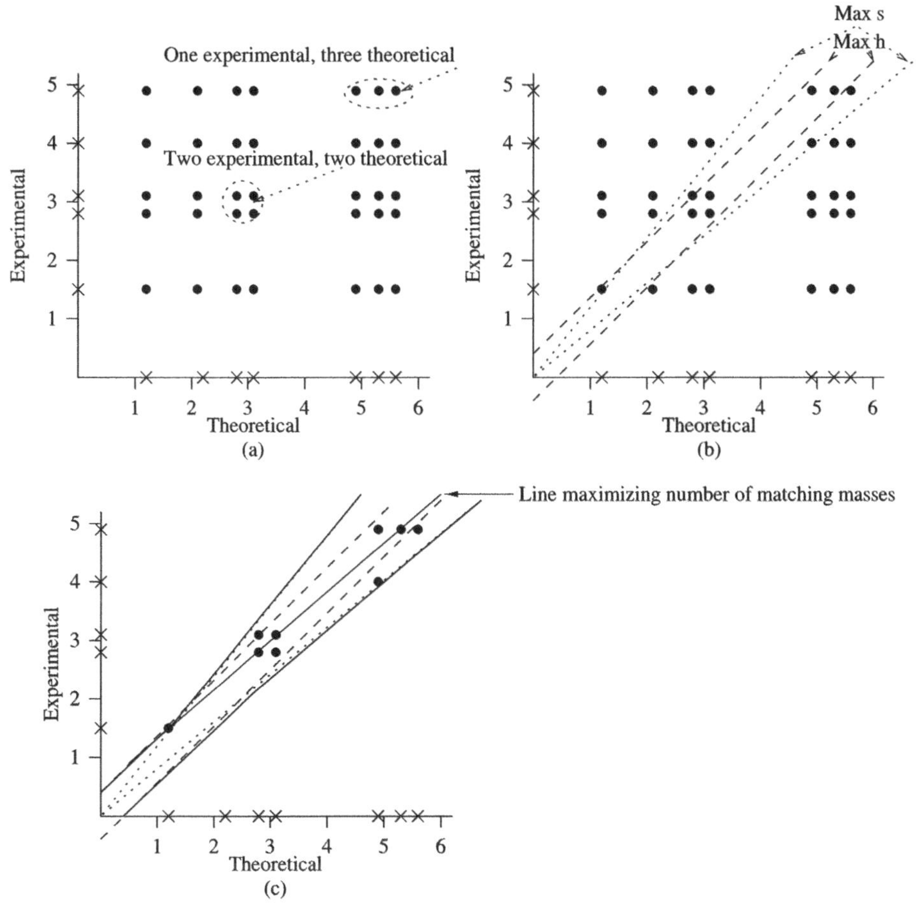

Figure 7.4 (a) The crosses on the axes indicate theoretical and experimental masses respectively, and bullets the possible matches. (b) The dashed line represents the maximum value for *h*, and the dotted line the maximum for *s*. Matches must lie inside the outermost boundaries (here nine bullets). (c) A line is then to be determined which maximizes the number of matching masses, inside the defined area, as the line drawn in the diagram shows. This line is found by using the Hough transformation

7.4 Score calculation

When a set of database sequences is found to have matches to the experimental peptide list, the sequences should be given a score to extract the sequence(s) that most likely correspond to the protein(s) from which the experimental peptide list is constructed. Ideally the spectrum should contain only the peptides derived from a single protein, but it is not uncommon that there are peptides from more proteins in a spectrum (as in a spot from a 2D gel).

The goal of a scoring scheme is to give the correct database sequences the best scores, and also to obtain a large difference between the correct sequence(s) and the wrong one(s).

The construction of good scoring schemes is not straightforward, and many different schemes have been proposed.

The score should be based on the similarity between the observed peaks and the theoretical peaks obtained when simulating an identical MS experiment on the database sequence. This similarity depends on a set of different components. The challenge is then to determine these components, how they should be measured, and how they should be combined in a formula with relative weights between them.

7.4.1 Score components

The components can be divided into two levels: the peptide level and the protein level. The components on the peptide level give scores to each individual peptide mass match. The following is a list of components that are (widely) used.

Peptide level

- The mass differences for the matching masses.
- The number of missed cleavages.
- The intensities of the experimental peaks.
- The expectation of the theoretical peptide's occurrence in a spectrum and intensity (based on its amino acids), see Section 6.1.3. This can include both theoretical peptide masses that are matched to experimental masses and masses that are not matched. For unmatched theoretical peptides one can calculate the probability that such a peptide should not be observed in a mass spectrometer.
- The number (and types) of modifications.

Protein level

- The number of experimental masses.
- The number of matching masses.
- The sequence coverage of the database sequence (%).
- The difference in experimental and theoretical protein masses.
- The difference in experimental and theoretical pI.
- The (estimated) number of proteins in the sample.
- The variation of the peptide matching errors. Large variation means that one should be more skeptical of the suggested sequence.

Determining the score coefficients

The problem with using several score components is determining the manner in which the components should be combined in a function, and how each component should be weighted. Simple scoring functions use the product of the component scores, or a linear function. Let the score of component i be s_i; then the sequence score for a linear function is $\sum_i c_i s_i$, where c_i are coefficients (weights) which have to be determined (or learned). Learning is typically done by considering a lot of manually verified identifications, and determining the coefficients such that 'good' matches get high scores.

7.4.2 *Scoring scheme examples*

As mentioned above, a number of different scoring schemes have been proposed. Briefly, they can be divided into non-probabilistic and probabilistic schemes. Non-probabilistic schemes typically combine a set of scoring components in some reasonable form, and probabilistic schemes try to calculate the probability that a given sequence is the origin of the spectrum. In this subsection we present some (rather diverse) scoring schemes.

Mascot–Mowse score

The exact score calculation used in Mascot is not fully available, but it is inspired from the Mowse score. The Mowse score is based on the following points:

1. Decreasing the score with increasing mass of the database sequence.
2. Increasing the score with increasing number of mass matches.
3. Decreasing the score (for a match) with increasing number of (theoretical) peptides in the database with similar masses to the experimental mass.

The protein sequence mass range is divided into intervals j of fixed length (10 kDa), and the peptide mass range into intervals i of another fixed length (100 Da). All theoretical peptides are placed in an (i, j) class, depending on the peptide mass and the mass of the protein sequence to which it belongs. Let $n_{i,j}$ be the number of theoretical peptides in the database belonging to class (i, j), and then define (for point 3)

$$f_{i,j} = \frac{n_{i,j}}{\max_i n_{i,j}}$$

Let a sequence d, with theoretical mass m_d, obtain r matches to the experimental masses and let c_k be the peptide class (i, j) of the kth match. Then the score of the sequence is

$$Score_d = \frac{50\,000}{m_d \prod_{k=1}^{r} f_{c_k}}$$

Example Consider two peptide mass intervals r, s (let for example r be the mass [800,899.9] Da) and one protein mass interval j (for example, the mass [20,29.99] kDa). Assume now that $n_{r,j} = 30$, $n_{s,j} = 60$, and that $\max_i n_{i,j} = 100$. Then $f_{r,j} = 0.3$, $f_{s,j} = 0.6$. Assume further that we have a database sequence d with theoretical peptides in intervals r and s. If we set $C = 50\,000/m_d$ the scoring for d is $C/0.3 \cdot 0.6 \approx 5.5C$.

△

The scoring scheme of Mascot is further described in Section 7.5.3.

ChemApplex

In the scoring scheme of the program ChemApplex each mass match is first given a chemical score ChemScore, which depends on the probability that this peptide would occur in a real spectrum. Thus knowledge about the protein digestion and the peptide ionization are used. ChemScore depends on the occurrence or absence of certain amino acids (`R,C,K`), and chemical modifications (on `C`, `M`, or `Q`). In addition, missed cleavages are taken into account, and the amino acids at the missed cleavage point (for example, a missed cleavage at `RD` will result in a higher score than a missed cleavage at `KV`, when all other conditions are equal). The final score of a peptide match is then

$$\frac{\text{Intensity} \times \text{ChemScore}}{\text{Masserror}}$$

where Intensity is the intensity of the (experimental) peak, and Masserror is the difference between the two masses in ppm. A simple formula for the final sequence score is the sum of the peptide scores. However, in order to remove a bias towards long sequences, the sequence length can also be taken into account in this score.

OLAV-PMF score

This is a probability-based scoring method. It tries to calculate the probability that a suggested protein sequence really is the origin of the experimental spectrum.

A commonly used formula for scoring schemes in bioinformatics is the log-odds formula, and this is the basis for OLAV-PMF scoring. In particular, the score of an event E is determined by first calculating (estimating) the probability of E to occur given two alternative hypotheses H_0 and H_1, namely $P(E|H_0)$ and $P(E|H_1)$. The score is then calculated as the log-odds

$$\log \frac{P(E|H_1)}{P(E|H_0)}$$

H_0 is often the 'random' hypothesis, such that $P(E|H_0)$ is the probability that E will occur just by chance. In the context we are discussing, E is the calculated value for some (peptide or protein) component's match. H_1 is the hypothesis that the considered match is a correct one, and H_0 that the match happens just by chance. $P(E|H_1)$ is then the probability that a correct match will result in the specified value for the given component. The notation used for $P(E|H_i)$ is $x_z^{H_i}$, where x specifies a frequency (or probability), z a component specified index, and i either 0 or 1, see below.

OLAV-PMF uses three components: sequence coverage v, amino acid composition c, and posttranslational modifications o. These three components are considered independent, and the score is a function of several component probabilities calculated for both of the hypotheses. For determining these probabilities, two training sets of peptides are used, T_1 for hypothesis H_1 and T_0 for hypothesis H_0. T_1 is constructed from correctly matched spectra, T_0 from a set of randomly generated sequences, which are then matched against the experimental spectra used for T_1. Distributions (or frequencies) are learned (determined) from these two sets, and are then used for estimating the probabilities for the different hypotheses.

Now let d be the database sequence we are calculating a score for, P the set of theoretical peptides from d that have a matching mass in the set of experimental masses, and p a member of P.

Sequence coverage score, S_v Sequence coverage distributions F are learned for both hypotheses, and are used for the score

$$S_v = \frac{f_v^{H_1}}{f_v^{H_0}}$$

$f_v^{H_1}$ is the probability that a correctly suggested sequence will have sequence coverage f_v, and $f_v^{H_0}$ that a random sequence will have the same sequence coverage. It was found that $f_v^{H_1}$ could be adequately fitted to a Gaussian distribution and $f_v^{H_0}$ to an exponential distribution.

Amino acid composition score, S_c The amino acid compositions of the peptides in the two training sets are compared, and the subset B of the amino acids that show a bias factor to one of the training sets is found. Let $u_b^{H_i}$ be the frequency of amino acid b in the training set T_i. Then a match $p \in P$ is scored as

$$S_c(p) = \prod_{b \in B} \left(\frac{u_b^{H_1}}{u_b^{H_0}} \right)^{n_b(p)}$$

where $n_b(p)$ is the number of b occurring in p. The final composition score for d is then

$$S_c = \prod_{p \in P} S_c(p)$$

Posttranslational modification score, S_o Let I be the set of modification types to consider in the search, with each modification type attached to only one amino acid (for example, oxidation of H). The probabilities of modifications at various positions in a peptide are considered independently, and the probability $p_i^{H_1}$, $i \in I$, for modification i to occur for hypothesis H_1 is found by using T_1. For H_0, $p_i^{H_0}$ is set to 0.5. For each (peptide) match there may be several possibilities for which modification types can be included (for example, one C has been alkylated by iodoacetamide, and one H and one M have been oxidized, resulting in a total of three modification types). If at least one amino acid is present in a higher number than the number of modifications of that type, the positions of the modifications are unknown. Let $M(p)$ be the set of all possible locations of the modifications, and $m \in M(p)$. The score of the modification m is

$$S_o(m) = \prod_{i \in I} \left(\frac{p_i^{H_1}}{p_i^{H_0}} \right)^{n_o^i} \left(\frac{1 - p_i^{H_1}}{1 - p_i^{H_0}} \right)^{n_t^i - n_o^i}$$

where n_t^i is the number of amino acids that could potentially be modified by i, and n_o^i is the number of these amino acids that really are modified. Note that the formula includes probabilities for modified and non-modified amino acids. The final modification score for p is then the average score of all the possibilities

$$S_o(p) = \frac{1}{|M(p)|} \sum_{m \in M(p)} S_o(m)$$

and the final modification score for d is

$$S_o = \prod_{p \in P} S_o(p)$$

The final total score of sequence d is then $S(d) = \log(S_v S_c S_o)$.

Example Suppose that the database sequence for which we want to calculate a score has a coverage of $v = 0.28$, that one of the matched peptides is $p =$ `HPDYSVVLWLRLAK`, and that the matching includes one oxidation. We will calculate the coverage score, and the composition and modification score for the peptide.

Coverage score Suppose that the exponential coverage distribution for H_0 is

$$f_v^{H_0} = \frac{e^{-\frac{v}{0.02}}}{0.02} = 0.000\,042$$

for $v = 0.28$. For H_1 let it be normally distributed, $(\mu, \rho^2) = (0.37, 0.14^2)$. Then

$$f_v^{H_1} = \frac{1}{\sqrt{2\pi\rho}} e^{-\frac{1}{2}(\frac{v-\mu}{\rho})^2} = 0.868$$

for $v = 0.28$. Thus we find for the coverage score $S_v = 0.868/0.000\,042 \approx 20\,000$.

Composition score The relative frequency ratios

$$\frac{u_b^{H_1}}{u_b^{H_0}}$$

for the amino acids found to be biased to one of the training sets are `A`: 0.89, `E`: 0.93, `F`: 1.12, `H`: 1.13, `M`: 0.72, `P`: 1.19, `R`: 1.22, `V`: 1.09, `Y`: 1.18. Thus the composition score of p is $S_c(p) = 1.13 \times 1.19 \times 1.18 \times 1.09^2 \times 1.22 \times 0.89 = 2.05$.

Modification score The oxidation can be at either the `H` or the `W`, thus presenting two possibilities. The modification probabilities for these oxidations are found to be $p_H^{H_1} = 0.02$, $p_W^{H_1} = 0.21$. Since there is only one of each of the actual amino acids, the formulas become quite simple:

- $S_o(\text{Oxidation at H}) = \frac{0.02}{0.5} \times \frac{0.79}{0.5} = 0.06$.
- $S_o(\text{Oxidation at W}) = \frac{0.21}{0.5} \times \frac{0.98}{0.5} = 0.82$.

The modification score of p is then $S_o(p) = (0.06 + 0.82)/2 = 0.44$.

△

7.4.3 Identification from a protein mixture

If there are reasons to believe (for example, from the search results) that the spectrum contains peptides derived from more than one protein, one can take this into account to potentially get a more efficient identification. One method is to slightly alter the search database, such that the two highest scoring sequences are 'combined', and then perform a new search. If the spectrum does indeed contain peptides derived from these two proteins, the score against the combined sequence should be high. Another approach is based on the *subtraction method.* The masses corresponding to the peptides from the sequence with the highest score are removed from the experimental masses, and a new search is performed with this reduced spectrum.

7.5 Statistical significance – the P-value

After a database search is performed, one sequence ends up with the highest score, S. However, this sequence may still not be the correct match, either because the experimental protein is not in the database, or because the correct sequence has received a lower score. One should also keep in mind that the scores from different experiments cannot generally be compared. A statistical significance for the score should therefore be determined: the probability that a sequence with such a high score is a correct identification. It is actually most convenient to calculate the complementary value: the probability that the identification is incorrect. This is commonly done by calculating the **P**-value, the probability of achieving a score of S or higher by chance. Another, but correlated measure is the **E**-value, the expected number of database sequences achieving such a high score by chance. For small values, the **P**-value and the **E**-value are equal. The final score is sometimes presented as $-C \log \mathbf{P}$ (where C is a constant), in order to give the 'best' matching sequence the highest score.

Calculating the **P**-value from theoretical considerations is only possible for rather simple scoring schemes.

7.5.1 *A priori probability for k matches*

Consider the number of matching peaks as a simple scoring scheme. We want to calculate the probability $Pr(d, k, n)$ that a spectrum R with n peaks has k matches to a *specific* sequence d in the database, just by chance. This means that R is considered a random match to d. The presentation is based on Eriksson and Fenyö (2002). Let us first assume that all masses in R have the same probability p for yielding a match in d, and that these probabilities are independent of each other. Then the probability of k matches follows the binomial distribution

$$Pr(d, n, k) = \binom{n}{k} p^k (1-p)^{n-k} \tag{7.1}$$

It is commonly known that p depends on the number of theoretical peptides in d. This number depends on the length (and therefore the mass) of d, and on the number of missed cleavages allowed in the *in silico* digestion. Therefore, p should be calculated for each sequence separately. Two other issues complicate the calculation further: (i) the number of peptides that have a specific mass depends on the mass; and (ii) the peptide masses occur in clusters, and the width of the peptide mass distribution increases with increasing mass of the clusters. This implies that p is not equal for all masses in R. To handle this the considered peptide mass region is divided into successive subregions $\{r_i = [m_i, m_{i+1}]\}$, where p is considered equal for all masses in a region. Let p_i be this value for region r_i.

Let n_i be the number of masses from R falling in region r_i, and k_i the number of these n_i masses that give a match to d. If q is the number of regions, we must therefore require $\sum_{i=1}^{q} k_i = k$. We see that k matches can be obtained for a number of different distributions of $\{k_i\}$ as long as the sum is k. Then from Equation 7.1 we get

$$Pr(d, n, k) = \sum_{(k_1, k_2, \ldots, k_q) | \sum_{i=1}^{q} k_i = k} Pr(d, n_1, k_1) Pr(d, n_2, k_2) \cdots Pr(d, n_q, k_q) \tag{7.2}$$

Determining p_i

p_i is the probability that a mass m in region r_i matches a theoretical mass in d. Let further:

- t be the number of theoretical peptides in d. t can be estimated from the mass of d and the maximum number of allowed missed cleavages.
- f_i be the frequency of the peptide masses falling in region r_i. f_i can be estimated from an *in silico* digestion of a sequence database.

Then we can calculate:

- $f_i t$ as the estimated number of theoretical peptide masses in d which are in region r_i.
- $p_i^* = f_i t/(m_{i+1} - m_i)$ as the probability that there is a mass in d in the same mass cluster as m. This follows from the fact that there are $m_{i+1} - m_i$ mass clusters in r_i (since the clusters are approximately 1 Da apart).
- In order to be considered a match, the two masses must not deviate by more than a given threshold, Δm.

Then p_i can be found by

$$p_i = p_i^* \delta(I, \Delta m)$$

where $\delta(I, \Delta m)$ is interpreted as a measure of the fraction of the peptide masses in a cluster (in region r_i) that falls within $\pm\Delta m$ from a random mass in the cluster. δ depends on the shape of the distribution of the masses in the cluster, and can be estimated from measuring peptide masses with the mass spectrometer used.

Note that $Pr(d, n, k)$ is not the **P**-value, since it calculates the probability for one *specific* database sequence to obtain exactly k matches. The **P**-value is the probability that at least one of the proteins in the database obtains k or more matches just by chance.

7.5.2 Simulation for determining the P*-value*

Since it is usually not possible to theoretically calculate the **P**-value, some more indirect methods must be used. The common way is to obtain a probability distribution of the highest scores that will be found when searching in the database with 'random' spectra.

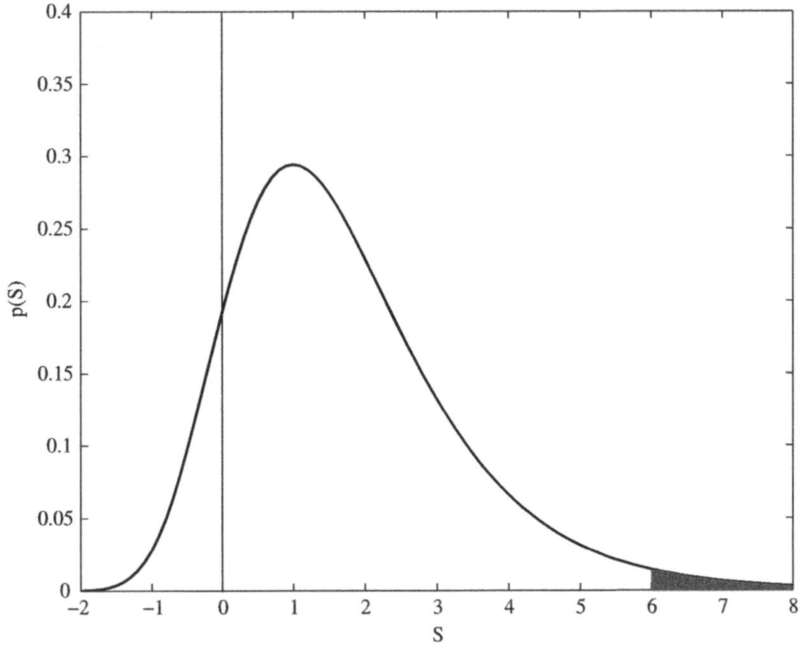

Figure 7.5 An example probability distribution function. The total area under the curve equals one, and the shaded area is 0.0181. This means that the probability for obtaining a value of $S \geq 6$ by chance is 0.0181

This distribution is supposed to be the distribution of scores that can be achieved just by chance. When a score S is found for a database sequence, the part of the distribution for scores higher than S is found. This is taken as the **P**-value of S, as illustrated in Figure 7.5.

Constructing the probability distribution

We describe three approaches for the construction of a probability distribution (Eriksson *et al.* (2000)).

Generate random spectra A set of independent *random* spectra (for example, 10 000) is generated. A search is performed with each of them against the same database, and a distribution of the highest scores is obtained. This is assumed to be a distribution of false positive (random) scores. This empirical distribution is then fitted to a proper probability distribution. For this procedure to yield reasonable results, some factors need to be taken into account:

- the random spectra must resemble possible experimental spectra;
- the spectra must be random in relation to the experimental context and the database used;
- the database search must be performed with the same search parameters as used in the search with the experimental spectra.

The most important task is to construct random spectra that could be generated by the experimental conditions used. In order to achieve this, two frequency distributions are constructed from the search database:

- the number of peptides (for the specific protease used) per protein, taking modifications and allowed missed cleavages into account;
- the number of peptides per mass unit.

From these two distributions a random spectrum can be constructed by drawing the number of peaks from the first distribution, and the masses for these peaks from the second.

Another, simpler way for generating a random spectrum is to randomly choose a protein in the database for each peak, and use a randomly chosen theoretical peptide mass from that protein for the peak mass. Experiments show no large differences for the usefulness of these two methods for generating random spectra.

Search with *ideal* spectra The distribution can also be obtained by first choosing a random set of proteins from the search database, and then constructing a theoretical spectrum for each of these. These theoretical spectra are in fact *ideal* spectra for those proteins. For each such spectrum, the *second* highest score is recorded when searching against the database. These second best scores are then considered false positive identification scores.

An alternative probability distribution Another technique compares the highest score to the second (or third) highest score, assuming that these latter scores are not correct identifications. A probability distribution of the score *difference* between the correct identification and the best incorrect identification can then be learned from experiments.

7.5.3 A simple Mascot search

As an illustration we perform some searches using Mascot. As mentioned above, the actual algorithm for calculating a Mascot score is not available, but what is known is summarized as follows:

- A probability P is calculated for the comparison between the experimental spectrum and a database sequence. P is the probability that the 'combined matches' constitute a random event. Note that P is independent of the database size.
- It is reasonable to assume that the database sequence that provides the 'best' match to the experimental spectrum should have the highest score. Therefore the score S is calculated as $S = -10 \cdot \log_{10}(P)$.
- From P and the size of the database, a score threshold T is calculated. T is determined such that a score that is equal to or greater than T can be considered random with a probability less than or equal to another threshold p. The default threshold used by Mascot is $p = 0.05$, but this can be adjusted by the user if desired.
- The **E**-value is calculated as $E = p \cdot 10^{\frac{T-S}{10}}$.

Consider entry P08253 (human matrix metalloproteinase-2, MMP2_HUMAN) in Swiss-Prot. This protein would yield 38 tryptic peptides in the mass range 600 to 3000 Da, if no missed cleavages are considered. Among them are four peptides with monoisotopic masses 611.28, 794.42, 1052.52, and 818.43 Da. We restrict the search first to the human (16 027) sequences, use ± 0.05 Da as the maximum allowed mass difference, and do not allow for missed cleavages. Database sequences with a score above 55 (T) are then significant at the 0.05 (p) level. First we search with the first two masses. P08253 receives the second highest score, and the highest score is 36. Then we search with three masses. P08253 now receives the highest score, this time 51. When searching with all four masses, P08253 receives a score of 70, which is significant at the 0.05 level. When searching in the whole Swiss-Prot database the score of 70 remains significant (T has now gone up to 67, due to increased database size). Note also that the score is reduced when the allowed mass error is broadened. For instance, searching with an allowed mass error of 0.15 Da in the human sequences gives a score of only 68 for the four matches, while T is still 55.

This experiment indicates that four masses at high accuracy could be enough for significantly (at the 0.05 level) identifying an average human sequence. We must, however, bear in mind that we have allowed no missed or unexpected cleavages, and that we have considered no modifications. Taking these items into account will increase the efficiency of the search (allowing us to identify more peptides), but could also increase the number of false positive peptide matches.

7.6 Characterization

Characterization in our context means discovering:

- the potential posttranslational modifications in the protein; and
- potential differences in the amino acid sequence between the sample protein and the sequence given in the database (conflicts or variants).

Since the identification procedure can take modifications into account, some of the modifications may have been found during this process. However, since a protein is usually identified with a sequence coverage of less than 50 %, nothing is known about the majority of the protein. Characterization is often performed by MS/MS analysis, but here we will briefly discuss how to approach this problem when only MS instruments are available.

Let us assume that the protein is identified, and thus known. The problem is then to match the experimental masses (peak list) to the known protein sequence. First, all conceivable contaminants are considered and removed from the peak list, and then peptide mass comparisons are performed as explained earlier. Since we now have only one sequence to consider, more modifications, or a combination of modifications, can be included in the search, as well as unexpected cleavages. These measures should then allow us to increase the coverage of the sequence.

By allowing for more modifications and cleavage variants, however, there is also a correspondingly higher chance for false positive matches. Specifically, an experimental mass can be found to match to several theoretical peptides, and several experimental masses can be matched to different (modified or cleaved) versions of the same peptide.

Example Assume that two of the masses in an experimental spectrum are $m_1 = 692.8$, $m_2 = 706.8$. Assume further that three theoretical peptides from a sequence are p_1=AGCVSTR, p_2=AASVTCR, and p_3=CQTWR. Then m_1 matches unmodified p_1 and unmodified p_3. Furthermore, m_2 matches unmodified p_2, but also p_1 and p_3 if the cysteines (C) in these two peptides are methylated. Note that although methylation could also occur on R, a tryptic cleavage would probably not occur in that case.

△

When several alternative theoretical peptides match a mass, a score can be calculated for each alternative. The score should be based on the mass difference and the probability that such a peptide would be ionized and detected in the spectrum, see Section 6.1.3.

The mass comparison will result in different types of data (also see Section 7.2.1):

- A set of experimental/theoretical mass matches, not necessarily one-to-one.
- A set of unmatched experimental masses. They can be caused by unconsidered contaminants, by the results of unexpected digestions, or by peptides that carry unconsidered modifications.
- Uncovered parts of the sequence for which no peptides were observed. These peptides may not have been ionized, or they could have eluded detection by not being recorded by the detector. The points mentioned above for unmatched masses may also explain this lack of coverage.

Since the experimental masses are expected to come from the sequence, it is possible to perform more comprehensive investigations:

- One could test for unexpected digestions by comparing the unmatched masses to every possible subsequence, or to semi-specific subsequences that satisfy the cleavage rule at only one end.
- One could consider more (unexpected) modifications.

Obviously, one could also perform multiple analyses of a single protein under different experimental conditions, and then compare the different results with respect to modifications. Ideally, such analyses should be undertaken in a common environment, with one common administrative unit. Some programs that can do this are listed in the bibliographic notes.

Exercises

7.1 Suppose the highest peak intensities of a spectrum are 29 621, 26 087, 19 699, 17 777, 9605, 7717, and that the total intensity is 253 422. Calculate the intensities normalized relative to highest value, and also by using cumulative intensity normalization.

7.2 Explain why recalibration can be dangerous.

7.3 Consider the Mowse score (Section 7.4.2). Assume the score for r matches a database sequence with mass m_d calculated to be S_d. Assume further that all f_{c_k} are equal to f. Derive an expression for the number of matches r^* required for a sequence with mass $2m_d$ to obtain the same score, S_d.

7.4 This exercise illustrates a rough calculation of the expected number of false positive matches. The number of different human sequences (entries) in Swiss-Prot is approximately 20 000. The average sequence has approximately 50 tryptic peptides, when no missed cleavage is allowed. About 45 % of these have masses in the range 500–1500 Da, so we assume that each sequence has 25 peptides with masses in the specified range. We know (Section 5.5) that the peptide masses occur in clusters, and, for the range considered here, there are approximately 1000 clusters. At this range the width of the clusters is around 0.3 Da. We now assume a uniform mass distribution over the range, that no two peptides from a sequence are in the same cluster, and that there are no modifications.

(a) Calculate the probability P that a specific sequence has a mass in cluster i.

(b) Find the expected number of sequences with a peptide mass in cluster i.

(c) Find the expected number of sequences with peptide masses in both cluster i and j.

(d) Find the expected number of sequences with peptide masses in all clusters i, j, and k.

(Note that the probabilities are calculated for *specific* clusters, since we search with given masses.)

7.5 Consider the Mascot score in Section 7.5.3. Assume that a score S is calculated from the probability P. Assume another score $S_1 = S + 10$.

(a) Show that the probability $P_1 = 10^{-1}P$.

(b) For the same database the **E**-value is proportional to the calculated values for P. Perform some of the searches described in Section 7.5.3 and show that this is correct.

Bibliographic notes

MSA	Egelhofer *et al.* (2000, 2002) Hough transformation; Duda and Hart (1972)
Aldente	Duda and Hart (1972); Tuloup *et al.* (2003) http://au.expasy.org/tools/aldente/

Scoring

Gras *et al.* (1999); Wool and Smilansky (2002); Zhang and Chait (2000)
Samuelsson *et al.* (2004)

Mascot http://www.matrixscience.com/help/scoring_help.html

ChemApplex	Parker (2002)
OLAV-PMF	Magnin *et al.* (2004)

Unmatched masses

Karty *et al.* (2002)

Identification from a protein mixture

Jensen *et al.* (1997); Zhang and Chait (2000)

Statistical significance

Eriksson and Fenyö (2002, 2003);Samuelsson *et al.* (2004)

Generating random sequences

Eriksson *et al.* (2000)

Programs for identification (and some characterization)

Expasy	Gasteiger *et al.* (2005)
Aldente	Tuloup *et al.* (2003) http://au.expasy.org/cgi-bin/aldente/help.pl
ChemApplex	Parker (2002)
GPMAW	Peri *et al.* (2001) http://welcome.to/gpmaw
MassSearch	James *et al.* (1994) http://www.cbrg.ethz.ch/services/MassSearch
Mascot	Perkins *et al.* (1999) http://www.matrixscience.com/search_form_select.html
MS-Fit	Clauser *et al.* (1999) http://prospector.ucsf.edu/
MultiIdent	Wilkins *et al.* (1999a, 1998) http://www.expasy.ch/tools/peptident.html
PeptIdent2	Gras *et al.* (1999)
PepMAPPER	http://wolf.bms.umist.ac.uk/mapper/
PeptideSearch	http://www.mann.embl-heidelberg.de/GroupPages/PageLink/ peptidesearchpage.html
Phenyx	http://www.phenyxms.com/
ProFound	Zhang and Chait (2000) http://proteometrics.com/prowl-cgi/ProFound.exe

Evaluation of identification algorithms

Chamrad *et al.* (2004)

Programs for characterization

FindMod	Wilkins *et al.* (1999b) http://au.expasy.org/tools/findmod
FindPept	Gattiker *et al.* (2002) http://au.expasy.org/tools/findpept.html
PeptideMass	Wilkins *et al.* (1997) http://au.expasy.org/tools/peptidemass.html
MassSorter	Barsnes *et al.* (2006) http://www.bioinfo.no/software/massSorter

8 Tandem MS or MS/MS analysis

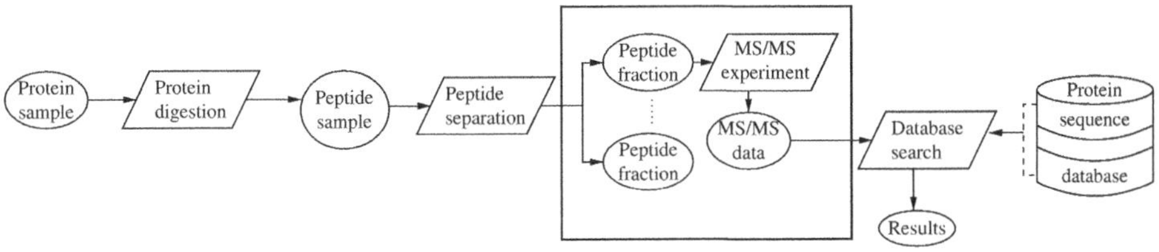

Identification of proteins by peptide masses (PMF) has some shortcomings:

- different peptides often have the same or similar masses, which necessitates extensive protein separation. However, this may even occur with peptides generated from a single protein, and thus other tools are needed to ensure the identification of such peptides.
- it is difficult to handle modifications efficiently as the exact location of the modification within the peptide sequence cannot be derived from the mass of the modified peptide alone if there are two or more positions in the peptide that can possess a modification of a certain mass.

A much better determinant for peptide identity (and indirectly for protein identification) is the peptide sequence itself. Although two peptides from different proteins may have identical sequences, such redundancy occurs far less frequently than at the peptide mass level. Usually one or two peptide sequences can already identify a protein. Additionally, if we can obtain mass information for each residue in the sequence, we can find the exact site of modification for modified peptides. This sequence-level information is obtained through tandem MS (or MS/MS) analysis.

The principle of MS/MS analysis is illustrated in Figure 8.1. This approach relies on two mass analyzers in tandem (hence the name). The first mass analyzer allows the selection of a particular m/z range, usually centered on the m/z of a peptide of interest. The selected peptide subsequently undergoes a *fragmentation* step with the second mass analyzer finally measuring the m/z of the resulting *fragment ions*. The resulting spectrum, which contains the m/z and ion intensity for each of the fragment ions, is called an *MS/MS spectrum*. A more standardized nomenclature refers to a fragmentation spectrum as a *product ion spectrum.*

Computational Methods for Mass Spectrometry Proteomics I. Eidhammer, K. Flikka, L. Martens and S.-O. Mikalsen
© 2007 John Wiley & Sons, Ltd

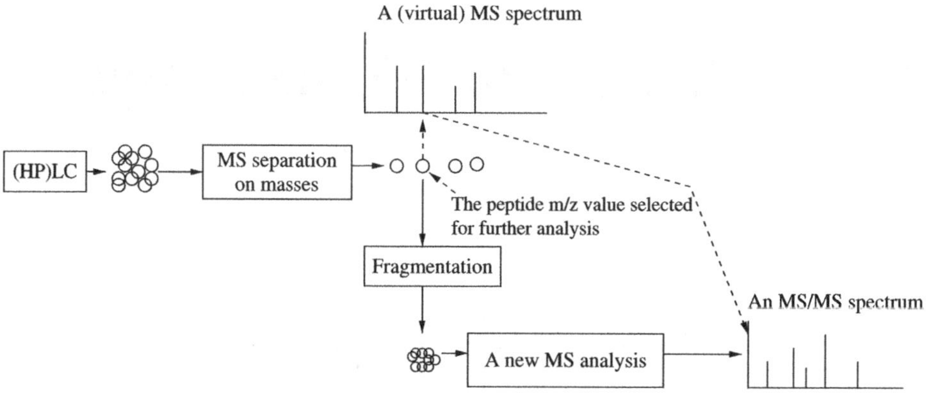

Figure 8.1 The principle for obtaining an MS/MS spectrum. Note that the MS spectrum does not need to be constructed

For identification and characterization the MS/MS spectrum is compared to subsequences in a database, by using one of the approaches described in Chapter 10.

The peptide ion chosen for fragmentation is called the *precursor ion* or *parent ion*, and the resulting fragment ions are likewise called *product ions* or *daughter ions*. Fragments (or fragment parts) that are neutral are called *neutral losses*. Note that the presence of these neutral fragments can only be inferred by observing the original fragment both with and without the loss. The neutral fragments themselves cannot be seen directly due to the lack of charge.

8.1 Peptide fragments

The peptides are mainly fragmented along the peptide backbone. Further fragmentation may occur, changing the produced fragments into other fragment types. Since the first mass analyzer selects ions that fall within a certain m/z range rather than a single m/z value, multiple ions can be selected simultaneously. Each of these selected precursors will yield fragments, of which those with sufficiently high abundances are shown as peaks in the MS/MS spectrum.

The precursor can carry one or more charges, depending on the ionization source and the peptide properties. When a singly charged precursor ion fragments, the single charge will necessarily be located on only a single fragment ion, effectively hiding the neutral sister fragment from the mass spectrometer. With multiply charged precursors, the different charges are usually distributed across the product ions. One advantage with multiply charged precursors is that instruments with lower maximum precursor m/z values can be used. Multiply charged ions also tend to fragment easier, and, as explained above, yield more fragment ions per fragmentation event.

The accepted nomenclature for the fragment ions is to denote each type by a letter. The most important fragment types are described below; for a complete list see the bibliographic notes.

Backbone fragments result from fragmentation along the peptide backbone. If a charge is retained on the N-terminal fragment, the fragment ion is classified as *a*, *b*, or *c*,

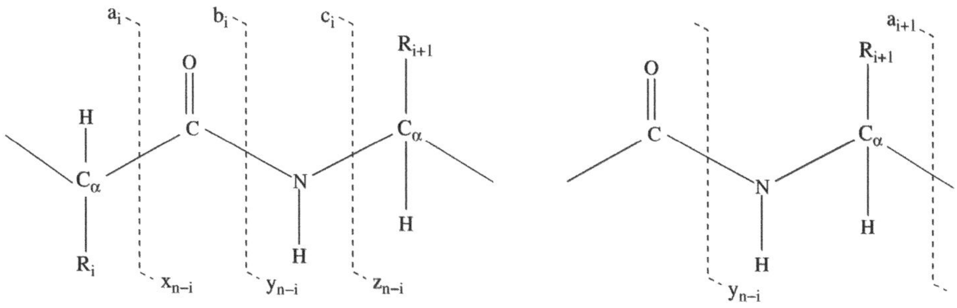

(a) (b)

Figure 8.2 (a) Nomenclature for different backbone fragments. *n* is the number of residues in the peptide, *i* refers to the *i*th residue of the peptide. (b) The basis of an immonium ion

depending on which bond is broken. If a charge is retained on the C-terminal fragment, the fragment type is either *x*, *y*, or *z*. The fragment ions are annotated by subscripts, indicating the number of residues in them (see Figure 8.2(a)).

Internal fragments result from double backbone fragmentation. Usually, these are formed by a combination of *b*-type and *y*-type fragments, and consist of five residues or less.

Immonium fragments are internal fragments composed of a single side chain formed by a combination of a *a*-type and *y*-type fragmentation. This means that both a C and O atom (−28 Da) are lost compared to the residue, see Figure 8.2(b). However, it abstracts an extra proton, hence an immonium ion for an amino acid is observed at an m/z of 27 Da less than the residue mass of the amino acid (assuming a single charge). A peak (with sufficient abundance) at the mass of an immonium ion (strongly) indicates the presence of the corresponding amino acid.

Water loss occurs when backbone fragments lose a water molecule (H_2O, −18.011 Da). The resulting fragments are denoted a°, b°, etc. Water is mainly lost from the side chains of serine, threonine, aspartic acid, or glutamic acid residues.

Ammonia loss occurs when fragments lose an ammonia molecule (NH_3, −17.027 Da). The resulting fragments are denoted a^*, b^*, etc. Ammonia is mainly lost from the side chains of arginine, lysine, asparagine, or glutamine residues.

Side chain fragmentation occurs when additional fragmentation of the side chain of a backbone fragment takes place. The most common are *d* ions from partial side chain fragmentation of *a* ions, *v* ions from complete side chain fragmentation of *y* ions, and *w* ions from partial side chain fragmentation of *z* ions. Side chain fragmentation can be used to differentiate between leucine and isoleucine, since the resulting fragments will have different masses.

Example Suppose the mass range around $m/z = 410 \pm 1$ is selected for fragmentation and further analysis. Suppose further that there is one peptide within this m/z interval,

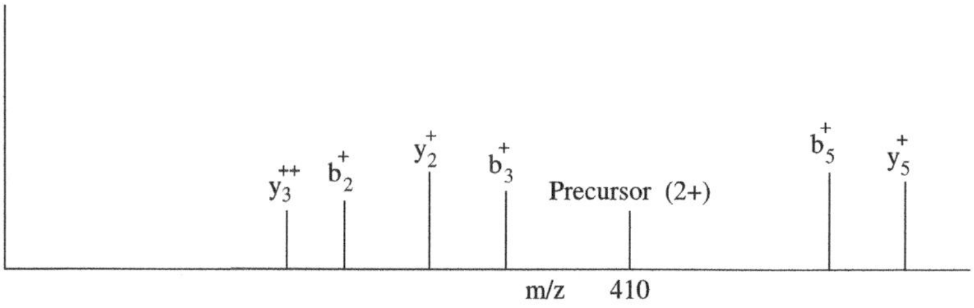

Figure 8.3 An example of an MS/MS spectrum. See text for explanation

the doubly charged peptide LICDVTR. For simplicity we assume only *b* and *y* ions, and that the following fragments are detected:

- the fragmentation after the second residue will produce detectable *b* and *y* ions, each with charge 1;
- the fragmentation after the third residue will produce detectable *b* ions with charge 1;
- the fragmentation after the fourth residue will produce detectable *y* ions with charge 2;
- the fragmentation after the fifth residue will produce detectable *b* and *y* ions, each with charge 1;
- the precursor is also detected.

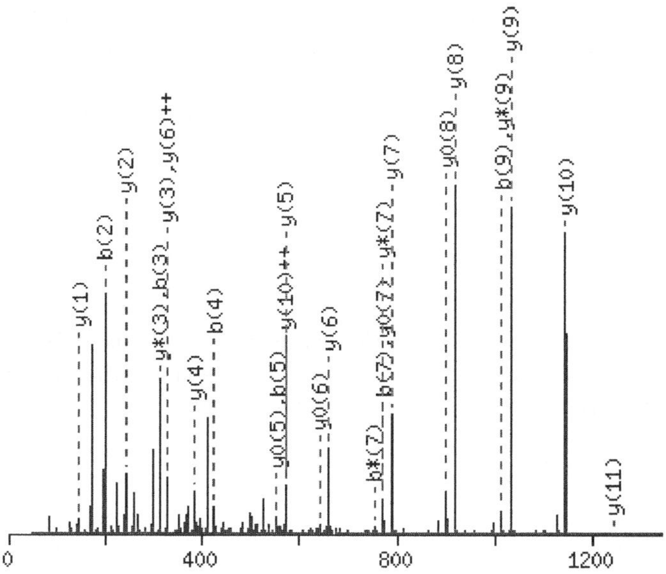

Figure 8.4 An example of an interpreted MS/MS spectrum as shown in Mascot output. Many fragment ions are observed, for example all singly charged *y* ions are recognized. Water loss (marked by 0) and ammonia loss (marked by *) are recognized for some fragments. Note that in practice b_1 ions are seldom are not observed, and b_2 ions can have a quite high intensity.

Then the MS/MS spectrum shown in Figure 8.3 is produced.

△

An example of an experimental spectrum is shown in Figure 8.4.

8.2 Fragmentation techniques

Fragmentation of molecules in mass spectrometers has been used for many years in order to obtain structure or sequence information. Historically, a broad distinction can be made based on the location of the fragmentation, yielding two types:

In-source decay, ISD means that fragmentation occurs 'simultaneously' with the ionization in the source (through the use of excess energy). This has historically been useful and was long considered the 'normal' form, hence the name 'normal ions' for the resulting fragment ions. Contrary to its applications for the study of small chemicals, however, ISD is considered a drawback when analyzing peptides. In order to prevent ISD, *soft ionization* techniques are therefore used for peptides (producing none or little fragmentation in the source).

Post-source decay, PSD implies fragmentation occurring after the source. These are ions that are not fragmented in the source, yet that either contain internally or receive externally sufficient excess energy to fragment in the analyzer. They are therefore sometimes called *metastable ions*.

The peptides most often do not have enough energy to fragment efficiently. A separate fragmentation unit is therefore commonly included in MS/MS instruments. Different techniques are used for fragmentation, yielding different types of fragment ions as the dominating types. These techniques can be divided into two broad categories, depending on whether they observe unimolecular or bimolecular decay. Four of the most used techniques are as follows:

Laser-induced dissociation, LID causes a unimolecular decay, that is the decay rate is only dependent on the number of molecules that are available to the decay. The decay rate k can be expressed mathematically as $d[A]/dt = -k[A]$, where A is the decaying ion. The laser used for ionization during the MALDI process will provide enough excess energy to the molecules to cause metastable ions. Decay occurs in the analyzer, classifying it as PSD. LID provides MS/MS spectra with mainly a, b, and y ions.

Collision-induced dissociation, CID is a bimolecular decay, as it depends on two colliding molecules. CID is the most commonly used fragmentation technique, primarily combined with ESI but sometimes also used in conjunction with a MALDI source. Fragmentation is performed in a special part of the mass spectrometer, called the *collision cell*. This cell contains an inert collision gas (for example, argon), and potential energy is built up in the precursor ions through repeated collisions with the gas molecules. A precursor molecule that reaches the energy threshold for fragmentation will fragment into product ions and/or neutral losses. As can be understood from the explanation above, the decay is dependent on both the number of decaying molecules, A, and the number of collision gas molecules, B. The decay rate is proportional to the rate at which the two

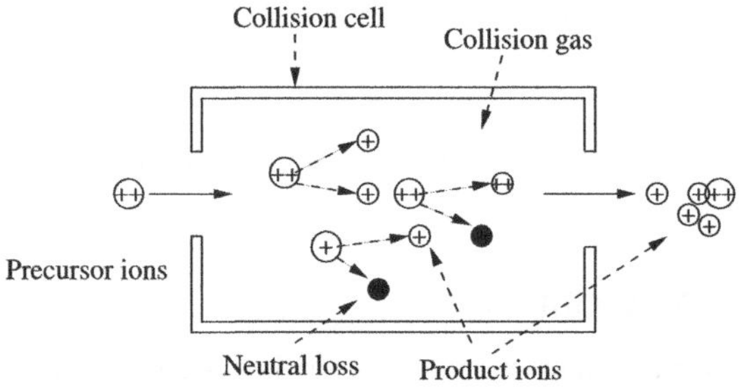

Figure 8.5 The principle of CID

molecules meet, which is in turn proportional to their concentration in the collision cell. Mathematically, a bimolecular decay rate is expressed as $d[A]/dt = -k[A][B]$ when the decay occurs as a single step.

Exactly which product ions and neutral losses are formed depends on the precursor and the energy involved. The principle of CID is shown in Figure 8.5. Commonly two levels of collision energy are used, either low (< 100 eV) or high (> 1000 eV) energy. High-energy CID can dissociate precursors of higher mass and produces more fragment ion types, but this makes the spectrum more difficult to interpret. For peptide fragmentation low energy is normally used, typically in the interval 25–70 eV, depending on the (expected) peptide size and charge.

Another term used for CID is CAD, Collisionally Activated Dissociation.

Electron capture dissociation, ECD occurs when a trapped precursor with more than one positive charge is bathed in low-energy electrons. Capture of such an electron will cause the precursor to fragment quickly, without allowing the energy surplus to spread across the different bonds. It preferentially produces variants of *c* and *z* ions, and occasionally *a* ions, making it complementary to CID which favors *y* and *b* ions. It can be used to fragment larger multiply charged peptides (or even intact proteins), but due to its rather low efficiency when compared to CID and experimental difficulties (a Fourier transform ion cyclotron resonance (FT-ICR) mass spectrometer is required for ECD), its use has remained limited to specialized topics. An advantage of the peculiarities of ECD fragmentation is that it can distinguish between leucine and isoleucine residues, as their distinct structure results in different fragments.

Electron transfer dissociation, ETD essentially brings the mechanism of ECD fragmentation to non-FT-ICR instruments (for example, 3D or 2D (linear) quadrupole ion trap instruments) by substituting free electrons with an anion species of sufficiently low electron affinity. The larger anions (in contrast to free electrons) can effectively be trapped long enough in the non-FT-ICR instruments to allow electron transfer to occur. The result is a fragmentation signature that is extremely similar to that of ECD.

8.3 MS/MS spectrometers

MS/MS-capable mass spectrometers in general consist of an ionization source, one or more analyzers, and a detector. When only a single analyzer is present, it is usually a trapping analyzer that can also eject unwanted ions (for example, a quadrupole ion trap) and act as a fragmentation cell. As explained earlier, the most common ionization sources for proteomic purposes are ESI and MALDI, described in Chapter 5. Remember that MALDI almost always results in singly charged precursor ions, but that double- or triple-charged ions are typical for ESI. However, other ionization sources can be combined with particular analyzers, for example *API* (Atmospheric Pressure Ionization).

8.3.1 Analyzers for MS/MS

Analyzers for MS/MS must perform two analyses: one for selecting the m/z range of interest, and one for measuring the m/z values and intensities of the fragment ions. These two analyses can be done in two different analyzers, called *in-space analyzers*, because the ions travel from one analyzer to the next, or they can be performed in the same analyzer but at different times, called *in-time analyzers*. In order to allow both MS and MS/MS analyses to take place on the same instrument, the analyzers can commonly operate in two different scanning modes. In *full-scan mode* all peptide ions from the source are analyzed and retained, allowing the recording of an MS spectrum. In *MS/MS mode*, however, the analyzer retains only ions that fall within the specified m/z range, ejecting or cutting off any ions that fall outside this range. After this selection step, the retained ions are subjected to a fragmentation step that yields daughter ions. These daughter ions are then measured by an analyzer operating in full-scan mode, yielding the final MS/MS spectrum. Mass spectrometers are often configured to switch continuously between these two modes, recording an MS spectrum in full-scan mode to select an interesting precursor ion, and subsequently switching to MS/MS mode in order to record the fragmentation spectrum of that precursor. When this continuous shifting is done automatically, it is usually referred to as *automated data acquisition*.

The mechanism by which the m/z values of the ions are measured depends on the type of analyzer, and can be based on the time of flight of the ions, their magnetic or electrostatic properties, orbital frequency, and path stability.

8.4 Different types of analyzers

The analyzers can be classified based on different properties, but here we have followed the more commonplace system that classifies them according to their mode of operation. Different types of analyzers can also be combined in so-called hybrid analyzers, to maximize the beneficial properties of each.

The main operations that the analyzers need to perform are the following:

1. Separate the peptides.
2. Select ions within an appropriate m/z range for subsequent fragmentation.
3. Fragment the selected precursor ion(s).
4. Measure the m/z of the resulting fragment ions.

The order in which points 2 and 3 are performed varies.

8.4.1 TOF/TOF

The TOF/TOF analyzer performs two TOF operations in sequence, and is therefore of the in-space type. MALDI is the most often used ionization source for a TOF/TOF, but some instruments also employ ESI. The part which enables peptide selection is called the *ion gate* (also called *mass gate*) or *timed ion selector, TIS*. An m/z range is selected by temporarily switching off a restrictive voltage on the ion gate when the ions of interest pass through. Ions that fall outside of the m/z range arrive before or after the voltage is removed, and are therefore deflected away by the electric field of the gate.

In the simplest case, PSD is the source of fragmentation, and this is illustrated in Figure 8.6. Since the fragment ions maintain the same velocity as the precursor, they also do not show any difference in their TOF when compared to an unfragmented precursor molecule. The TOF/TOF analyzer is therefore fitted with a reflectron that generates an electric field in which the distance that the molecules traverse depends on their m/z values. Thus, the larger the m/z of an ion, the more deeply it penetrates the reflectron field, and the ions are again separated in time based on their m/z value, with lower m/z ions (the fragments) arriving before the ions of higher m/z (the precursor).

There are several noteworthy disadvantages with PSD fragmentation: it takes a long time to produce spectra, fragmentation efficiency is typically low (on the order of 10 %), and unwanted fragmentation can occur at the reflector. In order to compensate for these shortcomings, a CID step often replaces PSD as the fragmentation method. A second source for ion acceleration can also be implemented, and the reflector can finally be used to reduce the spreading in time of the ions derived from the same molecule (or fragment). An outline of these additions is shown in Figure 8.7.

8.4.2 Triple quadrupole (triple quad)

A triple quadrupole instrument is constructed from three quadrupoles in series. The functioning of such a single quadrupole is explained next.

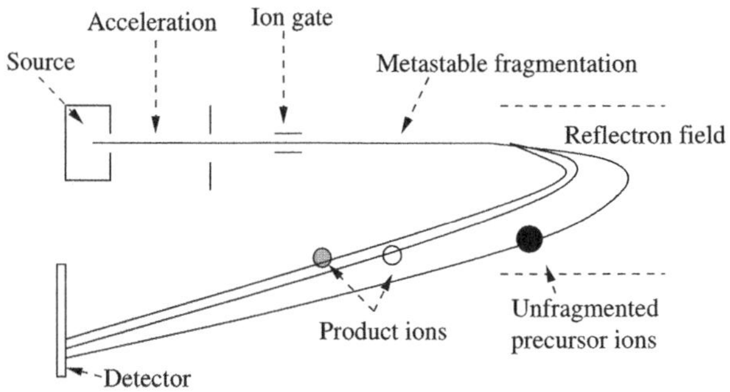

Figure 8.6 The basic principle of a TOF/TOF analyzer, here shown using PSD as the fragmentation method

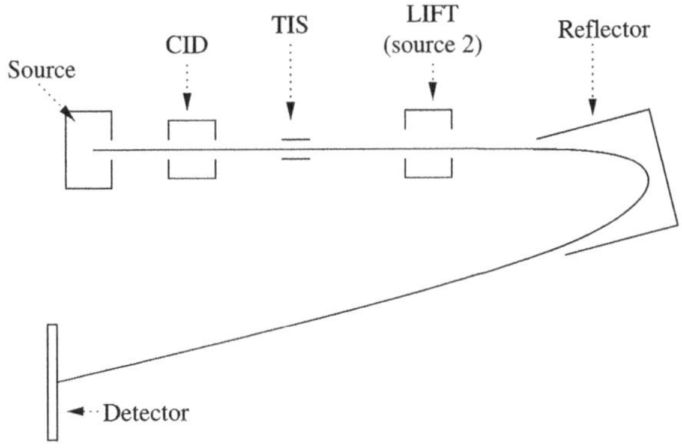

Figure 8.7 Illustration of a TOF/TOF instrument with fragmentation before peptide selection, from BRUKER DALTONICS. When a peptide fragments, it creates an 'ion family' with its product ions. All ions in a family have the same velocity, but different families have different velocities. Only one family therefore passes the TIS. The LIFT provides a lift in potential energy (hence the name), giving the different ions of the selected family different velocities, based on their m/z values

Quadrupole mass analyzer consists of four parallel metallic rods, as shown in Figure 8.8. The electric field between the rods is obtained by applying a superposition of a static direct current (DC) and a radio frequency (RF) alternating current (AC) voltage across the metallic rods. The combined voltage thus reads as $U + V \cdot \cos(\omega t)$. Ions passing through the field will follow a spiral trajectory. The radius of a particular ion's spiral depends on the m/z value of the ion and the offset voltage for the field. For a specific voltage, only ions of a particular m/z will reach the detector, while the others will be ejected from the quadrupole or collide with the rods. The m/z value of the ions that

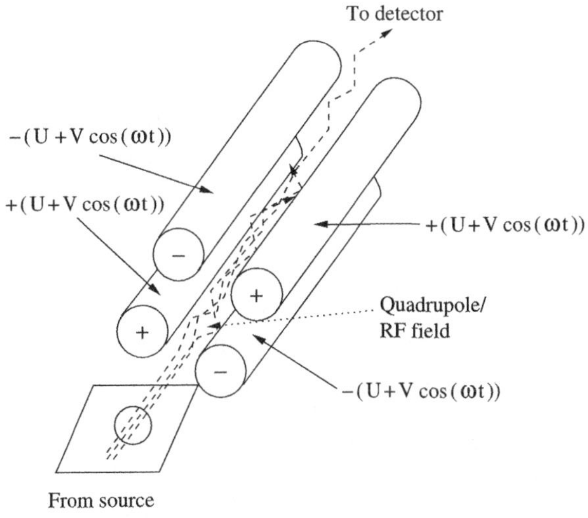

Figure 8.8 A quadrupole mass analyzer

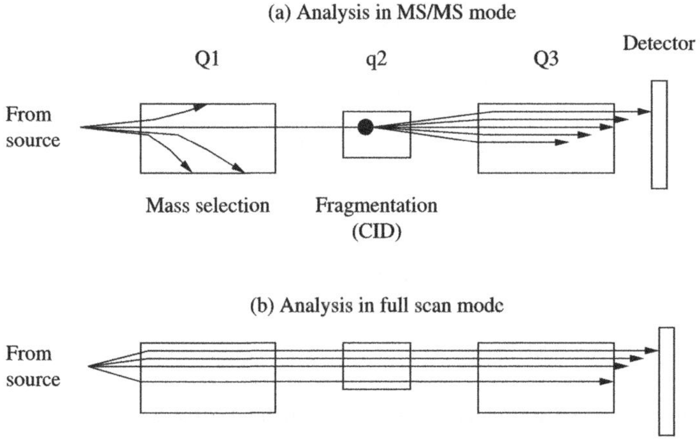

Figure 8.9 Schematic diagram of a triple quadrupole instrument in (a) MS/MS mode and (b) full-scan mode

will traverse the quadrupole safely can be calculated from the used offset voltage. By scanning over the voltage and observing the ion abundances for each voltage interval, a mass spectrum is obtained.

In a triple quadrupole instrument, the middle quadrupole is used for the fragmentation step (usually CID), while the first and last quadrupoles can be used as mass analyzers. This is shown schematically in Figure 8.9. The triple quadrupole is an in-space type of instrument and ESI is commonly used in the ion source.

8.4.3 Ion trap (IT)

Ion traps are analyzers that are able to trap ions in a confined space. MS and MS/MS analyses are both performed in the same unit, hence it is of in-time type. Since fragment ions remain trapped as well, it is possible to iterate the fragmentation process, forming MS^n spectra. We will briefly describe three types of ion traps.

3D ion trap

A 3D ion trap is based on a quadrupole and is therefore also called a quadrupole ion trap (QIT). Sometimes reference is made to the original inventor, in which case the name Paul trap is used. Figure 8.10 shows that the trap is built from three electrodes (two capping electrodes and one ring electrode between them).

We can think of a 3D ion trap as constructed from a quadrupole with two of the rods forming the endcap electrodes, a third bent into the ring electrode, and the fourth collapsed to a point in the middle. The ions are captured in the field by alternating compression and expansion along the x-axis (from source to detector), causing the cloud to expand radially and axially in turn.

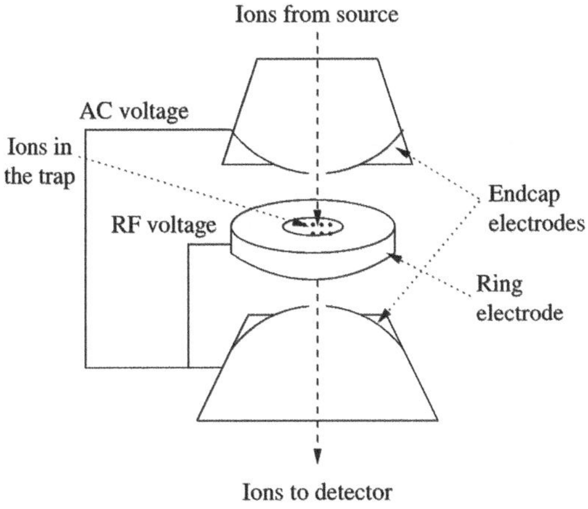

Figure 8.10 A simplified schematic diagram of a 3D ion trap

The principle of MS/MS mode operation of a 3D ion trap is relatively simple:

- the precursor ions enter the trap through the entrance endcap electrode;
- the m/z value of interest is selected, and ions of all other m/z values are ejected by correctly adjusting the voltages;
- collision gas is introduced into the trap, while the voltages are adjusted to increase the kinetic energy of the remaining ions, such that fragmentation can readily occur;
- the product ions are ejected out of the trap in the order of their m/z values by scanning over a voltage range, and are subsequently recorded by the detector.

Both ESI and MALDI are used for ionization, but this type of trap suffers from low resolution and poor accuracy. The size of the trap itself is about that of a lemon, and it has high sensitivity. However, the small size limits the number of ions that can be kept in the trap, limiting the dynamic mass range.

Linear ion trap (LIT)

The linear ion trap has been shown to remedy some of the disadvantages with 3D ion traps. Made of four parallel electrodes, it is simpler in construction. By using only two field components a larger number of ions can be captured, increasing the dynamic range beyond what can be achieved with a 3D ion trap. The signal-to-noise ratio also outperforms that of 3D ion traps. CID has typically been used for fragmentation, but newer LIT (and QIT) instruments can be equipped for ETD fragmentation.

Orbitrap

The Orbitrap consists of an outer and inner coaxial electrode, as shown in Figure 8.11, that form an electrostatic field. The ions form an orbitally harmonic oscillation along the axis of the electrostatic field. The frequency of a molecule's oscillation is inversely

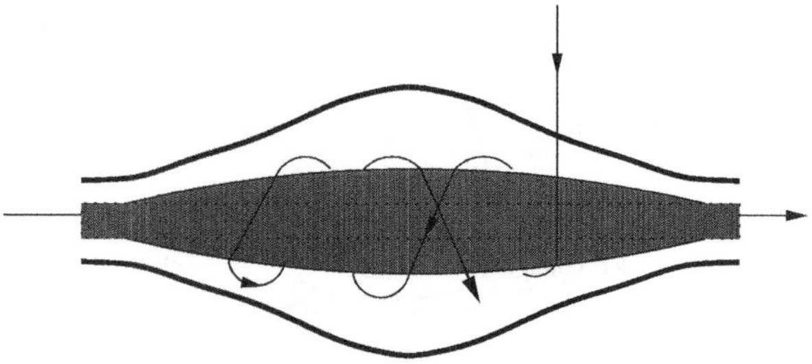

Figure 8.11 Schematic diagram of the Orbitrap. The ions are oscillating along the axis

proportional to its m/z value, and the m/z values of the individual molecules can therefore be calculated by (fast) Fourier transformation. Orbitrap analyzers have very high accuracy and resolution, and the design is patented by Thermo Scientific.

8.4.4 Fourier transform ion cyclotron resonance (FT-ICR)

An FT-ICR analyzer contains a cyclotron, a unit that accelerates charged particles to high energies. The cyclotron contains a (strong) magnetic field, and the ions are rotating around this magnetic field, due to an applied voltage, as illustrated in Figure 8.12.

The cyclotron frequency (or angular velocity) can in an approximated equation be related to m/z as $\omega = zeB/m$, where e is the charge of a proton and B is the magnetic field applied. The frequencies of several ions are determined simultaneously, resulting in a superposition of sine waves. The individual ion's frequency, and the corresponding intensity, must be extracted from this 'interweaved' sine diagram, and this is performed by applying a Fourier transformation. With the individual frequencies determined, it is then easy to calculate the corresponding m/z values.

The electric field allows the expulsion of all ions outside a particular m/z range. The retained ions can subsequently be fragmented by CID or ECD.

The ions are not recorded by a detector but by simply passing near some detection plates. The resolution and accuracy are extremely high, and the mass range lies between 25 Da and up to several kilodaltons. It is the most expensive of the instruments discussed

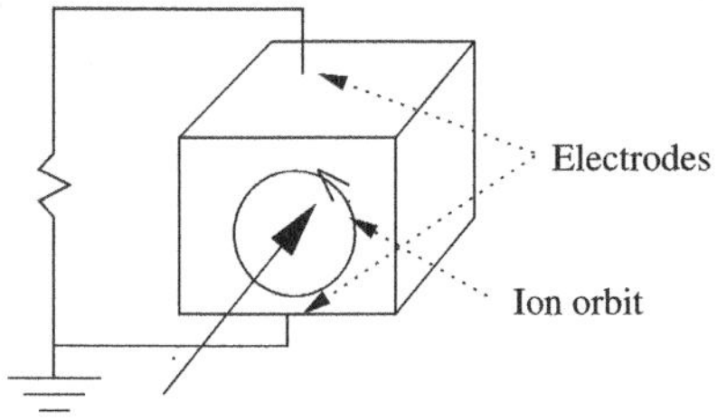

Figure 8.12 A simplified schematic diagram of an FT-ICR

here, mostly due to costs involving the magnet used (which generates fields of up to 9 tesla). FT-ICR spectrometry is sometimes denoted as FT-MS.

8.4.5 Combining quadrupole and time of flight – Q-TOF

This is a hybrid form of mass spectrometer, combining a quadrupole and TOF analyzer. Functionally it is identical to a triple quad (Figure 8.9), except that Q3 is replaced by a TOF analyzer. It allows for higher resolution as a TOF analyzer is used instead of a quadrupole in the final mass analysis step. It is also faster, since the voltage scanning in Q3 is avoided. Both MALDI and ESI are used for ionization.

8.4.6 Combining quadrupole and ion trap – Q-TRAP

This is a hybrid analyzer based on a triple quadrupole instrument, but with Q3 replaced by a linear ion trap. This setup is becoming quite popular in proteomics because of its ability to perform multiple reaction monitoring (MRM). Originally intended to selectively monitor products of chemical reactions, MRM is used in proteomics to substantially increase the selectivity of any analysis. We have discussed earlier that the quadrupole and ion trap mass selectors can only select an m/z range rather than an exact m/z value. As a result, the mass selectors may allow the transmission of several ions with narrowly spaced m/z values rather than a single ion. This degrades the signal-to-noise ratio and may interfere with identification. In the Q-TRAP, however, the first quadrupole can be used as the precursor selector, admitting only a narrow range of ions to proceed into the second quadrupole, which is used as the collision cell. The fragments of all the precursors subsequently enter the linear ion trap, where a second selection step can focus on a known (or predicted) fragment of the desired precursor. Distinct precursors are unlikely to have very similar fragment ions, which results in the highly selective isolation of a single fragment. Indeed, even though the linear ion trap can only select an m/z range, this range will most likely only be occupied by the fragment of interest. The single ion spectra obtained have a very high signal-to-noise ratio, are highly specific, and are ideal for quantitative measurements.

8.4.7 Combining TOF and ion trap

In these hybrid systems, an ion trap is followed by a TOF analyzer. The ion trap allows in-time separation and fragmentation of the ions, and the TOF subsequently supports the recording of high-mass-resolution spectra.

8.4.8 Combining linear ion trap with Orbitrap

A linear mass analyzer is followed by an Orbitrap, which together with an API source and a detector constitute the main components of the LTQ Orbitrap instrument from Thermo Electron Corporation. It has high accuracy and resolution, fast scanning time, and a wide mass range.

Table 8.1 Characteristics and performances of commonly used types of mass spectrometers. v indicate available, (v) indicate optional. +, ++, +++ indicate possible or moderate, good or high, and excellent or very high, respectively. LOD = Limit of Detection. Reproduced from Domon and Aebersold (2006) by permission of SCIENCE AAAS

	IT/LIT	Q-TOF	TOF/TOF	FT-ICR	Triple quad	Q-TRAP
Mass accuracy	Low	Good	Good	Excellent	Medium	Medium
Resolving power	Low	Good	High	Very high	Low	Low
Sensitivity (LOD)	Good		High	Medium	High	High
Dynamic range	Low	Medium	Medium	Medium	High	High
ESI	v	v		v	v	v
MALDI	(v)	(v)	v			
Identification	++	++	++	+++	+	+
Quantification	+	+++	++	++	+++	+++
Throughput	+++	++	+++	++	++	++
Detection of modifications	+	+	+	+		+++

8.4.9 *Characteristics and performances of some types of analyzers*

Table 8.1 outlines important properties for the analyzers discussed above. It is important to realize that these analyzers are continuously improved by the manufacturers and that the information given can therefore only be indicative.

8.5 Overview of the process for MS/MS analysis

Figure 8.13 shows the general principle for obtaining LC-MS/MS spectra. For ESI instruments, an MS spectrum is first (conceptually) constructed at given time points during the infusion of the sample. For MALDI instruments, the MS spectrum is obtained for the sample spot. From this MS spectrum, which often contains multiple peaks, the peaks for subsequent MS/MS processing are automatically selected. The criteria for this selection can be specified by the user, such as:

- the n most intense peaks, n normally between three and eight;
- the charge state;
- an inclusion list of specified m/z values to analyze;
- an exclusion list of m/z values not to be analyzed, for example certain contaminants likely to be present.

The inclusion list results in a targeted approach that is for instance used in MRM. The charge state of an MS peak is determined by the spacing between its isotope peaks, if possible (low-resolution analyzers such as ion traps often cannot distinguish isotopes, especially for higher charge states). In the figure the three most intense peaks are chosen for MS/MS, and these fragmentation spectra are recorded during the rest of the chromatographic time for the component under consideration (A). A final MS/MS spectrum

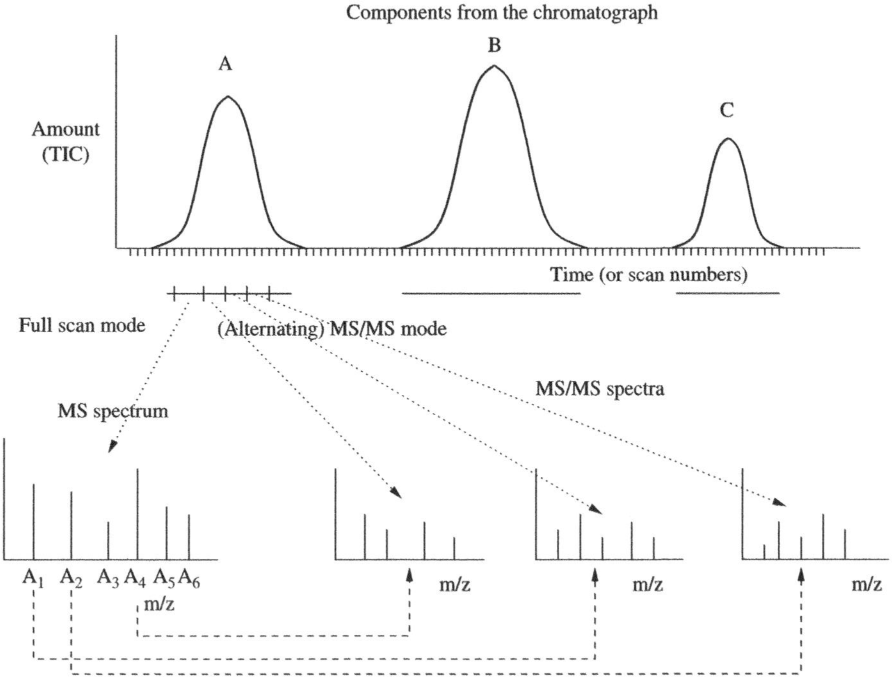

Figure 8.13 Illustration of the process used for recording LC-MS/MS spectra. The three highest peaks from the MS spectrum are selected for MS/MS. Recording of MS/MS spectra from a MALDI plate is similar, but the time dependency disappears there. The laser is then simply used to scan spot per spot. See text for a detailed explanation

for a precursor is typically made up of the accumulated spectra from multiple scans, and the accumulation time can usually be chosen by the user.

8.6 Fragment ion masses and residue masses

Interpretation of the spectra relies on knowing the relation between the mass of a fragment type and the sum of the residue masses of the amino acids in the fragment ion. This is illustrated in Figure 8.14 for fragment ions typically observed in CID. The equations below use the following notation:

$\langle N \rangle$ is the (nominal) mass of the N-terminal group, = 1 (H).

$\langle C \rangle$ is the (nominal) mass of the C-terminal group, = 17 (OH).

R_P is the sum of the residue masses of the peptide under consideration.

M_P is the neutral mass of the peptide under consideration ($= R_P + \langle N \rangle + \langle C \rangle$).

R_F is the sum of the residue masses of the fragment ion under consideration.

M_k is the mass of a fragment ion of type k under consideration (charge 1).

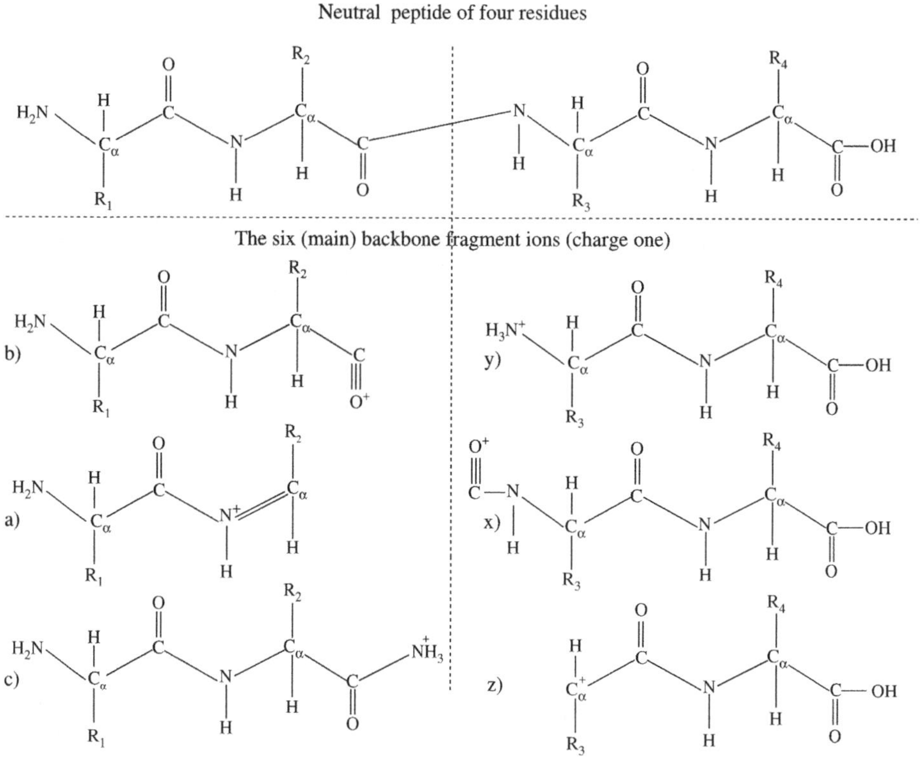

Figure 8.14 A neutral peptide, and a diagram for the six (main) backbone fragments with charge 1 for CID. Note that the *y* and *c* ions get two extra protons (H). The position of the charge can vary, with the illustrated positions indicating one possibility. The diagrams give sufficient detail for mass calculations

The masses for the singly charged fragment ion types can now be written as

$$M_b = R_F + \langle N \rangle \qquad M_y = R_F + \langle C \rangle + 2\mathrm{H}$$
$$M_a = R_F + \langle N \rangle - \mathrm{CO} \qquad M_x = R_F + \langle C \rangle + \mathrm{CO}$$
$$M_c = R_F + \langle N \rangle + \mathrm{N} + 3\mathrm{H} \qquad M_z = R_F + \langle C \rangle - \mathrm{NH}$$

Note that R_P is the sum of R_F for the two complementary fragment types, and that R_F is different for those two complementary types. Adding complementary masses results in

$$M_b + M_y = R_P + \langle N \rangle + \langle C \rangle + 2\mathrm{H}\ (= M_P + 2\mathrm{H})$$

$$M_a + M_x = R_P + \langle N \rangle - \mathrm{CO} + \langle C \rangle + \mathrm{CO}\ (= M_P)$$

$$M_c + M_z = R_P + \langle \mathrm{N} \rangle + \mathrm{N} + 3\mathrm{H} + \langle C \rangle - \mathrm{N} - \mathrm{H}\ (= M_P + 2\mathrm{H})$$

From these equations we see that the sum of single charged complementary *b* and *y* ions (and *c* and *z* ions) equals the mass of the double charged parent ion. The most frequently occurring fragment types by far for CID are the *b* and *y* ions and partly *a* ions, and many programs consider only these. We will also mainly consider these here for illustrative purposes.

8.7 Deisotoping and charge state deconvolution

The fundamental peptide-derived data used for database searches are the m/z value of the precursor and its MS/MS spectrum. Some preprocessing can often be useful to increase the quality of the identification. Peak detection and preprocessing as described for MS spectra in Section 6.2 can of course also be performed here. There are operations that are more specific for MS/MS spectra in that monoisotoping or deisotoping (depending on the resolution of the instrument) is combined with charge state deconvolution.

One goal of preprocessing is to obtain a spectrum where each fragment is represented by only one datum. This means that an isotope envelope is reduced to one peak (monoisotoping or deisotoping), and that a fragment that occurs in different charge states is also collated into one state (commonly to the singly charged ion, which is called charge state deconvolution). An m/z value for $z > 1$ is converted to $[((m/z) - 1)z + 1]$, and the intensities are summed for ions occurring with several charges. As explained in Section 5.4, these two tasks rely on the m/z differences between the peaks, and can therefore be done

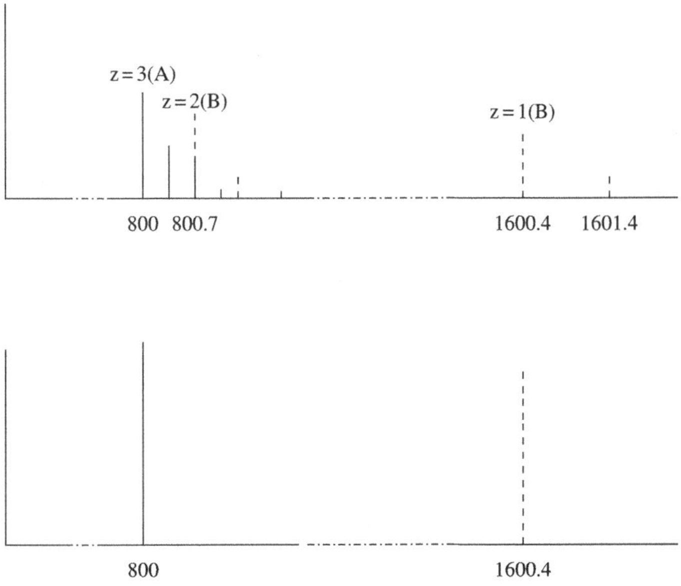

Figure 8.15 Illustration of deisotoping and charge state deconvolution. Two fragment ions are shown, one by the full line (A), and one by the dashed line (B). A occurs with charge 3, and has four isotopic peaks. B occurs both as charge 2 with three isotopic peaks and as charge 1 with two isotopic peaks. A and B have overlapping isotopic envelopes. A is monoisotoped to a single peak with charge 3, and B is deisotoped and deconvoluted to the peak with charge 1. The resulting intensities are simply the sum of the individual intensities of the isotopic peaks. As ion A has charge 3, its M_P = 2397

simultaneously. A serious problem is that isotopic envelopes can overlap, necessitating caution in estimating both the charge states and the total ion intensities for each of the overlapping ions. Figure 8.15 shows an example. Programs for charge state determination typically define a maximum limit for the charge (three or four for fragment ions). Furthermore, the charge cannot be higher than the charge of the precursor, and for each fragment it cannot be so large that the neutral mass of the fragment becomes larger than the mass of the precursor. For each peak, the occurrence of isotopes is checked against the predicted isotopic masses of the suggested charge. One can also take into account the expected number of isotopes for the mass of the peak, to see if these are present.

When it is found that a peak may belong to more than one isotopic envelope, the problem of distributing the intensity of that peak occurs. A reasonable way of solving this is to use models for the expected intensity distribution of the isotope envelopes included.

8.8 Precursor treatment

Precursor mass errors can occur when the wrong isotope has been selected as the monoisotopic peak, or if the charge is either not determinable, or incorrectly determined. Fortunately, precursor mass errors can be controlled for or estimated from their MS/MS spectrum.

8.8.1 *Precursor mass correction*

An error in the precursor mass or charge can be detrimental for the identification of the peptide, since such an error will produce systematic errors in derived data. By using complementary fragment ions it may, however, be possible to control for and correct wrong precursor masses. We show the principle assuming that the precursor charge is two. From Section 8.6 we have for the complementary ions (b, y) and (c, z) the equation $M_b + M_y = M_P + 2H$.

Let $R = r_1, \ldots, r_m$ be a spectrum. An *inversed* spectrum[1] is defined as $\bar{R} = \bar{r}_1, \ldots, \bar{r}_m$, where $\bar{r}_i = M_P + 2H - r_i$; then $b(c)$ ions in R should be translated to corresponding $y(z)$ ions in $\bar{R}$, and $y(z)$ ions to $b(c)$ ions. There should therefore be many common peaks in R and $\bar{R}$. If M_P is not correct, however, this correspondence will be low or not occur. M_P can therefore be considered an unknown x that is to be determined. Then define $\bar{R}(x)$ with $\bar{r}_i = x + 2H - r_i$. Let then $C(x) = C[R, \bar{R}(x)]$ be the number of common peaks (to a given accuracy) in R and $\bar{R}(x)$. The value of x that maximizes $C(x)$ should be chosen as the corrected precursor mass. This problem can be solved as a mixed integer programming problem, but also in polynomial time ($O(m^3)$) by Algorithm 8.8.1.

---

[1] Reversed spectrum is also used in the literature.

Algorithm 8.8.1 Estimating the precursor mass x for the MS/MS spectrum $R = \{r_1, \ldots, r_m\}$

var
$\{r_i\}$ the peaks of R
$\{\bar{r}_i\}$ the peaks of $\bar{R}$
i, j, k, l indices
n, n_1 number of coinciding masses in R and $\bar{R}$
x, x_1 proposed precursor masses
const
δ accuracy threshold
H the mass of H
begin
  $n := 0$
  for all $(i, j) \in [1, m]$ **do**
      calculate the precursor mass assuming that r_i and $\bar{r}_j$ are complementary
    $x_1 = r_i + r_j - 2H$
      calculate the inversed spectrum $\bar{R}(x_1)$
    for all $i \in [1, m]$ **do** $\bar{r}_i := x_1 + 2H - r_i$ **end**
    $n_1 :=$ the number of coincident peaks (r_k, r_l) such that $|r_k - \bar{r}_l| \le \delta$
    if $n_1 > n$ **then** $n := n_1$; $x := x_1$ **end**
  end
  x is the proposed precursor mass, and n the number of coinciding masses
end

Example Consider the peptide LICDVTR, and assume that the following (ideal nominal) MS/MS spectrum is constructed: $R = \{227(b_2), 276(y_2), 330(b_3), 375(y_3), 490(y_4), 544(b_5)\}$. We construct the inversed spectrum with $\bar{r}_i = x + 2H - r_i$. We find that the value $x = 818$ will maximize the number of equal values (four), $\bar{R} = \{593, 544, 490, 445, 330, 276\}$.

△

8.8.2 Estimating the charge state of the precursor

While MALDI rarely produce ions of charge state higher than one, multiple charge ions are commonly produced by ESI. However, fragmentation spectra from precursors with a charge higher than three are seldomly of good enough quality for subsequent identification.

The problem is therefore reduced to determining the charge as being one, two, three, or 'larger'. The charge state is commonly found by using the isotopic differences of the associated MS spectrum. However, especially for low-resolution instruments, the charge state assignment may be wrong or simply unknown. Knowing the actual charge state is important for performing a reasonable database search, all the more so since erroneous charge state assignments often lead to faulty results. An alternative to determine the correct charge state of the precursor is to perform the database search with several charge states, but this is time consuming, and it is not always possible to select the correct search results afterwards.

Several methods have therefore been developed for estimating the precursor charge state from MS/MS spectra. The easiest approach is to differentiate between charge of one and greater than one. If the charge is one, the (m/z) values of all the product ions are less than (m/z) of the precursor. For example, let T^+ be the total ion current for ions with (m/z) greater than that of the precursor, and T^- the rest of the total ion current. If the ratio (T^+/T^-) is small (for example, less than 0.1), the charge state can be estimated to one.

2to3 is a program that estimates the charge (2 or 3) of multiple charged precursors. It is supposed that neutral loss does not occur and that fragmentation of triply charged precursors into three fragment ions does not occur. Let $(m/z)_P, z_P$ be the (m/z) and charge of the precursor, and $(m/z)_b, z_b$ and $(m/z)_y, z_y$ be the corresponding two resulting complementary product ions. From $M_b + M_y = M_P + 2H$ (Section 8.6) we have for a precursor of charge 2

$$1(m/z)_b + 1(m/z)_y = 2(m/z)_P$$

since $(m/z)_P = (M_P + 2)/2$. If the charge of the precursor is three, and the b ion gets charge 2, we have

$$2\left(\frac{M_b+1}{2}\right) + \frac{M_y}{1} = M_b + M_y + 1 = 3\left(\frac{M_P+3}{3}\right)$$

hence

$$2(m/z)_b + 1(m/z)_y = 3(m/z)_P$$

We find an analogous equation if the y ion gets the extra charge. From this we get

$$z_P(m/z)_P = z_1(m/z)_1 + z_2(m/z)_2 \tag{8.1}$$

where 1 means b or y, and 2 means y or b. The idea is then to look at each pair of peaks (above a threshold intensity), and use Equation 8.1 to see if they can originate from complementary fragments when the charge of the precursor is two or three. Let the (m/z) values of two such peaks be Q_1 and Q_2. We then have:

if $Q_1 + Q_2 = 2(m/z)_P$ then it supports a doubly charged precursor;

if $2Q_1 + Q_2 = 3(m/z)_P$ then it supports a triply charged precursor;

if $Q_1 + 2Q_2 = 3(m/z)_P$ then it supports a triply charged precursor.

By considering all pairs of peaks, one can find the charge state with the most supporting evidence.

Different machine learning techniques are also used for the precursor charge estimation problem. One procedure is to utilize the observation that the (m/z) values of the product ions are in the interval $[0, z_P(m/z)_P]$. It is appropriate to divide the interval into subintervals $[(i-1)(m/z)_P, i(m/z)_P], i = 1 \ldots z_P$. The charge of the product ions will have different distributions in the different subintervals. For $z_P = 3$ the last subinterval will for example only contain product ions of charge 3. For each actual z_P the product ion patterns for each subinterval can be learned from known MS/MS spectra. The spectrum under consideration is then compared to these patterns, and given a score for each z_P. The value of z_P resulting in the highest score is used as the precursor charge estimate.

8.9 MS^3 spectra

Since the analysis in in-time analyzers uses the same analyzers for MS and MS/MS, one can actually repeat the process. Product ions from MS/MS can be selected for another fragmentation step and an MS^3 spectrum is thus produced. The process can be repeated even more, producing MS^n spectra for $n-1$ fragmentation steps. The number of ions available for fragmentation will decrease for each step, and a limit for producing reliable spectra from today's instruments is reached for an n of about 10. An interesting observation is that the additional selection steps will typically decrease the noise signal more than the signal of the actual fragment ions, increasing the signal-to-noise ratio. The MRM approach outlined above takes advantage of this property.

The fragment ions produced by the second fragmentation step are of three types (when considering only b and y ions as possible precursors):

1. y ions: these are produced from the y ions from the first fragmentation.
2. b ions: these are produced from the b ions from the first fragmentation.
3. Internal ions: these are produced as y ions from the b ions from the first fragmentation or as b ions from the y ions from the first fragmentation.

MS^3 spectra can be used to resolve ambiguities in an identification. For identification, the MS/MS spectra are compared to *segments* in a protein sequence database. A segment is a subsequence, in our context often a theoretical peptide. When more than two segments can fit in an MS/MS spectrum, the most intense peak in the spectrum can be selected for a second fragmentation step and the resulting MS^3 spectrum can then be compared to the two candidate segments. It can also be useful when no matching segment is found for the MS/MS spectrum, since the MS^3 spectrum is simpler and may be easier to reveal helpful information.

For the ion trap the ions have to be reloaded before each new step, which is in contrast to an FT-ICR instrument where no reloading is necessary, since the ions are detected as they are formed.

Exercises

8.1 Consider the peptide ACLHVR. Assume that there is a methylation modification at residue 4 (H). Construct a theoretical spectrum when the fragments have charge 1. Assume that the following ions are detected: $a_3, b_2, b_4, b_5, c_4, y_3, y_4, y_5, b_3^{\circ}, y_4^{*}$, and an immonium ion for V.

8.2 Assume that two peaks with nominal m/z values 172 and 343 are in a spectrum. Investigate if these can correspond to the same peptide, and determine in that case the mass (m) of the peptide.

8.3 In a spectrum there are peaks at m/z values $m, m+1, m+1.5, m+2$. Discuss how many different ions (fragments) these may correspond to, and the charges of the ions. What can you use to assess your suggestions?

8.4 Suppose an MS/MS spectrum is {72, 159, 175, 276, 287, 404}. Use Algorithm 8.8.1 to estimate the precursor mass.

8.5 Develop an algorithm that implements the 2to3 algorithm in an efficient way.

8.6 Consider the peptide ACLHVR. Construct a theoretical MS^3 spectrum of the b_4 fragment.

8.7 Assume two peaks with charge 1 from an MS/MS spectrum of a peptide where all residues are different, and have m/z values m_1 and m_2, where $m_1 < m_2$. Let us assume only b and y ions, that the number of residues in the peptide is l_P, and in the two fragments there are l_1 and l_2 residues respectively. The two peaks are both (separately) selected for a new fragmentation, also producing only b and y ions. Discuss how many peaks may be common in the two MS^3 spectra, using the lengths l_P, l_1, and l_2. Discuss this for each of the four possibilities for m_1, m_2 being b or y ions.

Bibliographic notes

Fragment nomenclature
Biemann (1988, 1990); Johnson *et al.* (1987); Roepstorff and Fohlman (1984)

Fragmentation pathways
Paizs and Suhai (2005)

Peptide mass correction
Dancik *et al.* (1999)

Precursor charge estimation
Hansen *et al.* (2001); Mann *et al.* (1989), (2to3) Sadygov *et al.* (2001)
Colinge *et al.* (2003a); Klammer *et al.* (2005)

Preprocessing of MS/MS spectra
Gentzel *et al.* (2003); Yates *et al.* (1998)

Characteristics of MS/MS analyzers
Domon and Aebersold (2006)

9 Fragmentation models

The analysis of MS/MS spectra is undertaken in order to allow protein identification and characterization. Several different kinds of analysis can be used, but all of them rely on predictions of the expected occurrences and intensities of ions from each fragment type. Imagine for example that you have a fragmentation spectrum R, and would like to verify if this could have been produced from a peptide with sequence S. In order to do so, you should have a model that predicts which fragments will be created and subsequently detected if an MS/MS experiment is performed on S. Ideally, this prediction would include intensities for the fragment ions as well. These expected fragment intensities could then be compared to the measured peaks in R and seen how similar they are. While complete knowledge of the actual fragmentation process would be very useful here, it is still not fully understood. Some theories and informal rules do exist, however, and we will be discussing some of these next.

The information we need can be delivered by a *fragmentation model*. This model takes a peptide sequence S as input, and returns a theoretical MS/MS spectrum.

There are two main approaches in the attempt to determine and/or understand the relationships between a peptide sequence and the produced MS/MS spectrum:

Chemical approach (bottom-up) relies on systematic investigation of the chemical processes involved in the fragmentation.

Statistical approach (top-down) is based on the investigation of large sets of MS/MS spectra to learn how the fragmentation depends on certain peptide properties, without necessarily understanding the fragmentation pathway.

The most promising investigations will likely be based on a combination of these two approaches.

9.1 Chemical approach

Central to the chemical approach is the determination of the location of the added proton(s). The proton(s) can be located at different *protonated sites*, such as a group in the side chain, the terminal amino group, and amide oxygen or nitrogen.

Computational Methods for Mass Spectrometry Proteomics I. Eidhammer, K. Flikka, L. Martens and S.-O. Mikalsen
© 2007 John Wiley & Sons, Ltd

9.1.1 The Mobile Proton Model (MPM)

A few models have been proposed to explain the fragmentation process, and one of the most popular is the *Mobile Proton Model*. This model states that as the dissociation energy increases the added proton(s) will move to a *protonation site*, if they are not sequestered by a basic amino acid side chain (R, K, or H).

The proton typically migrates to an atom at the amide bond (either the amide nitrogen or the amide oxygen), resulting in amide bond fragmentation (giving *b* and/or *y* ions) at the protonized atom. Peptides in which the proton(s) migrate freely are called *mobile peptides*, and this form of fragmentation is called *charge-directed fragmentation*.

Peptides in which the proton(s) are located at basic amino acids show very little proton mobilization. Recall that this follows directly from our definition of a 'base' according to the Brønsted–Lowry theory (see Section 1.3.3). Relocating the proton to another, energetically less favored site therefore requires additional energy, making such a protonation site unstable and unlikely. Other mechanisms for backbone dissociation can occur in such circumstances, for example *reactive intermediates*. This mechanism is commonly referred to as *charge-remote fragmentation*. It has been found that the loosely bound proton on the acidic side chain of glutamic acid (E), and especially of aspartic acid (D), can initiate fragmentation of the amide bond C-terminal to the acidic residue. Fragment ions derived by this cleavage thus dominate the MS/MS spectrum for such peptides.

9.2 Statistical approach

In a statistical approach, different statistical techniques (often relying on machine learning) are used to analyze MS/MS spectra in order to develop a fragmentation model, as illustrated in Figure 9.1. A model should be able to predict both the fragments that will occur and their intensities, given a peptide sequence.

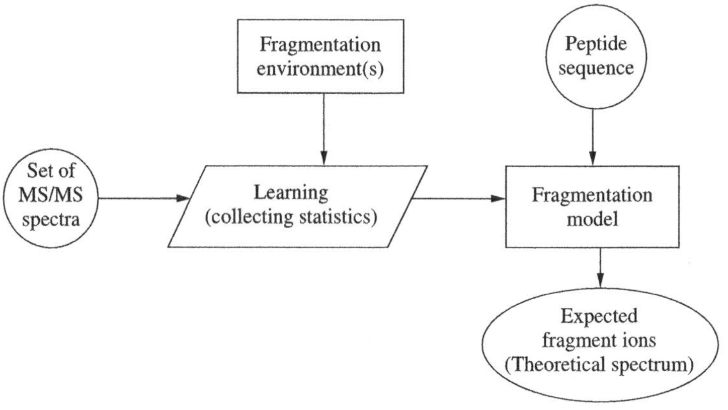

Figure 9.1 A fragmentation model is constructed from a set of spectra. The model will predict the fragment ions for a peptide sequence. The fragmentation environment determines how detailed the model should be

The complexity of the models depends on the detail of the collected data. A model can be thought of as being on one of three complexity levels:

Uniform level model (which does not need any statistics at all) will simply include all possible fragment ions, and assign a uniform intensity to each fragment.

Fragment level model makes a distinction between fragment types by assigning the same intensity to all fragment ions of the same type (all b ions get the same intensity, but this intensity is generally different from the one assigned to y ions). Another distinction can be made on the basis of fragment ion charge, so that b ions with a charge of one generally get a different intensity than those with a charge of two.

Residue level model describes how the fragmentation depends on certain fragment and peptide properties.

The statistics for a model must be collected from spectra produced in an experimental environment. Different fragmentation patterns can be expected for different MS/MS instruments. Also, different patterns can occur for different charge states of the precursor. It is therefore essential that the experimental environment in which the model is developed must be clearly specified.

9.2.1 Constructing the training set(s)

A large number of training spectra, from which the statistics are calculated, are needed for these analyses. Since different instruments and specific experimental conditions usually require distinct models to be developed, the training set(s) should ideally be constructed automatically. A training set should consist of (spectrum, segment) pairs, where the segments are identified by a search in a protein sequence database (remember that a segment here is part of a protein sequence). This is commonly done by searching the database with an 'accepted' search program, and considering those (spectrum, segment) pairs identified that receive a high score from the program. This is a slightly circular type of reasoning, since the input of the scoring model consists of those identifications that correspond to the model built into the program (that is, those that are confidently identified).

The training set should also be non-redundant, to avoid the introduction of bias towards some types of spectra. Avoiding redundancy here means removing fragmentation spectra derived from the same peptide sequence. Note that this means that reproducibility is implicitly assumed. A crucial point to consider when deciding on the type of model you want to construct is that there must first be sufficient data to derive statistical results to the desired detailed level.

Intensity normalization

The intensity normalization process should depend on the level at which the analysis is performed. On the fragment level, the normalization could be to the total intensity of the spectrum. At the residue level, the normalization should rather be to the most abundant peak of the fragment types considered. Distortions from other fragmentation pathways are thereby avoided, and spectra can have different signal-to-noise ratios without disturbing the analysis.

9.2.2 Spectral subsets

The fragmentation processes are likely to show different behavior for different peptides. The challenge is therefore to divide the set of training spectra into spectral subsets, in which the fragmentation is expected to show similar behavior. Each subset can then be used to train a specific instance of the learning algorithm. One possible partition is presented in Kapp *et al.* (2003). In accordance with the mobile proton hypothesis the peptides were divided into three groups:

- *mobile* peptides: defined as peptides where the number of basic residues (`R,K,H`) is less than the number of extra protons (the charge);
- *non-mobile* peptides: defined as peptides where the number of `R` residues is greater than or equal to the number of extra protons;
- *partially mobile* peptides: defined as peptides not classified as mobile or non-mobile.

A more detailed partition is performed in Huang *et al.* (2005), where subsets of the mobility groups were analyzed.

9.3 Learning (collecting statistics)

As explained previously, collecting statistics can be performed at two levels: fragment and residue level. Statistical analysis for the first level is fairly straightforward, although the counting can differ. In Huang *et al.* (2005), for example, two values are counted for each fragment type: frequency and normalized % intensity. The frequency is calculated as the number of spectra that contain a non-zero sum of normalized intensities for the fragment type, divided by the number of spectra in the considered training set. If half of the spectra contain at least one b^+ (b ion with charge 1), the frequency for b^+ would be 0.5. The intensity for a fragment type, on the other hand, is calculated directly from the spectra with a non-zero sum of normalized intensities for the considered fragment type. The median of these non-zero values is used as the predicted intensity.

In the next sections we mainly consider analysis at the residue level, but some methods are also valid for the fragment level. We define three terms here:

Fragmentation site is a position in the peptide where fragmentation can occur. If only fragmentation of the peptide bond is considered (that is, the formation of b and y ion pairs), there exist $n-1$ fragmentation sites for a peptide of length n.

Fragmentation environment is a set of fragment and peptide properties associated with a fragmentation site.

Fragmentation environment class is a class that a fragmentation environment belongs to.

A model must predict the tendency of fragmentation to occur for each environment in an environment class, relative to fragmentation to happen for all the environments in the class. The complexity (or detail) of a model is determined by the complexity of the fragmentation class it is going to serve.

Example A simple fragmentation environment class can consist of the two amino acid residues whose peptide bond is broken. If we do not take the position of the residues into account, this leads to 400 different fragmentation environments. Statistics have to be collected for each of the 400 possible pairs showing how strong the tendency is for the peptide bond to fragment between these two amino acids. A more complex environment class can include more neighboring residues, and/or the fragmentation position along the peptide sequence, and/or the hydrophobicity of the peptide or fragment etc.

△

Several ways of learning are employed, a few of which we will describe in the next subsections.

9.3.1 *Fragmentation intensity ratio (FIR)*

One way of learning is to calculate a fragmentation intensity ratio from a set of MS/MS spectra resulting from a set of peptides P for each environment of an environment class. Let $p \in P$, and n_p be the number of residues in p. Let f be the fragment ion type(s) under consideration, for example b ions, or both b and y ions. Then the average intensity over all fragmentation sites S_p in p is

$$\bar{I}(p, f) = \frac{1}{n_p - 1} \sum_{s \in S_p} I_s(f)$$

where $I_s(f)$ is the intensity of the f fragment(s) from fragmentation site s. The average intensity for all occurrences of an environment v is then

$$\bar{I}_v(p, f) = \frac{1}{|s \in S_p \wedge s = v|} \sum_{s \in S_p \wedge s = v} I_s(f)$$

The FIR of p for an environment v is defined as

$$FIR_v(p, f) = \frac{\bar{I}_v(p, f)}{\bar{I}(p, f)}$$

The FIR over all peptides is the average FIR. Let P_v be the peptides for which environment v exists. Then the final FIR is

$$FIR_v(P, f) = \frac{1}{|P_v|} \sum_{p \in P_v} FIR_v(p, f)$$

which is a measure for the intensities occurring in an environment v relative to the sum of all intensities.

Example Consider the peptide p=**ALCVLCR**, and let f be b ions. Suppose the observed intensities of the b ions are (from b_1 to b_6): (0, 12, 6, 8, 4, 3). Then $\bar{I}(p, b) = \frac{33}{6} = 5.5$. Let the fragmentation environment be the two residues flanking the fragmentation site, and consider $v = $ `(L,C)`. Then we get $\bar{I}_v(p, b) = \frac{16}{2} = 8$, and $FIR_v(p, b) = \frac{8}{5.5} \approx 1.5$, which shows a higher tendency for b ions at `(L,C)` than average (=1).

△

A possible extension of the environment in the above example could be to include the actual fragmentation site in either absolute or relative position, thus for example differentiate between `(L,C)` coming in third and fifth position, respectively. The question then is whether sufficient data are available to produce statistically significant results at this level of detail.

9.3.2 Linear models

A linear model can be constructed from a set of selected attributes. One such example is from Kapp *et al.* (2003). The linear model used was

$$\log_2(I_e) = I_B + t_{N\text{-}term}[e] + t_{C\text{-}term}[e] + k_1 b + k_2 b^2 + k_3 \log_2 L_p$$

where:

I_e is the expected intensity for fragmentation environment e (not differentiating between b and y ions).

I_B is a baseline fragmentation intensity term, representing the average intensity expected if neither of the neighboring residues have a special effect.

$t_{N\text{-}term}$ is a table with one entry for each amino acid. It represents an increase/decrease in the intensity that will be applied when this amino acid is the N-terminal flanking residue.

$t_{C\text{-}term}$ is an analog table for the C-terminal flanking residue.

b is the relative position of the fragmentation. An increased tendency of fragmentation is observed in the middle of the peptide.

L_p is the length of the peptide. This term corrects for the lower overall intensity on longer peptides due to the intensity normalization.

The fragmentation environment in this model consists of the amino acid at the N-terminal side of the fragmentation site, the amino acid at the C-terminal side, the position of the fragmentation within the peptide sequence, and the peptide length. A linear regression was performed to estimate the constants (k_1, k_2, k_3, and the 40 constants in the residue tables $t_{N\text{-}term}$ and $t_{C\text{-}term}$).

9.3.3 Use of decision trees

Elias *et al.* (2004) describe a learning method in which many fragment and peptide properties are investigated for their influence on the observed fragmentation pattern. The study relies on fragmentation spectra from doubly charged peptides, obtained on an ESI ion trap instrument. From the 1 000 000 candidate spectra, 27 000 (non-redundant) were chosen for the analysis. The SEQUEST software (Section 11.2.5) was used for finding matches in the sequence database. The identifications were limited to the confident assignments by applying rigorous filtering criteria.

Between 100 and 200 attributes were considered during the analysis, of which 63 were automatically selected as significant for the fragmentation. The 63 attributes can be classified into 10 attribute types:

1. The amino acid that occurs in certain positions, such as the N- and C-terminus of the peptide, and at positions x residues in N- and C-terminal direction from the fragmentation site.
2. Fraction and number of (`R,K,H`) residues in the peptide and the fragment.
3. The length of the peptide and the fragment, relative position of the fragmentation site, and distance from the fragmentation site to the termini.
4. Different attributes relating to gas phase basicity (both averaged over the fragment as well as for special residues).
5. Different attributes relating to helicity (both averaged over peptide and for special residues).
6. Different attributes relating to hydrophobicity (both averaged over peptide and fragment, and for special residues).
7. Ion type: b or y ion.
8. Number of missed cleavages.
9. m/z of the fragment and peptide, and the difference in m/z and mass between them.
10. Different attributes relating to isoelectric point (pI).

The result of the analysis was incorporated into two *decision trees*, which report a score for a spectrum and a candidate peptide sequence, indicating the probability that the spectrum is derived from the peptide. Apart from the end nodes, each node in a decision tree specifies a test of some attribute (here one of the 63 attributes), and each branch from the node corresponds to the result from the test. Although several branches can potentially extend from a node in a tree, in this case only two are present. When this is the case the tree is called a binary decision tree. Note that an end node in the tree represents a specific fragmentation environment. The trees were constructed by using Shannon entropy of intensity to select attributes and corresponding values for decision points (internal nodes).

Figure 9.2 shows part of a decision tree from Elias *et al.* (2004). The following attributes are used in this part:

POS is the fractional position of the fragmentation site;

ION is the fragment ion type;

M_Z is the m/z of the fragment;

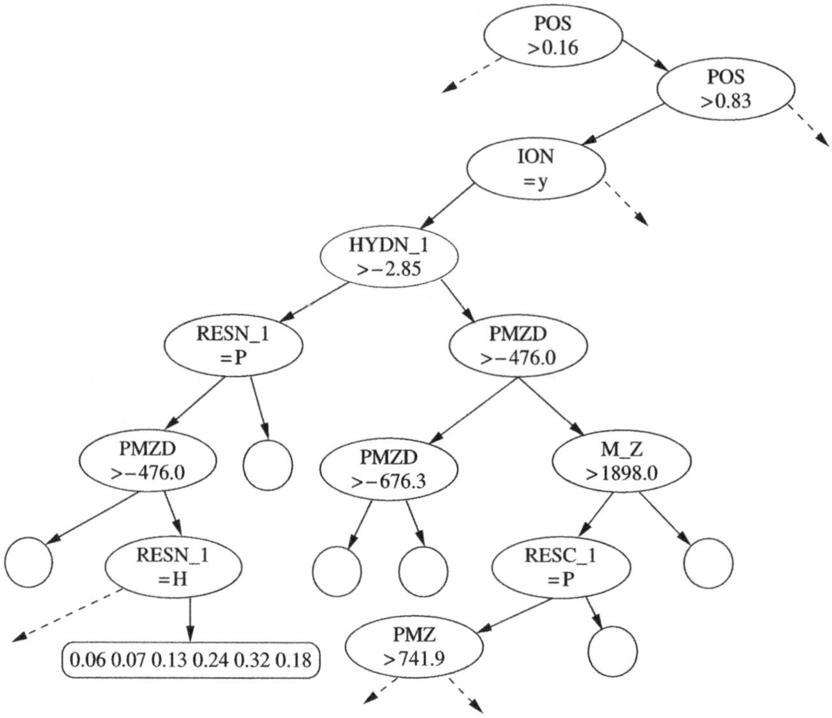

Figure 9.2 Part of a (binary) decision tree from Elias *et al.* (2004). The right branch is always selected if the test passes. This part concerns *b* fragments resulting from fragmentation in the middle of the peptide. The dashed lines are pointers to other nodes in other parts of the tree. Each end node contains an intensity distribution, but only one is presented. The intensity range is divided into six intensity bins. The end node shown here states that there is a probability of 0.18 that a fragment with this environment has an intensity of the highest intensity bin. (Reproduced by permission of Nature Publishing Group)

PMZD is the fragment m/z minus the precursor m/z (note that this is a negative number);

RESN_1 is the residue N-terminal to the fragmentation site;

RESC_1 is the residue C-terminal to the fragmentation site;

HYDN_1 is the hydrophobicity of the residue N-terminal to the fragmentation site.

To predict the intensity of a fragmentation environment, the tree is simply traversed to an end node, which contains an intensity distribution.

Example Consider a *b* ion from a fragmentation environment consisting (among others) of the following properties: ($POS = 0.48$, $HYDN\_1 = -3.2$, $RESN\_1 = \mathtt{H}$, $PMZD = -200$). Upon traversing the nodes, the end node shown in Figure 9.2 will be reached.

△

A mismatch tree is also constructed, and both trees are used for scoring. This is explained in more detail in Chapter 11.

9.4 The effect of amino acids on the fragmentation

The fragmentation learning procedures have greatly increased knowledge about the effects that the amino acids adjacent to the fragmentation site have on the fragmentation efficiency (and thus on the fragment intensities). Some of these results are briefly presented here.

Tabb *et al.* (2003c) investigated the preference for N- or C-terminal fragmentation for the different amino acids (N-bias fragmentation), using fragmentation spectra from doubly charged peptides from ESI ion trap instruments. The b ion N-bias of the amino acid a is defined as

$$N\_bias_b(a) = \frac{\sum_{i_a} I_b^N(i_a) - \sum_{i_a} I_b^C(i_a)}{\sum_{i_a} I_b^N(i_a) + \sum_{i_a} I_b^C(i_a)}$$

where $\sum_{i_a} I_b^N(i_a)$ is the sum of the intensities of all b ions from a fragmentation N-terminal to an a (i_a is an index running over all fragmentation sites with an a as the nearest C-terminal residue). The most b ion N-biased amino acid was found to be `P`, and less N-bias was observed for `G` and `S`. Hydrophobic amino acids such as `I`, `L`, and `V` show a C-bias fragmentation in y ions, and `H` shows a C-bias in b ions. A similar analysis is also included in the article by Kapp *et al.* (2003) on peptides with different charges, but also obtained from ESI ion trap instruments. For each of the mobility groups (Section 9.2.2) they analyzed amino acid specific fragmentation and position-dependent cleavage. The most important results are summarized below.

Mobile peptides:

- a strong positive N-bias was shown for `P`, and to a lesser degree for `G` and `W`;
- a positive C-bias was shown for `I` and `V`;
- a strong negative C-bias was shown for `P`, and to a lesser degree for `G` and `S`.

Partially mobile peptides:

- a strong positive N-bias was shown for `P`, and to a lesser degree for `G`, `H`, `K` and `W`;
- a positive C-bias was shown for `D`, `H`, `I`, `K` and `V`;
- a strong negative C-bias was shown for `P`, and to a lesser degree for `G` and `S`.

Non-mobile peptides:

- a positive N-bias was shown for `H`, `K`, and `P`;
- a (very) strong (11 times the average) positive C-bias was shown for `D`, and to a lesser degree for `E`, `K`, and `R`;
- a negative C-bias was shown for `G` and `P`.

9.4.1 Selective fragmentation

Selective fragmentation, meaning that the fragmentation of a peptide is strongly dominated by one fragmentation site, has been observed for years. It especially occurs when the number of arginines is higher than the number of protons.

Exercises

9.1 For a specific fragmentation technique we want to explore how the produced ions depend on:

- the residue N-terminal to the fragmentation site;
- whether the fragmentation site is nearest to the N-terminus of the peptide, the middle of the peptide, or the C-terminus of the peptide;
- whether the peptide is considered overall hydrophobic or hydrophilic.

How many fragmentation environments are there in this environment class?

9.2 Assume that the amino acids are grouped according to some properties. A fragmentation environment is defined by the group that the residue C-terminal to the fragmentation site belongs to. We will calculate the fragmentation intensity ratio for y ions, and consider the environment consisting of the aliphatic group $e = \{\texttt{I},\texttt{L},\texttt{V}\}$. We consider three peptides, and collect the following data:

	I		L		V		$\bar{I}(p_1, y)$
	No.	Ave. int.	No.	Ave. int.	No.	Ave. int.	
p_1	2	60	1	54	0		40
p_2	1	46	1	54	1	44	50
p_1	1	74	0		3	58	46

Calculate $FIR_e(P, y)$, where P is the set of the three peptides.

9.3 Consider the decision tree in Figure 9.2. Assume that the five end nodes in the figure numbered from the left are:

0.40 0.31 0.14 0.07 0.04 0.04
0.06 0.07 0.13 0.24 0.32 0.18 (the example in the figure)
0.36 0.32 0.18 0.07 0.04 0.03
0.33 0.25 0.17 0.13 0.08 0.04
0.17 0.28 0.22 0.16 0.11 0.06

A peak in an MS/MS spectrum is found with intensity in the second intensity bin. Use the decision tree to find the probability that this peak can be a b_4 ion from the peptide TSVFAVLR.

9.4 Consider six fragmentation sites. Assume that the intensities of the *y* ions resulting from the fragmented C-terminal to a V are 30, 40, 0, 45, 55, 80 and the corresponding ones from the N-terminal to a V are 30, 0, 20, 15, 0, 40. Explore whether V shows a C- or N-bias fragmentation for y ions.

Bibliographic notes

Fragmentation pathway and rules

Paizs and Suhai (2005), Papayannopoulos (1995); Wysocki *et al.* (2000)
Mann *et al.* (2001); Steen and Mann (2004)

Fragmentation statistics

Elias *et al.* (2004); Huang *et al.* (2002); Kapp *et al.* (2003);
Tabb *et al.* (2003c)
Huang *et al.* (2005)

Selective fragmentation

Huang *et al.* (2005)

10 Identification and characterization by MS/MS

The MS/MS identification problem can be formulated as: given a set of MS/MS spectra $\mathcal{R} = \{R_1, R_2, \ldots, R_n\}$, resulting from the peptides $\mathcal{P} = \{P_1, P_2, \ldots, P_n\}$ (generally with known m/z), and a set of protein database sequences $\mathcal{D} = \{D_1, D_2, \ldots, D_m\}$, find (identify) the sequences in $\mathcal{D}$ where the peptides $\mathcal{P}$ come from, if any. For each peptide we have four alternatives:

1. The peptide comes from a protein whose sequence is in $\mathcal{D}$.
2. The peptide comes from a protein whose sequence is homologous to a sequence in $\mathcal{D}$.
3. The peptide comes from a 'unique' unknown protein, with no homologue in $\mathcal{D}$.
4. Each of the cases above may include *modifications*; the spectrum may therefore come from a peptide that is modified.

In principle, the search is performed by comparing an experimental spectrum to each segment (theoretical peptide) in the database, and identifying the segment(s) that give the best match(es) to the spectrum. There may be millions of segments in a sequence database, however, and it is therefore often impractical to examine every segment. Some form of filtering is therefore used to isolate only those segments that constitute potential matches. The mass of the precursor is one such filter, and others are described in the following chapters.

One way of classifying the different approaches for protein identification is to arrange them by how the *basic comparison* (comparing an experimental spectrum R to a segment S) is performed. We chose to divide the methods into three approaches:

Spectral *compare spectra*: construct a theoretical spectrum T of the segment, and compare the experimental spectrum to the theoretical one.

Sequential *compare sequences*: perform *de novo* sequencing of the spectrum to obtain a *derived sequence* E. Then compare the derived sequence to the segment.

Threading *compare spectrum to segment*: either the spectrum is 'threaded' on the segment, or vice versa. This approach includes a few methods that do not fit into

Computational Methods for Mass Spectrometry Proteomics I. Eidhammer, K. Flikka, L. Martens and S.-O. Mikalsen
© 2007 John Wiley & Sons, Ltd

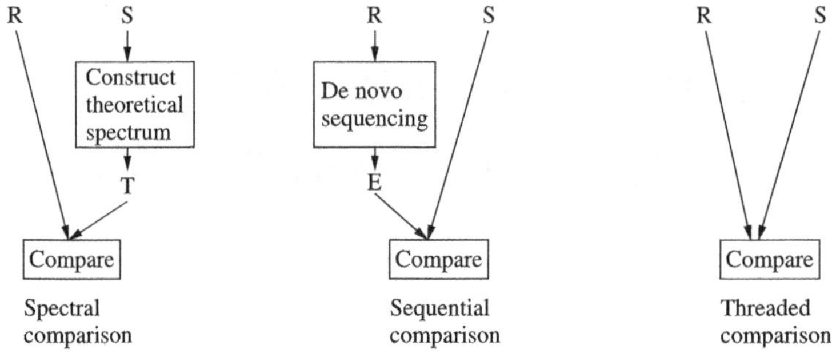

Figure 10.1 Illustration of the three different basic methods for spectrum-to-segment comparison

one of the other approaches. Some methods in this approach are briefly described in Section 13.4. It will, however, be clear when learning these methods that they have much in common with either the spectral ones or the sequential ones.

The different approaches are illustrated in Figure 10.1. For all approaches, *scoring* the matches is essential, and for MS/MS comparison the scoring scheme strongly depends on the matching approach.

The problem can now be formulated as: given a set of MS/MS spectra $\mathcal{R} = \{R_1, R_2, \ldots, R_n\}$ and a set of segments (theoretical peptides) $\mathcal{S} = \{S_1, S_2, \ldots, S_m\}$, find the segment(s) that give the best match(es) to each spectrum. We can consider two types of (spectrum, segment) comparisons: *straight* and *transformed* comparison. Transformed comparison means that possible transformations using *operations* are taken into account. The operations consist of types of modifications and the possible mutations (substitutions, insertions and deletions).

Straight comparison means finding the segment(s) in $\mathcal{S}$ that provide the best match(es) to a given spectrum when no transformations are considered.

Transformed comparison means finding the segment(s) that provide the best match(es) to a given spectrum, when at most k operations have been performed on the peptide that is the origin of the spectrum. The simplest case is when $k = 1$, and the mass of the operation is known.

10.1 Effect of operations (modifications, mutations) on spectra

A modification at residue i means that there is a mass shift in the b series of ions b_i to b_{n-1}, and in the y series of y_{n-i} to y_{n-1}, where n is the number of residues in the peptide. Consider a *complete* spectrum R_P of peptide P. Complete means that all ions of the considered fragment types are produced. Consider also another spectrum R_{PM} where P is modified at residue i. Then the b ions of R_P become equal to the b ions of R_{PM} by shifting $b_i, \ldots, b_{n-1}$ over a distance corresponding to the mass of the modification. A modification will therefore maintain the same number of peaks. A substitution will have the same effect as a modification, since it changes the mass of one residue. Insertions and

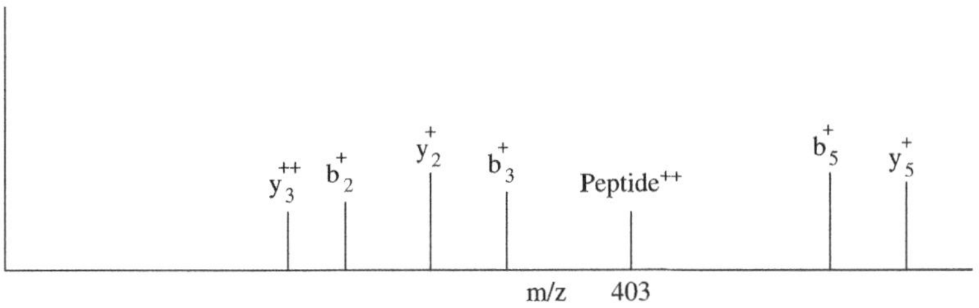

(a) Spectrum from unmodified peptide LICTVTR

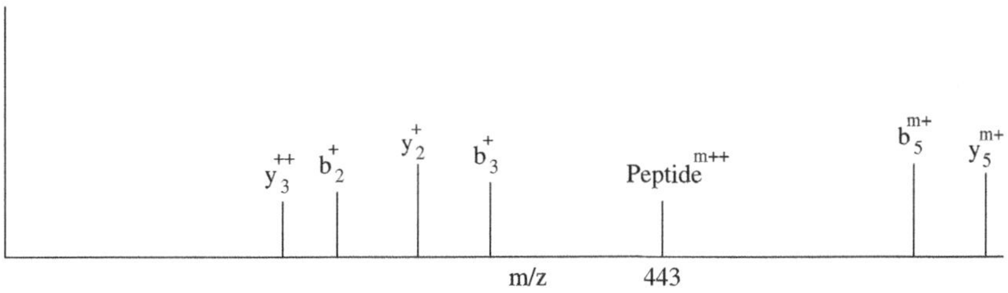

(b) Spectrum from peptide LICTVTR, phosphorylated at first T

Figure 10.2 (a) An example MS/MS spectrum (incomplete) of the unmodified peptide `LICTVTR`. (b) The same spectrum for the `LICTVTR` peptide, but now phosphorylated at the first `T` (80 Da)

deletions will also result in a shifting of peaks, but in addition to this, the total number of peaks is changed as well. This means that comparing spectra when taking modifications or mutations into account can be done by considering shifting some of the peaks, but also by considering the removal or insertion of peaks.

Example Consider the spectrum in Figure 10.2(a), derived from the unmodified peptide `LICTVTR`. If we assume a phosphorylation of the first `T`, then the peaks for b_5 and y_5 are shifted 80 units to the right, as shown in (b). Note that in this example, b_4, which also carries the modification, is not detected.

△

When we use the term *modifications* in the following chapters, it can also include mutations, depending on the context. This should not be confusing, however.

10.1.1 Comparison including modifications

Most of the identification methods where modifications or mutations are taken into account restrict the searching to include only a few specified types, and ignore the others. The searching can then be performed by (several) straight comparisons, by 'changing' the segments accordingly. If, for example, the possible methylation of cysteines needs to be taken into account, this can be done by comparing the experimental spectrum to the segment using the ordinary mass for cysteines, but also against that same segment with one or several of the cysteines carrying the increased mass of the modification. Such straightforward searching is, however, inefficient due to the possibility of a combinatorial explosion, and more intelligent search methods have therefore been developed.

In the last few years a couple of methods have been developed that are able to identify unanticipated modifications. The approach used by these methods is called *blind PTM identification*, since it operates in a blind mode, that is without any potential modifications specified before searching.

10.2 Filtering and organization of the database

In principle, each spectrum is compared to all segments in the database used, which in practice would imply too many comparisons. Some form of filtering is therefore necessary, meaning that only some of the segments are compared to the spectrum. This filtering needs to carefully balance two desires: not filtering out the correct segment (prevent false positive filtering), while filtering out as many of the incorrect segments as possible (prevent false negative filtering).

The most common filtering techniques use the precursor mass, and/or the specificity of the protease used for the digest. The mass filter can allow modifications or mutations, and either specific mass deviations are given, or a maximum allowed deviation is specified. The result is a set of masses, or an interval of masses, defining the mass constraints that segments must satisfy in order to be extracted for further comparisons. This is called *mass filtering*. It is important to realize that by using mass constraints, the ability to consider (unexpected) modifications will be limited.

Another filtering technique relies on the *de novo* extraction of short amino acid sequences from the spectrum. These sequences are typically two to four residues long and are called *sequence tags*. Only segments containing one of these sequence tags, sometimes including the correct flanking masses as well (essentially creating a combination of a sequence filter and a mass filter), are processed. This is called *sequence filtering*, and as we have seen above, it can be combined with mass filtering.

Example Consider a peptide with a neutral mass of 706 Da, and an observed MS/MS spectrum with peaks at {129, 246, 276, 345, 363, 432, 533, 579, 636}. If we expect that one residue may be phosphorylated, we have to consider two segments: one with a calculated peptide mass of 706 Da (in case there was no phosphorylation after all), and one with a calculated mass of 626 Da (which will yield the observed mass of 706 Da upon phosphorylation). When examining the mass differences between the fragment peaks we see that $345 - 246 = 99$, which is the residue mass of `V`, and $432 - 345 = 87$, which is the residue mass of `S`. It is then reasonable to believe that the peptide contains the subsequence `VS` (assuming we have been looking at consecutive b ions), or `SV` (if we have sequenced from the y ions).

△

In order to speed up the filtering step(s) the database can be indexed, and this can be done both for the mass filtering and for the sequence filtering. Advanced indexing organization can also include fragment masses, making it possible to use the fragment peaks of the MS/MS spectrum as entries in an index table. A disadvantage of using indices is that the index tables must be changed whenever the database is changed.

A second technique for speeding up the search is to organize the database as a suffix tree (see Section 13.4.1).

10.3 Scoring and statistical significance

Each search program uses some scheme for scoring the match between the segment and the spectrum, and the segment with the highest score is assumed to be the correct one. Many scoring schemes have been proposed, and the general discussion in Sections 7.4 and 7.5 concerning scoring schemes and statistical significance also holds here.

Exercises

10.1 Construct a theoretical spectrum of a peptide `ANCHIKR` using all *b* and *y* ions. Assume that in an experiment the peptide is modified by 14 Da on residue `C`, and that residue `I` is deleted. Construct also the spectrum for this case, and compare the two spectra by looking at the positions of corresponding ions and corresponding differences.

11 Spectral comparisons

As explained in Chapter 10, spectral comparison means that an experimental spectrum is compared to theoretical spectra constructed from segments from database sequences. The segments used are typically the theoretical peptides obtained from an *in silico* digestion. The comparison methods can be characterized by the following:

1. Which fragment ion types are considered.
2. How the intensities in the theoretical spectrum are calculated.
3. How the comparison is performed (the algorithm) and scored.
4. Whether modifications or mutations are taken into account, and, if so, how.

11.1 Constructing a theoretical spectrum

First one has to specify the fragment types expected in the experimental spectra, $\{\delta_i\}$. The segment is then subjected to a theoretical fragmentation, producing ions of the specified types (typically singly charged).

Peak construction For each fragmentation site one must decide for which of the fragment types $\Delta = \{\delta_i\}$ there should be peaks. The simplest case is to construct all. This results in a *complete spectrum*. An alternative approach is to use a *fragment type probability* $p(\delta_i)$, an estimate of the probability that an ion of type δ_i is produced at a fragmentation site. In addition, peaks can be produced due to noise. If q is the probability of the occurrence of a noise peak, then a peak should be constructed at the corresponding m/z value with probability $p(\delta_i) + [1 - p(\delta_i)]q$. The corresponding m/z values of the theoretical peaks are determined by using the equations in Section 8.6.

Peak intensities Determination of the peak intensities can be done at three levels, resulting in different types of theoretical spectra (or spectra at different levels), corresponding to the levels in Section 9.2:

UT spectra (Uniform Theoretical spectra), all peaks have the same height;

FT spectra (Fragment Theoretical spectra), the height of a peak depends on the fragment type, meaning for example that all b ions get equal height, and all y ions

Computational Methods for Mass Spectrometry Proteomics I. Eidhammer, K. Flikka, L. Martens and S.-O. Mikalsen
© 2007 John Wiley & Sons, Ltd

get equal height, but the height for the *b* ions is generally different from the height assigned to the *y* ions;

RT spectra (Residue Theoretical spectra), different heights are given to each peak of the same fragment type, depending on information about position, length, sequence, mass, etc., as described in Section 9.3.

These different types of spectra are reflected in different methods for spectral comparison and scoring:

- One can quickly construct UT spectra (or FT spectra) and compare these to the experimental spectrum. If the match gives a high score in this phase, one can then go on to a more sophisticated scoring, based on the same type of information as used for constructing RT spectra. This can therefore be considered a two-step procedure where the first step functions as a filter.
- One can directly construct RT spectra prior to starting the comparison, but this will cost more time.

Different schemes for scoring the comparison between an experimental spectrum *R* and a theoretical spectrum constructed from a segment *S* are used, and we divide them into non-probabilistic and probabilistic methods.

11.2 Non-probabilistic scoring

In this context, comparing spectra means comparing an experimental and a theoretical spectrum, although several of the scoring schemes were originally developed for comparing experimental spectra. Note, however, that there is a difference between comparing two experimental spectra (for example, to examine whether they are generated from the same peptide) and comparing an experimental and a theoretical spectrum, as all expected fragment types are included in the theoretical spectra. The comparison between two experimental spectra is treated in Section 14.6.

In the comparison two main methods are used:

- search for matching peaks;
- divide the m/z-axis into intervals, and compare the integral intensities in corresponding intervals.

The latter avoids the process of finding corresponding peaks, but carries the disadvantage of a loss of precision and suffers from problems when the ion's mass is on the border between two intervals.

Scoring schemes typically include a sum of the scores for each pair of matching peaks (or intervals), but often also a scoring component based on several matching peaks. Different variants of these methods are used, and they range from quite simple approaches to fairly advanced techniques.

11.2.1 Number and intensities of matching peaks or intervals

The simplest procedure for comparison is to process the spectra in parallel, counting the number of matching peaks. This can easily be extended to take intensities into account (either only from the experimental spectrum, or from both the experimental and theoretical spectra). One possibility is to calculate $\sum_{j=1}^{n} I_j^R I_j^T$, where I_j^K is the intensity of the jth matching peak in spectrum K, and n is the number of matching peaks. The scoring can also be weighted, for instance by introducing a factor that corresponds to the difference between the matching m/z values.

When interval division is used, the score of the comparison is, as above, $C = \sum_{j=1}^{m} I_j^R I_j^T$, where m is the number of intervals, and I_j^R is the total intensity in interval j of the experimental spectrum. It is clear that only intervals with non-zero intensities in both spectra affect the score.

The scoring scheme above has two components: the number of matching peaks or intervals, and the intensities of these matches. It assumes a linear increase in score as the number of matches increases. For instance, when all peaks have the same intensity, this means that a comparison with eight matches has a score that is twice as high as a comparison with only four matches. This is unreasonable, since some of the matches may occur simply by chance. Underlying probability functions for the number of matches occurring by chance are typically exponential, indicating an exponential increase in the score as a function of the number of matches. While this would not matter if all intensities were the same (the scores would arrange the segments in the same order), it has an effect when the intensities vary. Fenyö and Beavis (2003) have therefore proposed an alternative C' to the C score outlined above. This score is defined as $C' = Cn_b!n_y!$, where $n_b(n_y)$ is the number of matched $b(y)$ ions. Alternatively, they proposed $C'' = Ce^{n_b+n_y}$. They showed that both these nonlinear scoring functions outperformed the simpler, linear scores mentioned above.

11.2.2 Spectral contrast angle

A spectrum can be represented as an n-dimensional vector, where n is the number of considered m/z values (or intervals). The jth component of the experimental spectrum is then I_j^R. Two spectra can subsequently be compared by calculating (the cosine of) the angle between the vector representations, called the *spectral contrast angle*:

$$\cos\theta = \frac{\sum_j I_j^R I_j^T}{\sqrt{\sum_j (I_j^R)^2 \sum_j (I_j^T)^2}}$$

Two equal spectra have a contrast angle of zero, and an angle of 90° indicates maximal spectrum difference. The spectral contrast angle is mainly used to identify spectra produced by the same peptide, as discussed in Section 14.6.

11.2.3 Cross-correlation

A common function for calculating the correlation between any two signal series is the *cross-correlation* function. For the spectra we are considering, it can be formulated as

$C_\tau = \sum_{j=1}^{n} I_j^R I_{j+\tau}^T$, where τ is a relative displacement between the spectra. The simplest way to calculate the similarity (or correlation) is to use the correlation value for $\tau = 0$ (corresponding to the number of matching peaks when intensities are considered). It has (Powell and Hieftje (1978)) been found, however, that subtracting the mean of the cross-correlation function over a range $-k < \tau < k$ from the $\tau = 0$ value resulted in a better discrimination between similar spectra and that the resulting score was less sensitive to the (im)purity of the samples. In Eng *et al.* (1994) it was found empirically that 75 is a suitable value for k for MS/MS data.

In order to efficiently calculate this function for many τ, a Fourier transformation can be used. R and T are Fourier transformed, one of them is converted to its complex conjugate, and multiplication is performed. The result is then inverse Fourier transformed, to get the final value C_τ for many τ.

11.2.4 Rank-based scoring

Several rank-based scoring schemes have been developed. We will describe one from Searle *et al.* (2005). The experimental spectrum is first normalized by using a sliding window. After removing the precursor peak, the intensity of each peak is normalized by dividing the intensity by the average of all peak intensities in a window around the peak (± 100 units). The top 50 normalized peaks in the whole spectrum are ranked, with the highest ranked peak at position 50 and the lowest ranked one at position 1. The spectrum mass range is divided into unit intervals, and a vector I^R constructed such that I_i^R is the highest ranking value in interval i (which can be zero). The values I_i^R are then normalized by dividing by 50.

The intensities of the peaks in the theoretical spectrum are calculated in the range [1,50], depending on the fragment types (b, y, a, with water and ammonia loss). The b and y ions, for example, are assigned an intensity of 50. A vector I^T is then constructed for the theoretical spectrum, with I_i^T the intensity of the most intense fragment ion occurring in interval i. I^T is then normalized in the same way as I^R. The score is then

$$\sum_i (I_i^R - \overline{I^R})(I_i^T - \overline{I^T})$$

where $\overline{I^R}$ is the mean value over the values in I^R. Note that the expression is somewhat analogous to the spectral contrast angle.

The original scoring scheme contains a term which allows the inclusion of modifications, and was developed for assigning scores to *de novo* derived sequences (see Section 13.2.1).

11.2.5 SEQUEST scoring

Since several of the statistical and assessment analyses for MS/MS spectra use results from SEQUEST, we describe how SEQUEST scores its candidate segments when searching MS/MS spectra against a sequence database. SEQUEST employs two types of scores: a preliminary score and a final score. The preliminary score is computed in order to filter out segments that have only a very small probability for being the correct peptide. Those segments that achieve a sufficiently high preliminary score are then evaluated by the more computationally intensive final scoring scheme.

Preliminary scoring

The preliminary scoring uses *b*- and *y*-type continuity, and the presence of immonium ions.

The *b*-(*y*-)type continuity includes the number of matching peaks of type *b* ion (*y* ion) that also have a matching peak for the preceding *b* ion (*y* ion). If the total number of such peaks is C, a factor $1+C\beta$ is included in the scoring (with β=0.075).

The score for the immonium ions consists of calculating a value ρ_a for each of the amino acids a that are expected to yield observable immonium ions:

- if an immonium ion of a is not found in the experimental spectrum, then $\rho_a = 0$;
- if an immonium ion of a is found in the experimental spectrum, and the corresponding amino acid is present in the segment, then $\rho_a = 0.15$. If a is not present in the segment, then $\rho_a = -0.15$.

A factor $I_M = 1 + \sum_{a \in K} \rho_a$ is included, where K is the set of amino acids that are expected to yield observable immonium ions. The preliminary score then becomes

$$S_p = (\sum_{j=1}^{n} I_j) n (1 + C\beta) I_M / \eta_\tau$$

where I_j is the intensity in the experimental spectrum of matched peak j (matched within ± 1 unit), n is the number of matched peaks, and n_τ is the total number of predicted fragment ions.

Final scoring

The final SEQUEST score consists of several components, which include the cross-correlation score and the preliminary score, amongst others. The most important ones are as follows:

1. *XCorr*, the cross-correlation score.
2. *dCn*, the delta correlation value. The cross-correlation values are normalized such that the highest correlation value is one. *dCn* for a candidate match is one minus the normalized correlation value. It is of special interest to consider the *dCn* value between the first and second best match (first and second highest correlation values).
3. *Sp*, the preliminary score.
4. *RSp*, the rank that the segment under consideration got in the preliminary scoring.
5. *Ions*, the number of matched peaks divided by the number of peaks in the theoretical spectrum.
6. *dM*, the difference between the experimental precursor mass and the calculated mass of the segment under consideration.

Several programs exist for combining these different components into a single score. In Keller *et al.* (2002a,b) a method is described for combining different functions in a multifunctional scoring system, to achieve maximal discrimination between correct and incorrect identifications. The method is applied to SEQUEST scores. A *discriminant function F* is formed by a linear combination of the individual scoring components $\{v_1, \ldots, v_n\}$

$$F(v_1, \ldots, v_n) = c_0 + \sum_{i=1}^{n} c_i v_i$$

where the constants c_x are determined by techniques for discrimination analyses (see for example Section 14.5).

Treating modifications

SEQUEST is adapted for matching modified peptides. The initial segment selection procedure is changed so that segments are selected if they have a mass that is equal to or lower (to a given threshold) than the precursor mass. If the selected segment has a lower mass, the different possible combinations of the considered modifications are added to it, and the resulting predicted mass is compared to the observed precursor mass. If the masses are equal, the segment is retained for comparison, using the residue masses for the potentially modified residue(s).

11.3 Probabilistic scoring

A spectrum R is given, and a set of candidate segments $\{S\}$ from a database are found. We can then formulate a probabilistic scoring in two different ways:

- $Pr(S|R)$, the probability that the segment S is the origin of the spectrum R;
- $Pr(R|S)$, the probability that R is produced when performing MS/MS analysis on the segment S.

From Bayes' theorem

$$Pr(S|R) = \frac{Pr(R|S)Pr(S)}{Pr(R)}$$

we can see that the two formulas will rank the segments in the same order if the prior probabilities ($Pr(S)$) are equal for all S in the database. A theoretical spectrum (typically an RT spectrum) is produced either directly or indirectly, for calculating the necessary probabilities.

There are several probabilistic scoring methods. We will describe three that are fairly representative: a Bayesian method, and two different log-odds approaches. Only the main principles are described; the details can be found in the cited articles.

11.3.1 Bayesian method - SCOPE

SCOPE is a scoring scheme considering the experimental spectrum $R = \{r_1, \ldots, r_n\}$ and a database segment S.

Let $\mathcal{F}(S)$ consist of all fragment m/z values that are observed in a spectrum produced from S under the given experimental condition. Note that this means that the expected fragment types must be specified. Additionally, let F be a subset of $\mathcal{F}(S)$ such that $|F| \leq n$. F is then a potential set of observed fragments.

Example Let $R = \{177, 288, 312, 388, 489\}$ and $S =$ ASQTR. Assume that only b and y ions with charge 1 are expected, giving

$$\mathcal{F}(S) = \{72(b_1), 159(b_2), 175(y_1), 276(y_2), 287(b_3), 388(b_4), 404(y_3), 491(y_4)\}$$

Let us consider a fragmentation resulting in the fragments $\{y_1, b_3, b_4, y_4\}$, giving $F = \{175, 287, 388, 491\}$.

△

In order to calculate $Pr(R|S)$, one now – in principle – considers every potential fragment set F of S (under the given experimental conditions). For each F, calculate the probability for this fragment set to occur, $Pr(F|S)$. Then calculate the probability that the spectrum R is produced with fragment set F, $Pr(R|F, S)$. If we consider all fragment sets to be independent, we get

$$Pr(R|S) = \sum_{F \in \mathcal{F}(S)} Pr(R|F, S) Pr(F|S)$$

The calculation of these probabilities includes the assumption that each observed peak in R either must be assigned to a predicted peak in F, or must be considered noise. To achieve effective assignments, F is extended by noise peaks. A new fragment set is therefore defined: $\mathcal{F}'(S) = \mathcal{F} \cup R$. We now postulate $F \subseteq \mathcal{F}'(S)$ with the additional constraint that $|F| = n$. This means that F consists of fragments from S (those in both F and $\mathcal{F}(S)$) and noise (the remaining peaks in F but not in $\mathcal{F}(S)$).

Estimating $Pr(F|S)$

This is based on the probability for each fragment in the considered fragment set to occur. For the example above this means calculating the probability that only the ions $\{y_1, b_3, b_4, y_4\}$ are present. This calculation can be performed using empirically derived fragmentation statistics, see Section 9.3.

Estimating $Pr(R|F, S)$

Here we calculate the probability for 'R can be explained by F'. Since the number of peaks is equal in R and F, each can be ordered by increasing mass, and the desired probability can be estimated as

$$Pr(R|F, S) = Pr(R| \cap_{i=1}^{n} [r_i \hat{=} f_i], F, S)$$

where $r_i \hat{=} f_i$ denotes the event that peak r_i is a 'true' correspondence to f_i.

Example Consider the example above where $R = \{177, 288, 312, 388, 489\}$ and $\mathcal{F}'(S) = \{72, 159, 175, 177, 276, 287, 288, 312, 388, 404, 489, 491\}$. Let a theoretical spectrum be $F = \{175(y_1), 287(b_3), 312, 388(b_4), 491(y_4)\}$. Then R and F are compared, effectively estimating the probabilities that the peak at 177 is the y_1 ion, 288 the b_3 ion, 312 is a noise peak, 388 is the b_4 ion, and 489 is the y_4 ion.

△

The probabilities above are calculated by taking the accuracy and precision of the instruments into account, for example the likelihood that the y_4 ion is measured by a deviation of 2 units.

Implementation

The procedure above would require calculating the probabilities for a huge number of fragment sets (F). A dynamic programming procedure is developed that reduces the time complexity to $O(n|\mathcal{F}'(S)|)$.

A score corresponding to the most likely fragment set is

$$Pr'(R|S) = \max_{F \in \mathcal{F}'(S)} Pr(R|F, S)Pr(F|S)$$

that can be computed with the same time complexity. In practice $Pr(R|S)$ and $Pr'(R|S)$ produce values that are very similar, as almost all of the probabilities in $Pr(R|S)$ are also contained in the most likely fragment set.

11.3.2 Use of log-odds – OLAV

Many tools in bioinformatics are based on the log-odds principle. The advantage is that expected random results are taken into account. The probabilities of the result under two alternative hypothesis (H_1 and H_0) are calculated and divided, and finally the log taken. Typically the null hypothesis (H_0) is that the result corresponds to a random occurrence, and H_1 is the alternative hypothesis representing a correct match.

OLAV is a scheme calculating the score as

$$L = \log \frac{Pr(E|B, S, H_1)}{Pr(E|B, S, H_0)}$$

where E is the information extracted from an (R, S) comparison, and B represents available background information (fragmentation statistics etc.). The method is illustrated by letting E be the triplet (P, F, z):

- P is the difference between the theoretical mass and measured mass;
- F is the fragment mass matches;
- z is the charge of the precursor.

A model for statistical signal processing is built, assuming P and F, and P and z, to be independent. Using the general lemma

$$Pr(A, B|C) = Pr(A|B, C)Pr(B|C)$$

we get

$$Pr(E|B, S, H) = Pr(P, F, z|B, S, H) = Pr(F|P, z, B, S, H)Pr(P, z|B, S, H)$$

Since F and P are assumed independent, we get

$$Pr(F|P, z, B, S, H) = Pr(F|z, B, S, H)$$

Applying the lemma and the assumption of independence again, we derive

$$\begin{aligned} Pr(P, z|B, S, H) &= Pr(P|z, B, S, H)Pr(z|B, S, H) \\ &= Pr(P|B, S, H)Pr(z|B, S, H) \end{aligned}$$

This results in

$$Pr(E|B, S, H) = Pr(F|z, B, S, H)Pr(P|B, S, H)Pr(z|B, S, H)$$

Estimating the probabilities for H_1

For H_1 the individual probabilities are estimated as follows.

$Pr(P|B, S, H_1)$ is the probability that there is a difference of P between the observed precursor mass and the theoretical segment mass if S is the precursor sequence. This can be computed from an observed probability distribution of the precision of the instrument, which presumably follows a normal distribution.

$Pr(z|B, S, H_1)$ is the probability that a peptide with sequence S will get charge z. This can be found from a learned distribution. The distribution presumably follows an exponential function of the length of S. It should, however, be noted that the charge is also partly dependent on the amino acid sequence of the peptide, as basic amino acids will more readily attract additional protons.

$Pr(F|z, B, S, H_1)$ is the probability that a fragment set corresponds to an (R, S) match. Let $T = \{t_i\}$ be the theoretical spectrum constructed from S, based on the considered fragment ion types. Assuming independence between the fragment mass matches, we can formulate

$$Pr(F|z, B, S, H_1) \approx \prod_{j \in J} Pr_{t_j}(z) \prod_{j \in I-J} (1 - Pr_{t_j}(z))$$

where I is the set of indices corresponding to every theoretical fragment mass, and J is the set of matched theoretical masses. $Pr_{t_j}(z)$ is the probability of the actual fragmentation producing t_j when the precursor charge is z, and again can be learned as described in Chapter 9. Note that unmatched experimental peaks are not included in the probability calculation.

Example Consider an experimental spectrum $R = \{177, 288, 312, 388, 489\}$ with precursor charge $z = 2$ and $m/z = 283$, and compare it to the segment S=ASQTR. Let $F = \{175(y_1), 287(b_3), 388(b_4), 491(y_4)\}$. The calculated m/z ($z = 2$) for S is 281.5. $Pr(P|B, S, H_1)$ is then the probability of having a 1.5 unit difference from the correct value at a measurement of 283, for the equipment used. $Pr(z|B, S, H_1)$ is the probability that S acquires a double charge with the used ionization. Comparing R and F, allowing for a 2 unit measurement error, finds matching peaks for $\{y_1, b_3, b_4, y_4\}$. For $Pr(F|z, B, S, H_1)$ we must then calculate the probability for each of these four ions to occur, and the probabilities for each of the other possible fragment ions to be absent from the spectrum.

△

Estimating the probabilities for H_0

$Pr(P|B, S, H_0)$ is given by a uniform distribution.

$Pr(z|B, S, H_0)$ is taken to be constant or is learned from random sequences.

$Pr(F|z, B, S, H_0)$ is estimated by a formula as for H_1, but the fragment probabilities are calculated independently from z.

Implementation

Several of the independence assumptions are not valid – for example, that every fragment match is independent. In reality, if residue i in the peptide is protonated, then every b ion of length $\geq i$ should (ideally) be detected, and a corresponding argumentation holds true for y ions. Moreover, as the locations of the protonation sites are not constant for every ion for that peptide, it should preferably be modeled as a probability distribution. In order to accommodate this more complex approach, a hidden Markov model is constructed that is able to favor consecutive matches in each ion series, while allowing one or two gaps.

A general scoring scheme is also developed, which makes it possible to include several additional components besides the ones used above.

11.3.3 Log-odds decision trees

Decision tree modeling as described in Section 9.3.3 is used to define a scoring scheme including both a match and a mismatch decision tree.

To construct a mismatch decision tree the second best ranked segments in database searches were used. These second best matches were chosen over randomly selected peptides so that alternative matches giving a similar score would be discriminated from one another. Such discrimination represents a dilemma often faced in the interpretation of MS/MS spectra.

When a match is found between a theoretical and experimental spectrum using SEQUEST, each peak t_i of the theoretical spectrum is scored against its corresponding peak r_i in the experimental spectrum (note, however, that probably some of the t_i do not have any corresponding peak):

- By using the fragment/peptide properties of t_i, the match tree is traversed to an end (terminal) node. This node contains a probability distribution for the intensity expected for a peak with those fragment/peptide properties. From this the probability of the intensity observed at r_i, $I(r_i)$ is found: $Pr_{\text{match}}(I(r_i)|$ fragment/peptide properties of $t_i)$.

- The same procedure is used for the mismatch tree, determining $Pr_{\text{mismatch}}(I(r_i)|$ fragment/peptide properties of $t_i)$.
- The log-odds of these two probabilities is then calculated:

$$\text{lod}_{t_i} = \log_{10}\left(\frac{Pr_{\text{match}}(I(r_i)|\text{fragment/peptide properties of } t_i)}{Pr_{\text{mismatch}}(I(r_i)|\text{fragment/peptide properties of } t_i)}\right)$$

- The final score is found by summing the log-odds scores over all peaks t_i of the theoretical spectrum

$$SC = \sum_{t_i \in T} \text{lod}_{t_i}$$

This log-odds score has an advantage over using the probabilistic score from the match tree only, since the latter score would include a product of probabilities, and would therefore unfairly penalize long segments.

11.4 Comparison with modifications

As the straightforward method of including modifications in the identification (constructing several theoretical spectra) is inefficient, other methods have been developed. We will describe two of them.

11.4.1 Zone modification searching

A method called *zone modification searching* is used in the program ProID of Interrogator. In this context, a zone is defined by dividing the peptide mass into k equal parts, each part being a zone. Let the mass difference between the precursor and the considered database segment be δm, and the tasks are (i) to examine if there is a high probability that this δm is derived from a modification, and (ii) to determine, if possible, the position of the modification. The modification site is localized by dividing the mass range $[0, M_p]$ into k ($k = 6$ as standard) zones, and by calculating a score for each zone. The modification is then supposed to be in the highest scoring zone.

Scoring the zones

Assume that the length of the precursor peptide of an experimental spectrum is n, and that a modification is present at residue p. If the experimental spectrum is compared to a theoretical spectrum of b and y ions constructed from the correct segment, experimental peaks from $b_1 \ldots b_{p-1}$ and $y_1 \ldots y_{n-p}$ will match. Peaks from $b_p \ldots b_{n-1}$ and $y_{n-p+1} \cdots y_{n-1}$ will match to an inversed theoretical spectrum (constructed from the complementary masses). Assume that the modified residue at p falls in zone h. Then we have the following:

- every experimental b ion in zone $1 \ldots h-1$ will match a theoretical b ion;
- every complement of an experimental b ion in zone $h+1 \ldots k$ will match a theoretical y ion;
- every experimental b ion in zone h will match a theoretical b ion, or its complement will match a theoretical y ion;
- every experimental y ion in zone $h+1 \ldots k$ will match a theoretical y ion;
- every complement of an experimental y ion in zone $1 \ldots h-1$ will match a theoretical b ion;
- every experimental y ion in zone h will match a theoretical y ion, or its complement will match a theoretical b ion.

The score for a zone h is then calculated by counting the number of peaks in the spectrum that can be matched, assuming that the modification is present in that zone h. All peaks in the spectrum are first considered b ions and these are matched, and subsequently they are considered y ions and matched again. Since the assumption is that the modification is found in zone h, either the original peak or its complement will actually be matched, depending on the ion series under consideration and the relative position of the peak with regards to the zone h. The total number of matches, summed over the b ion and y ion assumptions, is counted as the score. This is clarified further in the next example.

Example Suppose a spectrum is produced from the modified peptide $P = \texttt{AGC}^m\texttt{VSTR}$, where m means methylated (corresponding to a nominal mass increase of 14 Da). The precursor mass is $M_P = 706$. Let the peaks of an experimental spectrum be as shown in the first two rows of Table 11.1. We define four zones, (0–176.5, 176.5–353, 353–529.5, 529.5–706). The table also shows the complementary masses and the zone number in which the masses will fall if they are assumed to be b or y ions.

The calculated theoretical peaks (without modifications) are at

b ions:	72,	129,	232,	331,	418,	519
y ions:	175,	276,	363,	462,	565,	622

The score for zone 2 (thus assuming the modification is in zone 2) is calculated as follows.

Table 11.1 Data for the spectrum used in the example

Peaks	129	246	276	345	363	432	533	579	636
Ions	b_2^+	$b_3^{m,+}$	y_2^+	$b_4^{m,+}$	y_3^+	$b_5^{m,+}$	$b_6^{m,+}$	$y_5^{m,+}$	$y_6^{m,+}$
Compl. masses	579	462	432	363	345	276	175	129	72
Zone if									
b ions	1	2	2	2	3	3	4	4	4
y ions	4	3	3	3	2	2	1	1	1

Assume b ions:

Zone 1: 129 matches (theoretical) b_2, the original peak is used since the modification is assumed to be in zone 2.

Zone 2: complement of 246 matches y_4, complement of 345 matches y_3, note that complements of peaks are used from here on, as the modification now needs to be taken into account.

Zone 3: complement of 432 matches y_2.

Zone 4: complement of 533 matches y_1.

Assume y ions:

Zone 1: complement of 636 matches b_1, complement of 579 matches b_2.

Zone 2: 363 matches y_3.

Zone 3: 276 matches y_2.

Zone 4: no match.

The total number of matches is nine (all the experimental peaks), thus the score of zone 2 is nine. The other zones are scored in a similar way, resulting in lower scores. The modification is therefore likely to be in zone 2, corresponding to residues 3 or 4.

△

Database indexing

The zone modification searching includes numerous comparisons, some of which should not match (which may result in erroneous matches due to noise). For effective use in large databases, additional intelligent search tools must be available. In ProID the zone modification searching is combined with a comprehensive indexing organization of the database. The sequences are theoretically digested (by trypsin or another selected protease), and the theoretical peptides (segments) are collected into peptide bins of similar masses, such that each bin contains at most a specified number of theoretical peptides (for example, 100 000). All the theoretical peptides in a bin thus have the same zone division. A second indexing is based on the theoretical fragment masses (b and y ions).

Additionally, masses including possible modifications are precalculated for each segment. For the interpretation of the spectrum, the user defines a range $[0, maxmm]$, for the possible modification masses. In order for a segment to be further investigated there must be a possible modification mass $t \in [0, maxmm]$ for which $|M_S + t - M_P| \leq \epsilon$, where M_S is the theoretical peptide mass, M_P is the measured precursor mass, and ϵ is the chosen accuracy.

By using the fragment indices, and the parallel processing of the remaining theoretical peptides from a bin, reasonable execution time is achieved for zone modification database search.

11.4.2 Spectral convolution and spectral alignment

Spectral convolution and spectral alignment make it possible to include a *blind search* for modifications in the identification process. A blind search means that nothing is assumed about possible modifications before searching.

We recall that cross-correlation for $\tau \neq 0$ measures the similarity when one of the spectra is displaced τ units relative to the other. Suppose now that we have a peptide P of length n and let Q be the same peptide but modified at residue i by mass m. Suppose for simplicity only b ions; then the cross-correlation of noise-free and complete spectra of P and Q for $\tau = 0$ will have $i-1$ peaks in common. For $\tau = m$ they will have $n-i$ peaks in common, and adding the number of common peaks for these two correlations will result in $n-1$ peaks. Such numbers of common peaks furnish one of the ideas behind *spectral convolution*, which for two spectra R, T is defined as

$$R\Theta T = \{r - t | r \in R, t \in T\}$$

Furthermore, $(R\Theta T)(x)$ is the number of pairs $r \in R, t \in T$ such that $r - t = x$. In some sense x can therefore be compared to τ for cross-correlation. If T is a complete b ion spectrum from P, and R from Q, then $(R\Theta T)(0) = i-1$, $(R\Theta T)(m) = n-i$, and 0 and m will be local maximum values for $(R\Theta T)(x)$.

Example Consider the (modified) peptide P=AGC^{m}VSTR, and the experimental spectrum

$$R = \{129(b_2^+), 182(y_3^{++}), 246(b_3^{m,+}), 276(y_2^+), 432(b_5^{m,+}), 462(y_4^+), 533(b_6^{m,+}), 636(y_6^{m,+})\}$$

m means methylated (corresponding to a nominal mass increase of 14 Da). A theoretical spectrum T of P with unmodified b and y ions larger than one and with charge 1 would be $\{129, 232, 276, 331, 363, 418, 462, 519, 565, 622\}$. We then find for the spectral convolution $(R\Theta T)(0) = 3$, $(R\Theta T)(14) = 4$.

△

Consider now two operations (modifications or substitutions) for P, at positions r_1, r_2 with masses m_1 and $m - m_1$. Let R again be an experimental complete spectrum of modified P, and T a theoretical complete unmodified spectrum. With b_i^K the ith b ion for a spectrum K, we then have

b ion:

- $b_i^R = b_i^T$ for $i < r_1$
- $b_i^R = b_i^T + m_1$ for $r_1 \leq i < r_2$
- $b_i^R = b_i^T + m$ for $r_2 \leq i$

y ion:

- $y_i^R = y_i^T$ for $i \leq n - r_2$
- $y_i^R = y_i^T + (m - m_1)$ for $n - r_2 < i \leq n - r_1$
- $y_i^R = y_i^T + m$ for $n - r_1 < i$

From this we see that $(R\Theta T)(x)$ will have (local) maximum values at $x = 0$, $m_1, m - m_1, m$. For a random peptide with uniform distribution of the modifications over the peptide, the ratio of the expected heights of these maximum values is 2:1:1:2 (the first and last maximum values are effected by both b and y ions).

Example Consider again the (modified) peptide P=AGC^{m}VS^{p}TR, and the experimental spectrum

$$R = \{129(b_2^+), 222(y_3^{p,++}), 246(b_3^{m,+}), 276(y_2^+), 512(b_5^{mp,+}), 542(y_4^{p,+}), 613(b_6^{mp,+}), 702(y_6^{mp,+})\}$$

p means phosphorylated (corresponding with a nominal mass increase of 80 Da). We then find $(R\Theta T)(0) = 2$, $(R\Theta T)(14) = 1$, $(R\Theta T)(80) = 1$, $(R\Theta T)(94) = 3$. Note, however, that there are many values for x where $(R\Theta T)(x) = 1$.

△

We have seen that modifications will have a tendency to turn up as local maximum values in the spectral convolution, but finding these maximum values requires calculation of the function for many values. Spectral convolution, however, has an even more serious drawback: the set of local maxima does not necessarily need to correspond to possible modifications.

Example Consider two spectra, $T = \{10, 15, 23, 32, 37, 42\}$ and $R = \{10, 17, 25, 32, 37, 44\}$, and suppose again for simplicity only one ion type. The spectral convolutions for $x = 0$ and $x = 2$ both yield 3, indicating a modification with mass 2, but there is no consistent mass shift in T (corresponding to a modification) that would explain both maximum values.

△

The example shows that one needs to constrain the way in which the number of matching peaks for different displacements is calculated. This can be obtained by some sort of alignment between the spectra, called *spectral alignment*.

Spectral alignment

The key fact to take into account is that a true modification results in a consistent shift of all succeeding peaks in R relative to T.[1] The spectral alignment addresses the problem of finding the maximum number of matching peaks when at most k shifts are allowed, corresponding to k modifications.

---

[1] This is not always correct; in fact it is only correct for each fragment type separately, see the following example.

Table 11.2 Matching of R and T, and spectra resulting from a shift in T, explained in the example

	b_2^+	y_3^{++}		$b_3^{m,+}$	y_2^+						
R	129	182		246	276						
T	129		232		276		331		363		
T'	129			246		290		345		377	
T''	129		232		276		331		363		
		$b_5^{m,+}$	y_4^+			$b_6^{m,+}$				$y_6^{m,+}$	
R		432	462			533				636	
T	418		462		519		565		622		3
T'		432		476		533		579		636	5
T''		432		476		533		579		636	5

Example Consider the first example in this subsection. A comparison between R and T and shifts in T are shown in Table 11.2. The last column contains the number of matching peaks. The number of matching peaks for no shift (T) is three. T' is the theoretical spectrum with one shift (14 Da) from 232 onwards, resulting in five matching peaks. Note, however, that y_2 and y_4 are now wrongly shifted. T'' is the spectrum with one shift from 418 onwards, again resulting in five matching peaks, but this time with y_4 wrongly shifted. The two solutions for the maximum number of matching peaks with one shift (one modification) will localize the modification at different residues, with only one of them being the correct one.

△

The maximum number of matching peaks in the example should ideally be seven (that is, all singly charged peaks in R). The reason that only five are found lies in the fact that R is a mixture of b and y ions, and that unmodified y ions have larger masses than modified b ions.

The number of matching peaks can be extended if we also compare T to the inversed spectrum of R (see Section 8.8), in practice comparing T to a combination of R and R inversed. For the example above two new peaks would be matched, $y_5^{m,+}$ and $b_4^{m,+}$. Note, however, that for experimental spectra, using the inversed spectrum will increase the number of noise peaks.

The maximum number of matching peaks for different numbers of shifts can be found by a dynamic programming (DP) procedure. Assume that T contains m peaks $\{t_1, \ldots, t_m\}$ and R contains n peaks $\{r_1, \ldots, r_n\}$. Then define a two-dimensional matrix A_{mn}, such that row i corresponds to t_i and column j to r_j.[2] Let $A_{ij}(k)$ be the maximum number of matching peaks between the i first peaks of T and the first j peaks of R, under the

[2] The presentation here is slightly different from the one in the original article by Pevzner *et al.* (2000).

assumption that t_i and r_j match and that there are at most k shifts. Further define two cells $(i'j')$ and (ij) in A as *codiagonal* if $t_i - t_{i'} = r_j - r_{j'}$.[3]

Let $i' < i$ and $j' < j$. Then we have

if $(i'j')$ and (ij) are codiagonal **then** $A_{ij}(k) \geq A_{i'j'}(k) + 1$
else $A_{ij}(k) \geq A_{i'j'}(k-1) + 1$

This can be used to define the recursion of the dynamic programming procedure:

$$A_{ij}(k) = \max_{i'<i,j'<j} \begin{cases} A_{i'j'}(k) + 1, & \text{if } (i', j') \text{ and } (i, j) \text{ are codiagonal} \\ A_{i'j'}(k-1) + 1, & \text{otherwise} \end{cases}$$

For the initial condition define $A_{00}(k) = 0$. This procedure searches for each k for the 'path' with the maximum number of cells when at most k shifts are allowed. The recursion algorithm considers only shifts that do not change the order of the peak values, which means that a shift of t_i must be larger than $t_{i-1} - t_i$.

Example Let $T = \{7, 11, 15, 18, 21, 24, 30, 38, 43\}$ and $R = \{7, 11, 13, 19, 22, 25, 31, 33, 38\}$. The recurrence for the spectral alignment for $k = 0, 1, 2$ is shown in Table 11.3, where only the cells used for finding the best alignments are filled in. $k = 0$ means no shifts, hence values for $k = 0$ are only shown for cells with the same values in both spectra. The maximum number of matching peaks for $k = 0$ is three, and for $k = 1$ this is six. Two alignments with one shift result in this value, shown underscored and in bold face, respectively. The maximum number of matching peaks for $k = 2$ is eight, which is shown as an extension of the alignment in bold face.

△

The dynamic programming as described here has a running time complexity of $O(n^4k)$ for spectra of length n. A more effective procedure has been developed, as follows schematically:

- For calculating $\max A_{i'j'}(k)$ let $u(i, j)$ be a pointer to the $A_{i'j'}$ which is $\max A_{i'j'}(k)$ over all (i', j'), $i' < i$, $j' < j$, and $t_i - t_{i'} = r_j - r_{j'}$. These pointers can be constructed in $O(n^2)$ time if one constructs and uses a lookup table of length $t_n + r_m$.
- For calculating $\max A_{i'j'}(k-1)$ one can use another two-dimensional matrix M, and require that $M_{i-1,j-1}(k-1)$ at the time of calculation contains $\max_{i'<i,j'<j} A_{i'j'}(k-1)$. By a little reformulation this can be written as $M_{ij}(k) = \max_{i'\leq i,j'\leq j} A_{i'j'}(k)$.
- This means that the recursion becomes

$$A_{ij} = \max(A_{u(i,j)} + 1, M_{i-1,j-1}(k-1) + 1)$$

- The recurrence for M is then calculating the maximum of three values, $M_{ij}(k) = \max(A_{ij}(k), M_{i-1,j}(k), M_{i,j-1}(k))$.

---

[3] Note that the cells do not need to be codiagonal in the strict understanding of the word, but would be in a matrix defined over all integers constrained by t_m and r_n.

Table 11.3 Example of a spectral alignment. Two alignments are shown, as underscore and in bold face. See the text for explanation

$T\backslash R$		0	7	11	13	19	22	25	31	33	38
	$k=0$	**$\underline{0}$**									
0	$k=1$	0									
	$k=2$	0									
	$k=0$		**$\underline{1}$**								
7	$k=1$		1	1							
	$k=2$		1	1							
	$k=0$			2							
11	$k=1$		1	**$\underline{2}$**	2						
	$k=2$		1	2	2						
	$k=0$										
15	$k=1$			2	3	3					
	$k=2$			2	3	3					
	$k=0$										
18	$k=1$				$\underline{3}$	**3**	4	4			
	$k=2$				3	4	4	4			
	$k=0$										
21	$k=1$					4	**4**	5			
	$k=2$					4	5	5			
	$k=0$										
24	$k=1$					$\underline{4}$	5	**5**	5		
	$k=2$					4	5	6	6		
	$k=0$										
30	$k=1$							4	**6**	3	
	$k=2$							6	7	6	
	$k=0$										3
38	$k=1$								3	$\underline{5}$	3
	$k=2$								6	**7**	7
	$k=0$										
43	$k=1$									3	$\underline{6}$
	$k=2$									7	**8**

Exercises

11.1 Assume that an experimental spectrum is compared to two theoretical spectra T_1 and T_2, constructed from two segments. All intensities in the theoretical spectra are set to 50. Assume that three matching peaks with intensities 40, 20, and 50 are found when comparing to T_1, and four with intensities 20, 10, 25, and 25 when comparing to T_1. Calculate the scores C when correlation is used (Section 11.2.1), and the adjusted score C', assuming two b ions, and one and two y ions, respectively.

11.2 Calculate the spectral contrast angle for the spectra in Exercise 11.1.

11.3 A rank-based score is to be calculated. For illustration we suppose eight intervals, and we rank the six highest intervals. Suppose these six ranks in the experimental spectrum are {5, 0, {1,4}, 2, 6, 0, 3, 0}, and in the theoretical spectrum {0, {2,5}, 4, 0, 3, 1, 6, 0}. (This means that there are two peaks observed in the third interval of the experimental spectrum.) Calculate the rank-based scoring.

11.4 A scoring using OLAV is to be calculated, under the following assumptions:

- The peptide mass difference P is 1 Da. Let the probability function for this difference follow the standard normal distribution.
- Assume the charge $z = 2$, and the peptide length $n = 10$. Let for simplicity the charge probability for $z = 2$ follow the function $P_{z=2}(n) = e^{\frac{n-20}{20}}$.
- For the theoretical fragments we ignore those of length 1 and $n-1$. Assume that there are matches to five b ions and four y ions, and that we use a probability of 0.6 for a potential b ion to occur, and 0.5 for a potential y ion to occur (for $z = 2$).
- Let $Pr(P|B, S, H_0)$ be uniformly distributed in the nominal masses 400–3000.
- Let $Pr(z|B, S, H_0)$ be $\frac{1}{3}$ for each of $z = 1, 2, 3$.
- Let the probability for b ions and y ions to occur in random peptides be 0.4 for both.

11.5 Assume we have an experimental spectrum with peaks at (129, 232, 276, 331, 377, 432, 533, 579, 636) and with precursor mass $M_P = 706$. We have reason to believe that the sequence of the precursor is AGCVSTR, and that there is a modification. Use zone modification searching to try to identify where a modification may be. Use the zones (0–176.5, 176.5–353, 353–529.5, 529.5–706).

11.6 Assume an experimental spectrum (182, 209, 276, 312, 476, 512, 565, 613, 702). We have reason to believe that the sequence of the precursor is AGCVSTR, and that there are one or two modifications.

(a) Try to find the maximum number of matches when one and two shifts are allowed. Masses for theoretical peaks can be found from examples in Section 11.4.2.

(b) Perform the dynamic programming procedure on the same dataset.

Bibliographic notes

Cross-correlation
Eng *et al.* (1994); Owens (1992)

Rank-based scoring
Searle *et al.* (2005)

SEQUEST
Eng *et al.* (1994); Gatlin *et al.* (2000); Keller *et al.* (2002a,b); Yates *et al.* (1995)

SCOPE
Bafna and Edwards (2001)

OLAV
Colinge *et al.* (2004, 2003c)

Decision trees
Elias *et al.* (2004)

Other scoring schemes
Dancik *et al.* (1999); Havilio *et al.* (2003); Matthiesen *et al.* (2004)
Tanner *et al.* (2005); Zhang *et al.* (2002)

Zone modification searching
Tang *et al.* (2005)

Spectral convolution and spectral alignment
Pevzner *et al.* (2000, 2001),Tsur *et al.* (2005)

Blind search
Lie *et al.* (2006a); Yan *et al.* (2006)

12 Sequential comparison – *de novo* sequencing

De novo sequencing means trying to derive the original peptide sequence from an MS/MS spectrum without any sequence knowledge beforehand. (*De novo* is Latin and means anew or over again).

De novo sequencing was primarily developed for analyzing spectra that were not identified by other search methods, indicating that no homologous sequences exist in the database, or only sequences with weak homology. However, *de novo* sequencing followed by a database search has also proved to be able to compete with other approaches.

The *de novo* peptide sequencing problem via tandem mass spectrometry can be formulated as: *for an experimental spectrum R, a peptide mass M_P, and a set of fragment ion types Δ, derive a sequence (or set of sequences) with mass M_P that gives the best match to spectrum R*. Posttranslational modifications can also be taken into account during *de novo* sequencing. Choosing the best match implies that a score scheme must exist for scoring the derived sequence candidates against the spectrum.

The approach taken by nearly all methods for performing *de novo* sequencing of an experimental spectrum R consists of a two- or three-step procedure:

Step 1: Derive a set of sequence candidates $\{E_i\}$, usually ranked by some scoring scheme.

Step 2: Score each of the candidates against R (another scoring scheme).

Step 3: (optional) Search for the highest scoring candidates in a protein sequence database.

The straightforward way to solve Step 1 is to generate all possible sequences corresponding to the peptide mass. However, in the general case this implies generating too many sequences to be practically useful. Therefore other approaches must be used, and several different methods have been developed. Here we focus on one approach, sequencing by using *spectrum graphs*, and only briefly mention some others. We focus on spectrum graphs because this approach is well investigated and implemented in several algorithms, it is intuitive, and it clarifies most of the issues concerning *de novo* sequencing.

Several methods for performing a subsequent database search have been developed, and these are described in Chapter 13.

Computational Methods for Mass Spectrometry Proteomics I. Eidhammer, K. Flikka, L. Martens and S.-O. Mikalsen
© 2007 John Wiley & Sons, Ltd

In the remaining part of this chapter we assume singly charged fragment ions, if not otherwise stated.

12.1 Spectrum graphs

The ideal situation for *de novo* sequencing would be that only one fragment type was produced, that fragmentation occurred at every single fragmentation site along the peptide backbone, and that all fragments had the same charge.

Example Assume a peptide with neutral mass $M_P = 692$. Only singly charged b ions are produced (and this at every fragmentation site), resulting in a spectrum with peaks at $\{72, 129, 232, 331, 418, 519\}$. Knowing that the first peak is the b_1 ion, we find that the mass of the first residue is $72 - \langle N \rangle$, where $\langle N \rangle$ is one, corresponding to the mass of the extra H at the N-terminus. Hence the first amino acid is A. The mass differences between succeeding peaks correspond to the sequence of the residues; hence they are G, C, V, S, T. The last residue mass is $M_P - \langle N \rangle - \langle C \rangle - (519 - 1) = 156$, corresponding to the residue mass of the C-terminal R.

△

One way of visualizing the method shown in the example is to use a *graph*. A graph consists of a set of *nodes* and a set of *edges*. An edge is a connection between two nodes, and it can be undirected or directed. If an edge between nodes n_1 and n_2 is directed, it means that the connection is from n_1 to n_2. An undirected edge is commonly represented by a line, and a directed edge by an arrow. Both the nodes and the edges can contain values.

In order to use graphs in *de novo* sequencing we define a node for each peak of the spectrum, and arrange the nodes by increasing mass. Then we define a directed edge between any two nodes that have a mass difference that corresponds to the mass of an amino acid residue (or corresponding to the combined mass of several amino acid residues if necessary). An edge always goes from a lower mass to a higher mass, and is labeled with the corresponding amino acid. Such a graph is called a *spectrum graph*. Figure 12.1 shows the spectrum graph for the ideal situation (only b ions) described in the preceding example. For completeness there is also a first node at mass $\langle N \rangle$, and one last node at mass $M_P - \langle N \rangle - \langle C \rangle + 1$ (for the 'b ions graph').

In reality, the spectrum is never as simple as this example indicates. There will be additional peaks of other fragment types and noise peaks. Furthermore, usually several b ions will be missing from the spectrum, and modifications add extra complications to the spectrum. (Bear in mind that when constructing the spectrum graph we have no knowledge about which types of ions the peaks correspond to.)

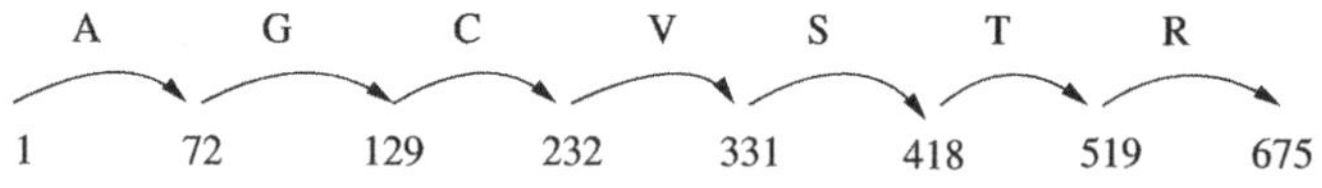

Figure 12.1 A spectrum graph for the spectrum $\{72, 129, 232, 331, 418, 519\}$ and peptide mass $M_P = 692$. The path through the graph corresponds to the derived sequence. The mass difference between subsequent numbers corresponds to the residue mass

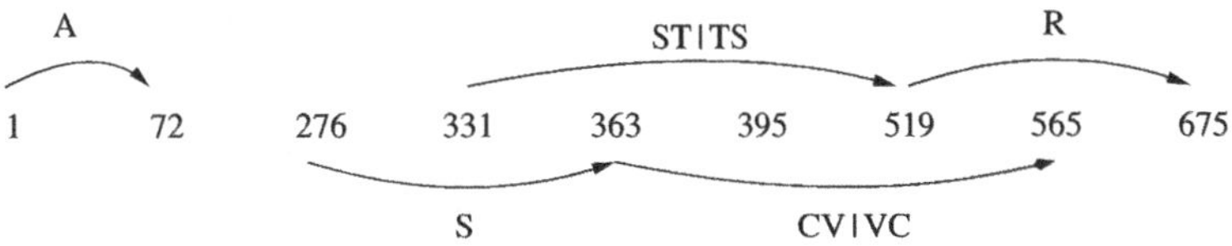

Figure 12.2 A 'b ion' spectrum graph for the spectrum $\{72, 276, 331, 363, 395, 519, 565\}$ and $M_P = 692$

Example We have a spectrum $\{72, 276, 331, 363, 395, 519, 565\}$ and peptide mass $M_P = 692$. A spectrum graph assuming b ions is shown in Figure 12.2. Here an edge is drawn between two nodes when the mass difference corresponds to one or two amino acids. Note that the order of the two amino acids in an edge spanning a mass difference of two residues cannot be determined.

△

There is no path through the whole graph in this example, and a complete sequence cannot be derived. The reason is that there is a lack of b ions. However, there can be some y ions, and the complementary properties between b and y ions can be utilized. The idea is for each peak in R to first assume that it is a b ion, and then assume it is a y ion. This assumption is used to construct a new spectrum, a *sequence spectrum* Q. For each peak $r \in R$ with ion mass m_r there will be a peak in Q at m_r, and in addition there will be a peak at $M_P - m_r - 2H$, which is the mass of the complementary b ion if r is a y ion. Thus Q is the union of R and R inversed. The number of peaks in Q will be between n and $2n$ when n is the number of peaks in R. By using this approach we hopefully increase the number of b ion masses, although some of them are found indirectly by calculating the complements of y ion masses.

Example The inversed spectrum of the spectrum in the example and in Figure 12.2 is $\{129, 175, 299, 331, 363, 418, 622\}$. We see that two of the peaks, 331 and 363, are common to the spectrum and its inversed spectrum. Since the sum of these two masses is $M_P + 2$, they probably come from a fragmentation producing both b and y ions. The sequence spectrum then becomes $\{72, 129, 175, 276, 299, 331, 363, 395, 418, 519, 565, 622\}$and the spectrum graph is shown in Figure 12.3. From the derived sequence `AG(CV|VC)STR` we find that R contains $\{b_1, b_4, b_6\}$, and $\{b_2, b_5\}$ are constructed in the inversed (complementary) spectrum.

△

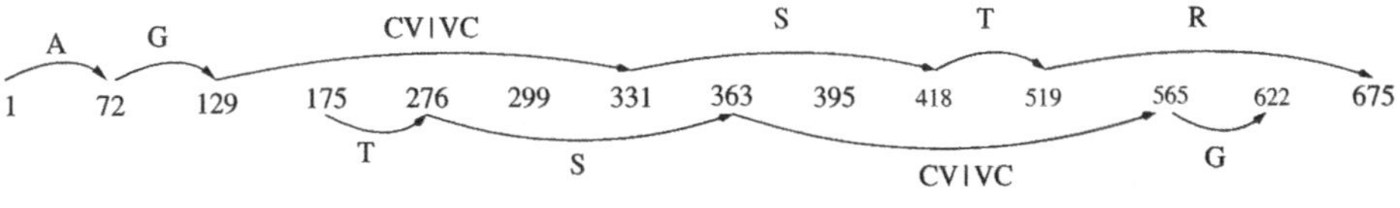

Figure 12.3 The spectrum graph for the addition of the inversed spectrum. We see that one path indicates that the original sequence is AG(CV| VC)STR

12.1.1 A general spectrum graph

b and y ions are the most frequently observed ion types when fragmenting with low-energy CID, but other types can also occur to a greater or lesser extent. In order to take all possible fragment types into account, we will describe a general spectrum graph. The spectrum graph is as before normally constructed through a sequence spectrum.

A sequence spectrum – the nodes of a spectrum graph

The procedure for transforming an experimental spectrum to a sequence spectrum is as follows:

1. Choose the set of fragment ion types Δ, to be considered.
2. Choose one fragment type as the *basis type*, usually b ions.
3. For each peak in the spectrum, for each fragment type $i \in \Delta$, calculate the corresponding mass of the basis type ion if the peak was of type i. Let this calculated mass represent a peak in the sequence spectrum. For each peak in the original spectrum there will be in the sequence spectrum as many peaks as there are members in the ion set, $|\Delta|$.
4. Assign a score to each peak in the sequence spectrum. This can for example be based on a weight for each fragment type in Δ, showing how likely it is that the ion type is present.
5. Calculate a height (intensity) for each peak, based on the score found in the preceding step.

Note that a peak in the sequence spectrum can come from several peaks in the original spectrum, up to $|\Delta|$ if all fragment types are produced for a fragmentation site.

Using the notation from Section 8.6, let b ions be the basis, and assume only singly charged ions. Suppose there is a peak in the original spectrum at M_o. Denote the m/z value for the constructed peak as $M_b^*(i)$, when the peak at M_o is assumed to be of fragment type i. We calculate $M_b^*(i)$ for i being backbone fragments. With b ions as the basis, $M_b^*(b) = M_o$. This means that $M_b^*(a)$ is the mass a corresponding b ion will have if M_o is an a ion. For the N-terminal ions a and c, we find

$$M_b^*(a) = M_o + CO \ (28\,\text{Da}), \ \ M_b^*(c) = M_o - NH_3 \ (17\,\text{Da})$$

For the C-terminal ions we rely on the relationship $M_b = M_P + 2H - M_y$, and then find

$$M_b^*(x) = M_P - M_o + CO \ (28\,\text{Da}), \ \ M_b^*(y) = M_P - M_o + 2H \ (2\,\text{Da}),$$
$$M_b^*(z) = M_P - M_o - NH \ (15\,\text{Da})$$

Note that the complementary ion masses depend on the mass of the precursor. It is therefore very important that this mass is correct.

Example Let a peptide mass be $M_P = 774$, and have an experimental spectrum as shown in Figure 12.4. The peaks constructed in the sequence spectrum from the experimental

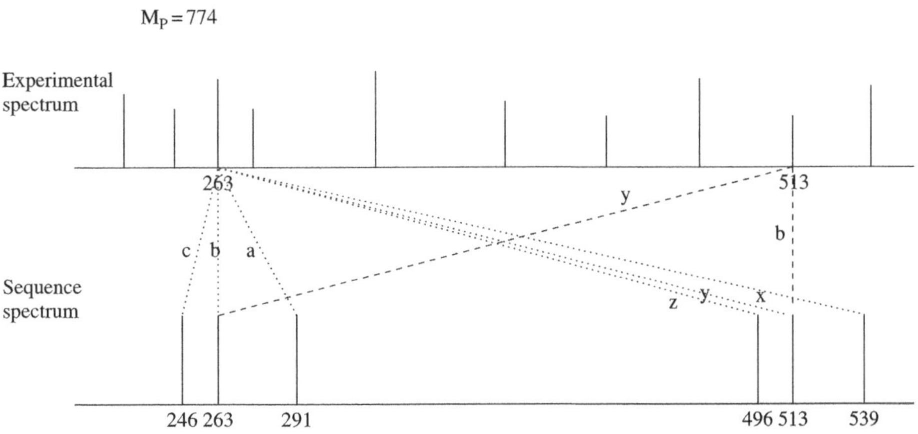

Figure 12.4 An experimental spectrum and some peaks from the corresponding sequence spectrum. Note that 'a' in the figure means that the mass is found for the corresponding *b* ion if the mass in the experimental spectrum was from an *a* ion

peak at mass 263 are shown, and also two of the constructed peaks from the experimental peak at mass 513. We see that the constructed peak when the experimental peak at 263 is assumed to be a *b* ion, and the constructed peak when the experimental peak at 513 is assumed to be a *y* ion, have equal masses.

△

Since many fragment types may be considered, the spectrum graph can quickly become quite complex. In order to limit this complexity, most methods only consider *b*, *y*, and possibly *a* ions.

12.2 Preprocessing

A preprocessing step should try to transform the experimental spectrum to as near the ideal situation as possible. General preprocessing steps such as deisotoping, charge state deconvolution, and precursor mass correction as explained in Chapter 8 should be done. Additionally, peaks resulting from neutral losses from the precursor (water −18 Da, ammonia −17 Da) can be removed. Neutral losses from product ions should also be identified and the original product ion rewarded. This means that if there is a peak at mass −17 or −18 Da from a peak *r*, there is more reason to believe that *r* is not a noise peak. Note that several other types of neutral losses may occur, but loss of water and ammonia are the most common ones.

Selecting the correct number of peaks for constructing the graph is essential for achieving an optimal solution. Too few peaks give incomplete paths, and too many result in noisy nodes that will easily confound the analysis. A straightforward method for reducing the numbers of nodes in the graph is to use an intensity filter for the peaks used to create the spectrum graph. The idea relies on the assumption that the *b* and *y* ions generally have the highest intensities among the fragment ions. A sliding window is often used, selecting the k most intense peaks in the window. Examples of values are $k = 4$ and a window size of 60 Da. If it is possible to justify that it is unlikely that a specific peak r in the experimental spectrum is of type i, then it is not necessary to make a basis peak of r using i.

Trying to classify the peaks into different fragment types is an ambitious task. Machine learning techniques are used for learning the typical 'pattern' of each fragment type using spectra from known peptides. The neighborhood of a peak in an experimental spectrum is then compared to the different patterns and the probabilities of its being a b or y ion, or any other considered type, are estimated.

Errors in the measured masses are a source of complications in this approach. Peaks corresponding to the same fragmentation site in the peptide should ideally have one of the constructed peaks in Q in common. Due to inaccuracies in the mass measurements this is not always the case, however. A greedy algorithm could be used for merging two peaks, provided that the mass difference between them is less than ϵ, starting with the peaks having the smallest mass difference. ϵ will depend on the accuracy of the instrument. After merging, several methods can be used for determining the intensity and mass of the merged peak, for example the mean of the masses of the merged peaks could be used for the mass, and the geometric mean of the original intensities could be used for the intensity of the merged peak. Because the mass of a merged peak need not correspond with any of the original peaks, the merging is repeated iteratively until all peaks are at least ϵ apart.

12.3 Node scores

A spectrum graph is compared against existing sequences, or is used for *de novo* sequencing. Programs for these tasks will generally produce numerous potential solutions. In order to rank them, the different proposed sequence solutions should be scored, based on how good they match the experimental spectrum. Since the inferred spectrum graph is used, and not the original experimental spectrum, the different components (nodes, edges) of the graph should also have a score showing how much support they have from the experimental spectrum. Let r be a peak in the experimental spectrum, and q be a peak in the sequence spectrum such that if r is of fragment type i, then q is the corresponding basis type peak. We say that peak r *supports* peak q through fragment type i. Let the support for q be denoted Γ_q. A simple method for scoring is to give each fragment type i a weight ρ_i, reflecting how often ions of this type are present in spectra produced by the instrument used, under the actual experimental conditions (Chapter 9).

Example Suppose $\Delta = \{a, b, c, x, y, z\}$ $(\rho_a = 8, \rho_b = 12, \rho_c = 3, \rho_x = 10, \rho_y = 12, \rho_z = 9)$. Let the basis be b ions. Take $M_P = 774$, and the following masses for the spectrum peaks: (263, 400, 485, 513, 610, 668, 676, 694). There will be a peak at mass 263 in the sequence spectrum that is supported by the original peak at 263 (assuming it is a b ion, since $M_b^*(b) = M_o$). The original peak at 513 also supports the sequence spectrum peak at 263 if 513 is assumed to be a y ion. Both of the supporting peaks carry weight 12, so the score of the sequence spectrum peak at 263 becomes 24.

△

A more advanced method relies on ion type probability while taking into account that peaks can occur at a given position by chance in the experimental spectrum (noise). Assume that peaks of type $\{i\}$ are produced independently with probabilities $\{\rho_i\}$ (the ion probabilities can be learned from a set of spectra for the actual instrument). Then there is a peak in R that supports a peak at position q through ion type i, with probability $h_i = \rho_i + (1 - \rho_i)h_R$, where h_R is the probability of a peak being generated by noise. A scoring formula can then use 'a bonus for explained ions' and 'a penalty for unexplained ions'. This is normalized by h_R, giving the following formula for the score of a peak q:

$$\prod_{i \in \Gamma_q} \frac{h_i}{h_R} \prod_{i \in \Delta - \Gamma_q} \frac{1 - h_i}{1 - h_R} \tag{12.1}$$

Log-odds may also be used to calculate the probability that the observed peak intensity is derived from a real fragmentation event, as opposed to simply being a noise.

12.4 Constructing the spectrum graph

The peaks of the sequence spectrum become the nodes of a spectrum graph, and in addition two end nodes are constructed, as illustrated in Figure 12.3.

Depending on how the nodes are assigned a score, one can get different types of graphs. Referring to the classification of theoretical spectra (Section 11.1) we get:

U-graph, the nodes have uniform scores.

F-graph, the scores are calculated from the fragment types of the supporting peaks in the experimental spectrum. This is the graph type that is most commonly used.

R-graph, this is more artificial, as advanced analysis can only be performed if residue-specific information is available (that is, we must know the fragmentation site of the ion).

A directed edge is drawn between two nodes if the mass difference corresponds to one or several residue masses. The number of edges can be constrained by how large an edge mass can be, or how many residues an edge can represent. Modifications can also be taken into account in the determination of the edges, by simply allowing edges to be drawn between nodes with mass differences corresponding to modified residues.

Scores (or weights) can be attached to the edges. If an edge corresponds to the mass of one amino acid a, and the basis is an N-terminal fragment type, the start node n of the edge corresponds to a fragmentation site xa, where x means any amino acid. In order to provide additional support for the occurrence of n, the edge can then be weighted by the probability of such a fragmentation occurring for the considered fragment types.

Chen *et al.* (2001) describe an efficient algorithm for determining the edges of a spectrum graph. Let the maximum mass corresponding to an edge be M, and let the precision of (the sum of) the amino acid masses be δ. Then there are M/δ different masses. Define an array A with an entry for each of the different masses, and let $A[m] = 1$ if the mass m corresponds to the (added) mass of some amino acids (otherwise it is zero). A can be determined for increasing m in that $A[m] = 1$ if m corresponds to an amino acid mass, or if $A[m - m_a] = 1$ for an amino acid mass m_a. The edges of G can then be found easily by direct lookup in A by the node mass differences. There is an edge between nodes (n_i, n_j) if and only if $(0 < m(n_j) - m(n_i) < M)$ *and* $(A[m(n_j) - m(n_i)] = 1)$, where $m(n_i)$ means the mass of n_i.

12.5 The sequencing procedure using spectrum graphs

The general procedure for *de novo* sequencing using spectrum graphs can be formulated as follows:

1. An MS/MS spectrum R is experimentally produced from a real peptide P.
2. From R a spectrum graph G is constructed. This is possible via a sequence spectrum Q.
3. From G there are derived sequences $\{E_i\}$ corresponding to the peptide mass M_P:

$$P \to R \to Q \to G \to E$$

The goal is to achieve an E with the best match to P. P is unknown, however, so we must search for an E with the best match to R, according to some scoring scheme. The general procedure is to first find a set of sequences that give a high score against G, and then compare these against R, for example by the methods described in Chapter 11.

Deriving E from G means finding high-scoring paths through G. The score for each path can be calculated from the node and edge scores included in the path. A dynamic edge scoring can also be calculated as follows. When entering the node n one can record the amino acid b corresponding to the incoming edge, and if a is the amino acid on the selected outgoing edge, the amino acid of the fragmentation site becomes ba, and a probability associated with this can be used.

The mass similarities between some of the amino acids present problems for both practical and theoretical handling of MS/MS data. These mass similarities include:

- the mass equality of `L` and `I`;
- the mass similarity of `Q` and `K`;
- the similarity of the mass of some large amino acids to the sum of the masses of two small amino acids, for example the mass of `W` is equal to the sum of the masses of `E` and `G`;
- ambiguous pairs or tuples of amino acids, for instance the sum of `S` and `F` is similar to the sum of `P` and `H`;
- the mass of oxidized `M` (`M` readily oxidizes *in vitro*) is similar to that of `F`.

12.5.1 *Searching the graph*

A crucial point in searching for paths through the graph is that a path corresponding to a sequence should not include two nodes resulting from the same fragmentation site of P.

This can be indirectly controlled by assuring that a path cannot include two nodes originated from the same peak in R, but remember that there can also be complementary peaks in R.

The straightforward method for sequencing is to traverse the graph from the start node (N-terminus), and this can be done in either a depth-first or a breadth-first manner. Depth first means that subsequences are extended, one at a time, until the end node (C-terminus) is reached, or until it is found that the subsequence cannot be extended to a complete path. Breadth first means that one traverses the graph one node at the time, assuring that when visiting a node, this node has a data structure with all subpaths matching the graph from the start node to this node. This data structure is populated by scanning one node ahead each time to find all potential edges.

Example Given a spectrum R={72, 175, 303, 406} from a precursor with mass $M_P =$ 604. Assuming a single charge, the complementary peaks become {534, 431, 303, 200}, with the final spectrum graph shown in Figure 12.5. An edge corresponds to the mass of one or two amino acids. The depth-first approach will follow a path for as long as it can, for example along the nodes (a, b, c, e, f, h).

Note that a choice of path must be made at some of the nodes (b, e). When the path is at a dead end (h), backtracking is done to the last node where a choice was made (e), and the alternative path is followed to nodes g and i. A path is then found from start to end, corresponding to the sequence alternatives `AC({GA}|K|Q)({GA}|K|Q)R`. Then backtrack to b (for finding alternative paths), and visit for example (d, e, f, h), and afterwards also visit (g, i) from e. Some form of intelligence could, however, be included, such that f is not visited this second time, since it could be recorded the first time that this path through f resulted in a dead end. Breadth first means that the nodes are visited in the order (a, b, c, d, e, f, g, h, i). When visiting b, for example, scanning ahead for possible edges reveals a path from a to d along (a, b, d). When d is then visited, after c, this path is known.

Heuristics can be used for pruning in both approaches: subsequences with little likelihood of being extended to a sequence matching the graph are abandoned. This likelihood is calculated based on scoring of subsequences.

△

Since the nodes (and possible edges) have weights, and the scores of the sequences are calculated from these weights, one can consider using general algorithms for finding the path(s) with the highest score in a graph, for which there exist efficient solutions. These algorithms cannot, however, be directly used, since complementary nodes must be avoided. This is discussed in Dancik *et al.* (1999) for the program SHERENGA, for the case in which only b and y ions are considered. For a graph G define a set T of

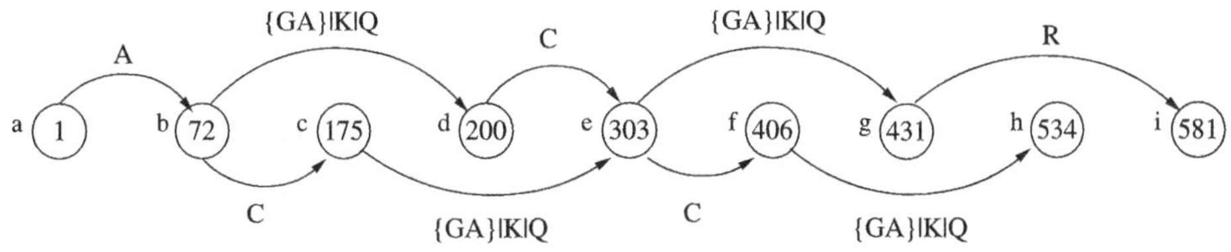

Figure 12.5 The spectrum graph corresponding to the example in Section 12.5.1

forbidden pairs in G. Two nodes in G are forbidden if they are complementary. A path in G is *antisymmetric* if it contains at most one node from every forbidden pair. The sequencing problem can then be formulated as finding the antisymmetric path with the highest score in a graph G with a set of forbidden pairs T. Unfortunately, solving this problem has exponential time complexity (NP-hard). Using the special characteristics of spectral graphs, however, an efficient algorithm can be constructed, as follows:

- The graph is acyclic, since the mass of the end node of an edge is always larger than the mass of its start node.
- Every two forbidden pairs of complementary nodes $(b_i, y_{i'})$ and $(b_j, y_{j'})$ are *non-interleaved*. This can be shown by assuming, without loss of generality, that $b_i < b_j$. Then we have three possibilities for placing $y_{i'}$, which also determine the placements of $y_{j'}$ (remember that $m(b_i) + m(y_{i'}) = m(b_j) + m(y_{j'})$, when m means mass). Then we have the following for the forbidden pairs:

  $y_{i'} < b_i$ implies $y_{j'} < y_{i'}$;

  $b_i < y_{i'} < b_j$ implies $y_{j'} < b_i$;

  $b_j < y_{i'}$ implies $b_i < y_{j'} < y_{i'}$.

A graph with a set of forbidden pairs is called *proper* if every two forbidden pairs of nodes are non-interleaving. There are efficient algorithms for finding the antisymmetric maximum path in a proper graph.

Chen *et al.* (2001) describe a method for *de novo* sequencing using dynamic programming for finding the maximal path through a spectrum graph. They consider only b and y ions. A possible path of the graph $G = (V, E)$ can be found in $O(|V||E|)$ time and $O(|V|^2)$ space complexity.

The method is extended to sequencing where one of the residues in the peptide is modified. Chen *et al.* first assume that a solution to the original unmodified problem is not found. The modified problem, given $G = (V, E)$, is solved by asking for two nodes v_i and v_j, such that there does not exist an edge (v_i, v_j), but adding this edge to G creates a (plausible) solution that contains this edge. If the mass difference between v_j and v_i corresponds to the mass of a modified amino acid, then the found path corresponds to a sequence containing one modification. The algorithm is of $O(\max(|V|^2, |V||E|))$ time and $O(|V|^2)$ space complexity.

Partition of the spectrum graph

If the fragment types of the peaks in R were known, then the sequencing problem would become much easier. Methods for trying to classify the peaks into separate fragment types have therefore been developed. However, Bern and Goldberg (2005) assert that such classification is unnecessary, and propose a method for trying to divide the peaks of the spectrum into two classes, one containing the b ions and the other the y ions, without knowing which are which. This is done by constructing a graph consisting of only the peaks of R, and using several rules for constructing the edges. Partition of the nodes of the graph into two parts is formulated as an eigenvalue problem. Each part can then be sequenced separately.

Sequencing from incomplete spectra

A complete spectrum is a spectrum that contains at least one fragment type for each fragmentation site, resulting in a spectrum graph with a node for all possible fragments of the basis type. Incomplete spectra are quite common, however, and are generally dealt with by allowing an edge to represent the sum of several amino acid masses. Another procedure is to start at several places in the spectrum graph, and thereby produce several subsequences. One can for example find `ACD` at the lower end of the spectrum and `PPT` at the higher end. If the gap between them corresponds to a set of amino acids, for example `LFR`, a proposal for the complete peptide becomes `ACD{LFR}PPT`.

12.5.2 *Scoring the derived sequences against the spectrum*

The highest scoring sequences derived from G are subsequently scored against R. The amount of sequences that need to be compared to R depends on how advanced G is: whether G is a U-, F-, or R-graph. More advanced graphs result in less sequences, assuming that the scoring is already fairly restrictive.

Since the score is assigned to a match between a spectrum and a sequence, the scoring methods described in Chapter 11 can be used.

12.6 Combined spectra to improve *de novo* sequencing

The largest problem of *de novo* sequencing is incomplete fragmentation of the peptide bonds, and a low signal-to-noise ratio. To increase the fragment coverage, and improve the reliability of the annotated peaks, at least two techniques are proposed for collecting more information before attempting the sequencing. One is using several spectra from the same peptide, where some of the spectra preferably come from a posttranslational form of the peptide. This is described in Section 14.6.4. The other method is the use of two fragmentation techniques, and is described next.

12.6.1 *Use of two fragmentation techniques*

The methods explained so far are all based on using a single fragmentation technique, CID. The reliability of the derived sequences depends on finding complementary fragments. This can be taken to another level by combining two fragmentation techniques, CID and ECD, as in Savitski *et al.* (2005a,b). Ions derived from both CID and ECD spectra then furnish highly reliable data.

CID mainly produces b and y ions, and ECD preferentially gives variants of c and z ions, and occasionally a ions. The fragment diagram presented in Figure 8.14 (Section 8.6) refers to ions formed with CID. Using the notation a^E for the a ions (and similar for c and z ions) produced by ECD, we have

$$a^E = a + H, \ c^E = c, \ z^E = z + H$$

The idea is then to try to relate the following pair of ions in the two spectra: $(b, c^E), (y, z^E), (y, c^E), (b, z^E)$. Using the formulas in Section 8.6 we get the following equations for the masses (remember that the ion masses include a proton for the charge, and that M_P is the neutral peptide mass):

1. $b - c^E = -17$;
2. $y - z^E = 16$;
3. $y + c^E - 19 = M_P$;
4. $b + z^E + 14 = M_P$.

Nodes are constructed in the spectrum graph for the masses that satisfy one of these equations and that are part of a complementary pair in the CID spectrum as well. This does not commonly cover the whole spectrum, so progressively less reliable masses are iteratively added until a graph covering the full spectrum is obtained, making it possible to derive a full peptide sequence.

Example Consider a precursor mass of $M_P = 692$, producing a CID spectrum $R_{CID} = \{129, 232, 276, 363, 418, 462, 519, 565\}$ and an ECD spectrum $R_{ECD} = \{146, 159, 206, 347, 348, 446\}$. Then the following peaks in R_{CID} are complementary (129, 565) (232, 462) (276, 418). We also find the following pairs from R_{CID} and R_{ECD} that satisfy one of the equations given above (129, 146) (equation 1), (363, 347) (equation 2), (462, 446) (equation 2), (519, 159) (equation 4). Four pairs are therefore found between the two spectra, but only two of these are also part of a complementary pair in R_{CID}. From the matching pairs we also suppose that 129 is a b ion, 363 and 462 are y ions, and that 519 is a b ion.

△

Exercises

12.1 Propose a data structure for quickly assigning amino acids to mass differences, both single and tuples.

12.2 An MS/MS spectrum {72, 175, 258, 333, 387, 444, 535} is obtained from a peptide with mass $M_P = 718$.

(a) Construct a sequence spectrum for this (assuming b and y ions). Note that the spectrum may contain noise peaks.

(b) Try to derive the sequence of the peptide from the sequence spectrum.

12.3 Consider the example in Section 12.3. Assume that a peak has support from b, c, and y ions, and that the probability that this peak is a noise peak is 0.08. Use Equation 12.1 to calculate the score for this peak.

12.4 Find which fragment ions the masses in the spectral graph in Figure 12.5 correspond to.

12.5 Show that two forbidden pairs of nodes are non-interleaved, as explained in Section 12.5.1.

12.6 Derive the four equations in Section 12.6.1.

Bibliographic and additional notes

Spectrum graph

The term *sequence spectrum* was first used by Bartels (1990), and *idealized spectrum* was used by Hines *et al.* (1992) and Scarberry *et al.* (1995). The greedy algorithm for the merging of calculated peaks is described in Dancik *et al.* (1999).

Preprocessing and peak reductions are described in Frank and Pevzner (2005), Grossmann *et al.* (2005), and Zhang (2004). Lubeck *et al.* (2002) try to classify the fragments into b and y ions (program PepSUMS), as do Zhong and Li (2005).

Simple methods for peak scorings are described in Bartels (1990) and Fernández-de-Cossío *et al.* (1995). A more advanced method is described in Dancik *et al.* (1999). A comprehensive analysis for calculating ion probabilities (ρ_i) is described in Scarberry *et al.* (1995). Node score using log-odds is in Frank and Pevzner (2005).

The method using the combined fragmentation technique is described in Horn *et al.* (2000) and Savitski *et al.* (2005a,b).

Other approaches

A few other methods not using sequence graphs have been developed.

Sequencing by constraint satisfaction

A method using constraint satisfaction for deriving a sequence E is presented in Bruni *et al.* (2004). A set of variables $\{x_{aj}\}$ is defined, where $a \in \mathcal{A}$, the set of all amino acids, and $j \in N$, the residue positions of E. $x_{aj} = 1$ if amino acid a is in position j, 0 otherwise. The task is then to determine values to $\{x_{aj}\}$ and $|N|$ under the following constraints:

- Compatibility of the peptide mass: $\sum_{a,j} x_{aj} r_a + \langle N \rangle + \langle C \rangle = M_P$, where r_a is the residue mass of amino acid a.
- Each peak in the spectrum R should correspond to a fragment of P (that is, a substring of E). This means that each peak $k \in R$ with mass p_k must satisfy an equation $d_k + \sum_{a,j} x_{aj} r_a = p_k$, where d_k is one of the residue mass deviations for one of the actual fragment types, that is 1 for b ions, −27 for a ions, etc.
- E must have exactly one amino acid for each position, $\sum_a x_{aj} = 1; \forall j \in N$.

Branch and bound techniques are used to reduce the search space.

Sequencing by genetic algorithms

De novo sequencing can be looked upon as an optimization problem, with general search techniques used. A method using genetic algorithms is described in Heredia-Langner *et al.* (2004). The initial individuals (possible sequences) are found by SEQUEST, hence the method can be looked upon as 'pseudo sequential'. The individuals must be evaluated in each generation, both to (i) test if a satisfying solution is found (the *score*), and (ii) decide which individuals to use in reproduction (the *fitness*). The score indicates how well an individual sequence matches the experimental spectrum.

Sequencing by divide and conquer

Zhang (2004) describes a method using divide-and-conquer techniques. Such techniques work by breaking a larger problem into two or more subproblems, which are similar to the original problem but of smaller size. The subproblems are then solved recursively. For *de novo* sequencing the spectrum is divided into smaller and smaller subspectra until a subspectrum is small enough to allow an exhaustive search of all possible sequences to be performed. A subspectrum is defined by the demarcating peaks b_{start} and b_{end}. If the difference between these two masses is larger than a preset threshold, it is divided in two new subsequences by choosing a pivot ion b_{pivot} between them. In order to be reasonably sure that an actual b ion is found to act as the pivot, several (3–10) pivot ions are used, selected on the basis of their intensities. The obtained subsequences are ultimately combined into sequences for the whole spectrum.

Some programs for de novo sequencing

Lutefisk	Taylor and Johnson (1997, 2001)
SHERENGA	Dancik *et al.* (1999)
PEDANTA	Pevzner *et al.* (2000)
	Chen *et al.* (2001); Lu and Chen (2003a)
SeqMS	Fernández-de-Cossío *et al.* (1998, 2000)
PepSUMS	Lubeck *et al.* (2002)
PEAKS	Ma *et al.* (2003)
DACSIM	Zhang (2004)
AUDENS	Grossmann *et al.* (2005)
	Zhong and Li (2005)
PepNovo	Frank and Pevzner (2005)

13 Database searching for *de novo* sequences

When *de novo* sequencing of a peptide is performed, it is commonly followed by a sequence database search. However, the spectra are rarely of such a high quality that a unique sequence is derived for the complete peptide. There are also uncertainties in the sequencing due to the similarity of amino acid residue masses, depending on the accuracy of the measurements. One way of presenting a derived sequence to a search program is to use a notation that allows for ambiguity. It should be possible to specify the different results of the sequencing, such as:

- unambiguously derived amino acids;
- ambiguous amino acids (either single amino acids or pairs, very rarely also of larger tuples);
- unknown order (normally of two residues);
- masses that could not be assigned.

Example An example of a derived sequence specification can be represented as CD[AS]QT[FA|CD]<69.1>[I|L]R. The notation indicates the following: the unambiguously identified sequence CD, then a mass difference corresponding to the residue masses of AS (note that this implicitly means that the order is unknown). The next two amino acids QT are again clearly identified, followed by a mass corresponding to either FA or CD. The sequence stretch is then interrupted by an unassigned mass of 69.1 Da. The sequence finally documents the occurrence of either I or L before ending on an unambiguous R. Note the different notation for the two occurrences of CD. In the first instance, both amino acids could be unambiguously identified due to the presence of a peak separating C and D in the spectrum, whereas in the second instance the sum of their masses corresponds to the difference between two peaks. Additionally, in the second occurrence of CD, the observed mass difference between the peaks can also be explained by another pair of amino acids (FA).

Computational Methods for Mass Spectrometry Proteomics I. Eidhammer, K. Flikka, L. Martens and S.-O. Mikalsen
© 2007 John Wiley & Sons, Ltd

If the *de novo* sequencing had considered modifications, it would have discovered that ⟨69.1⟩ corresponds to dehydration (neutral loss of water) of serine, and (using slightly more sophistication) that the mass of `FA` (218.1 Da) also corresponds to the mass of `CT` if `C` is considered methylated.

△

A database search program should be able to deal with such ambiguities and unassigned masses as mentioned above, but the actual representation will vary depending on how the search program works. This is discussed below when the different methods are treated. A general strategy is to search 'simultaneously' with several derived sequences from a single sample, and taking into account that they may all come from the same protein. Also, if several (high-scoring) sequences, which are different enough that they cannot readily be described together in the notation used, are derived for a spectrum, each of them should be included in the search.

Another issue with consequences for the score calculation is whether the peptides in the sample are assumed to come from a single protein, or from several proteins. If all peptides come from a single protein, the score for a database sequence should increase as the number of matching derived sequences increases. This correlation between number of matching sequences and total database entry score has less validity when several proteins are included. In general, the score against a database sequence is calculated through a combination of (i) how many derived peptides match that sequence, and (ii) how good the individual matches are. It is important that overlapping matches must be avoided when applying this scoring scheme, since they do not contribute additional sequence information and therefore should not increase the score. In practice, this means that if there are several identical derived sequences, which all match a database sequence, only one of them should take part in the scoring.

Regardless of the actual method used for searching, there may be a huge number of spectra/segment comparisons, resulting in unsatisfactory search time. Filtering techniques are therefore used for skipping those segments that are unlikely to be correct.

The search programs should be able to deal with errors in the sequencing (*sequencing error tolerant search*), mutations in the precursor peptide in relation to database segments (*homology tolerant search*), and preferably also modifications.

The task can be formulated as follows given a set of spectra $\mathcal{R} = \{R_1, \ldots, R_i, \ldots, R_n\}$, each R_i is first sequenced, producing derived sequences $\mathcal{E} = \{E_1, \ldots, E_i, \ldots, E_n\} = \{\ldots\{E_i^1, \ldots, E_i^{i_m}\}\ldots\}$. The sequences from $\mathcal{E}$ are subsequently searched against a sequence database in order to identify the protein(s) of origin and/or homologous proteins. Advanced methods will allow for the possibility that the peptides are modified.

13.1 Using general sequence search programs

There are several high-performance programs available that are intended for finding homologous sequences in protein sequence databases. It is a reasonable idea to use one of them for MS/MS-derived peptide sequences. However, these programs typically require a complete, unambiguous protein sequence as input, whereas the input in the context of *de novo* sequencing after mass spectrometry consists of a set of (ambiguous) small peptide sequences. These programs must therefore be adapted to accept this kind of input data, but their main strategy of operation should be kept (otherwise new programs could just as well be developed). Most of the homology searching programs rely on a scoring matrix (for example, a PAM or BLOSUM matrix), which provides the 'similarity' scores for the different amino acids. By changing these score matrices in accordance with the representation for the derived sequences, the programs can be adapted to handle the ambiguities in the derived sequences.

In the rest of this chapter we also use the term *query sequences* for derived sequences from MS/MS experiments, since this is the terminology most often used in such programs. The most popular sequence database search programs are FASTA and BLAST, and we will start by briefly sketching the main operational principle of these algorithms.

13.1.1 The main principle of FASTA and BLAST

FASTA and BLAST build on the same main principle that can be briefly outlined as follows:

1. Preprocess the query sequence E, by dividing E into (overlapping) words (BLAST) or k-tuples (FASTA), and store these in a fast lookup table or state diagram.
2. Search along the database sequence D for similar words or k-tuples to those in E.
3. Extend the alignments from the similar word pairs found to higher scoring *local alignments* without gaps. In BLAST such alignments are called HSP (High-Scoring Segment Pairs).
4. Perform dynamic programming around the high scoring (sub)alignments found. It is possible that gaps are introduced at this stage.

For dealing with derived sequences, Step 3 is changed, and Step 4 is usually skipped. Note that this means that gaps are not allowed.

FASTA is the starting point for the programs CIDentify and FASTS, MS BLAST builds on BLAST. CIDentify assumes that all peptides are coming from one protein while the other two accept peptides derived from several proteins.

13.1.2 Changing the operation of FASTA/BLAST

The main changes for adapting these programs to MS/MS-derived sequences are:

- the program must accept uncertainties in the query, and also mass specifications;
- the scoring matrix must be changed in accordance with the input representation;
- the original programs initially searched for *local* alignments, while now the alignments should be global for the queries (the whole derived sequence), but remain local for the database sequence;
- the scoring scheme of the alignments can be kept, but calculation of statistical significance has to be changed.

A more detailed discussion is provided as follows:

- The mass equality of leucine and isoleucine. `L` can for example be used in the queries for both of them, and the scoring for `(L,L)` and `(L,I)` in the scoring matrix can be changed to reflect the average of the identity score for `L` and `I`. Average values can also be used for the scoring of `L` to the other amino acids.
- The mass similarity of glutamine and lysine. How this is handled depends on whether information about the cleavage protease is taken into account in the search. For instance, if trypsin is used, `K` can be used if the residue is at the C-terminus of the derived sequence, and `Q` elsewhere. Otherwise `K` and `Q` can be handled in the same way as leucine and isoleucine.
- Ambiguous order and/or assignment of residues. An example of such an ambiguous mass assignment is a mass delta of 186.1 Da, which can be explained by `W`, `EG`, `GE`, `AD`, or `DA`. Such ambiguities must be specified, and a search in the database sequence should cover all possibilities.
- Unidentified mass. This can be represented by an `X`, corresponding to one, two, or several residues, depending on how advanced the sequencing procedure is. An alternative is to use the mass value directly.
- If trypsin is used for digestion, `B` (trypsin cleavage symbol representing either `K` or `R`) can be inserted in front of the query sequence (but this prefixing by `B` should obviously be ignored when matching the query against the N-terminus of a database sequence). Also, (soft) constraints for `B` at the C-terminus can be included. Similar end letters can be used to reflect the specificities of other proteases.

Example One illustrating example is shown below, comparing a query *Q* to a database sequence *D*:

```
Q:            A C D A S V 186.1 X G E R
                | |     | |\          |
                | |     | | \         |
D:    . . V K D C D S A V A   D   L W   R I . .
```

`CD` is recognized in both sequences in Step 2 of the search, and an extension of the alignment follows. `(AS, SA)` is not considered a match here, since the order in the query *Q* is determined from the sequencing. `AD` is considered a match to 186.1, and the score can be calculated as the sum of the identity scores of `A` and `D`. `W` in the database sequence *D* is not considered a match to `GE`, although they have the same mass. It could, however, be interpreted as an indication that the peak used for recognizing `G` and `E` is potentially a noise peak.

△

The example illustrates that we should appreciate the possibility of alternative interpretations of the derived sequence, and that we should be open for errors in the experimental sequencing. Other possibilities for differences between the derived sequence and the database sequence are mutations (polymorphisms) and alternative mRNA splicing that was not previously registered in the sequence databases. The node scores of the spectrum graph could contribute valuable information in such situations.

A two-step procedure presents a reasonable strategy for the comparison between the query and the database. Such a method is used by CIDentify, in which the segments in the database that give a high scoring match to the unambiguous residues of the query are identified first, and only then are the unknown mass specifications aligned against these segments.

An important difference in searching with derived sequences, as opposed to a regular search, is that in the former case there is a set of (small) query sequences, and several of these can match the same database sequence. The search can be performed separately for each query, or for all queries together. The latter approach is used in MS BLAST, where the query sequences are concatenated into one large query sequence, with a minus sign (−) inserted between each individual query (a high negative score is assigned to the minus sign in order to prevent matches that include (parts of) two derived sequences). MS BLAST also allows the inclusion of several alternative sequences derived from the same spectrum.

When several queries match the same protein sequence from the database, they should be joined to produce the best overall alignment, taking care to use only one derived sequence from each spectrum.

13.1.3 Scoring and statistical significance

The final match to a database sequence often consists of several individual query matches. Calculating a score for each query match is normally done simply by adding the scores (from the scoring matrix) of each residue pair match, and the final score for the database sequence can be the sum of the scores of each query match. In this way the database sequences can be ranked by decreasing score. However, since such scores do not provide any information about the likelihood of the matched sequence(s) really corresponding to the protein(s) originally in the sample, a statistical significance should also be calculated. As explained in Chapter 7, this is usually done by calculating the **P**-value. Note that the **P**-value (defining a *probability*) is related to the **E**-value (defining an *expectation*) in the following way: $P = 1 - e^{-E}$. The calculation, or estimation, of the **P**-value becomes more difficult in this context since not all the spectra will find a match, and they can match in any order along the database sequence. This makes it hard to calculate a scoring distribution for random sequences, and computer simulation is often used for performing an (informal) statistical analysis.

Example We describe how significance is estimated in MS BLAST. The significance of a score should depend on how many sequences there are in the combined query, and how many of these match (as HSP) the considered database sequence. Then searches with 10, 20, and 50 (concatenated) non-redundant, randomly generated sequences of length 11 are carried out, and this is repeated 1000 times for each peptide set size. The total score and the number of HSPs used in the score are then recorded for the best match in each search. For each of the three set sizes (10, 20, and 50) the searches are divided

Table 13.1 Threshold scores determined via simulation experiments when a modified PAM-30 scoring matrix was used; 99 % of the top hits that matched the specified number of HSPs had a score below the presented thresholds. After Shevchenko *et al.* (2001)

	No. of unique peptides in the query		
No. of reported HSPs	10	20	50
1	68	72	75
2	102	106	111
3	143	146	153
4	...		

into categories depending on the number of HSPs that were used in the scoring. For each individual category, the threshold is fixed at the point where 99 % of the best matches had scores below this threshold, as shown in Table 13.1.

Suppose a search with 18 queries resulted in three HSPs that match a database sequence, yielding a total score of 148. From Table 13.1 this can be considered to be significant at the 1 % level.

△

Changing existing homology database search programs has some shortcomings. They are restricted in the number of isobaric equivalences (similar masses) that can be used, they are not well suited for spectra of poor quality, and the possibilities for posttranslational modifications are not explicitly taken care of. They also do not explicitly consider sequencing errors.

These programs are thus not suited for application in a high-throughput environment, and specialized search programs for these more demanding conditions have been developed.

13.2 Specialized search programs

Below we describe two programs which are specifically developed for searching sequence databases with derived sequences, using different approaches.

13.2.1 OpenSea

OpenSea deals with derived sequences from all MS/MS spectra from a given experiment. It is based on the fact that a *de novo* derived sequence consists of a series of uniquely derived subsequences, separated by unidentified masses (gaps), for example `AGFR<243.1>VD`. OpenSea identifies unique sequence tags of a defined length. It then searches the database sequences with these sequence tags, and when one is found the remaining part of this specific database sequence is aligned to the (ambiguously or erroneously) derived sequence. (Note the difference between OpenSea and the use of peptide sequence tags as described in Section 13.3 of this chapter. Those methods only extract sequence tags from the spectrum, while OpenSea uses the whole sequence.)

The alignment is done in steps of short local alignments of length 1, 2 or 3 residues. A gap of length 1 is allowed in the short alignments (aligning for example three database residues to two residues in the derived sequences). The procedure is illustrated in the next example. Note that a requirement for OpenSea to find the correct segment is that the derived sequence must contain at least one correct sequence tag.

Example We illustrate the main working procedure of OpenSea with an example. Suppose we have a derived sequence and a database sequence as follows:

```
Database sequence   ...VKT  CT    MPDAMF  TG    RY...
                         6  5     4  1    2     3
Derived sequence         T<218.0>EVDAMS<158.1>R
```

(1) A sequence tag match (DAM) is found between the derived sequence and the database sequence, illustrated by the number 1. Five additional short local alignments were then performed in order to achieve the final alignment: (2) F is first compared to S and then to S<158.1>, both without yielding a match. FT is then (in principle) compared to S, to S<158.1>, and finally to S<158.1>R, again without finding a match. Then FTG is compared to S<158.1>, finding a match (TG to <158.1>) with a substitution (F to S). (3) The matching residue R is found. (4) After some trial comparisons MP is aligned to EV, since they have similar masses. This indicates that the derived sequence may contain an error. (5, 6) After some trials the common T is found, and a (mass) mismatch between CT and <218.0>.

△

The individual local alignments are given a score separately using a scoring matrix, where an alignment between an amino acid a and a mass specification is scored as the average non-identity value for a. The scores of the local alignments are added by a linear combination to give the final (segment) alignment score. The linearity comes from the fact that the local alignment scores are weighted, depending on the type of matches (one-to-one, one-to-many, or many-to-many). In the example above, alignments 1, 2, and 4 are all many-to-many and therefore grouped together and given a specific weight; 5 is an example of one-to-many. This scoring mechanism is called OSAS.

Determination of substitutions and modifications

Consecutive mass mismatches are grouped into one mismatch (since modifications or substitutions of consecutive residues are less likely). Each mass mismatch is then explored to determine if it can be explained by substitutions or modifications. The mass difference is used as an index for a substitution lookup table, or a modification lookup table. These tables contain a log-odds score of how likely the corresponding substitution/modification is for each mass difference. If there are different explanations for a mass difference, the one with the highest log-odds score is chosen.

Example Consider the mass mismatch in the example above, of 14.0 between CT and <218.0>. This corresponds to a substitution of T into D. If methylation is included in the lookup table, this will also be detected as a possible explanation. The log-odds scores are then used to choose the most probable of these two explanations.

△

Final scoring

The final score between the spectrum and the detected database segment consists of two (sub)scores:

1. A measure of mass-based sequence homology, by the OSAS score explained above.
2. A measure of the similarity between the experimental (normalized) MS/MS spectrum and a theoretical spectrum constructed from the database segment. This is performed by a rank-based score, as explained in Section 11.2.4. This expression is extended by a value depending on the number (n) of assigned modifications and substitutions; this value is 1 if $n = 0$, or $1 - (n-1)^2$ otherwise. The result is that segments that have many substitutions and/or modifications are penalized.

The two scores are then combined in a linear fashion (with weights 0.9 and 6.0, respectively).

If there are several possible locations for a modification (for example, a cysteine methylation in a subsequence CTC), the position is chosen by using the second (sub)score. Finally, mass mismatches that are not explained are explicitly specified.

The total protein sequence score is then calculated as the sum of the individual segment scores.

13.2.2 SPIDER

SPIDER uses an interesting approach in that it considers the sequencing error tolerant search and the homology tolerant search separately. Consider the process

$$P(\text{peptide}) \to R(\text{spectrum}) \to E(\text{sequence}) \to S(\text{segment})$$

Define $f(E, P)$ as a measure of the sequencing error, and $g(P, S)$ as the homology difference between the original peptide and a database segment. The homology difference is measured by the *edit distance*. The edit distance between two sequences is the minimum number of operations for transforming one of the sequences to the other, where the operations are substitution, deletion, and insertion of single symbols.

The task is to find the S in the database that best suits the derived sequence E, when the sequencing error and homology difference is considered. A measure for the distance between E and S is therefore needed, and a distance score that takes the sequencing error and homology difference into account is $f(E, P) + g(P, S)$. But P is (usually) unknown, so this formula cannot be used directly. Instead, the SPIDER distance between S and E is defined as

$$d(E, S) = \min_P (f(E, P) + g(P, S))$$

When f and g are defined, the problem becomes finding the P that minimizes this expression.

Defining f and g

f measures the distance (cost) between a derived sequence (E) and a peptide (P). These two can be aligned, and a block of the alignment can be defined as a column of equal amino acids, or conflicts due to sequencing error. The cost $f^*(a)$ of the first depends on the amino acid a, and the last $f'(m)$ on the mass of the conflicting amino acids, m. Then the total distance f can be calculated as the sum of the costs of all blocks.

g measures the distance between a peptide and a segment (S), and a best alignment with cost g can be found in a traditional dynamic programming approach using a scoring matrix $g'(a, b)$ and a gap cost $\overline{g}$.

When E, P, and S are known, the alignments for (E, P) and (P, S) can be found, and then combined in an alignment for the three of them.

Example Let E=`AQSFVLR`, P=`AQPHVIR`, S=`ASQHVR`. Then the following alignments can be constructed:

```
     (E,P)         (P,S)      (E,P,S)
E  AQ[SF]VLR                AQ[SF]VLR
P  AQ[PH]VLR    AQPHVLR     AQ[PH]VLR
S               ASQHV-R     AS[QH]V-R
```

where [] denotes a block with different amino acids but similar masses. The different costs are then $f(E, P) = f^*(A) + f^*(Q) + f'(234.1) + f^*(V) + f^*(L) + f^*(R)$ and $g(P, S) = g'(A, A) + g'(Q, S) + g'(P, Q) + g'(H, H) + g'(V, V) + \overline{g} + g'(R, R)$.

△

The SPIDER distance between E and S can therefore be found by making alignments, calculating the distances for all possible P, and recording the minimum distance. This is of course not realistic, but the authors have developed a polynomial dynamic programming procedure to construct the sequence for P giving the minimum distance. Thus each segment in the database can in principle be compared to the derived sequence.

13.3 Peptide sequence tags

Trying to derive the complete sequences from a lot of spectra can be time consuming, and an alternative is to derive only small tags. When such a tag is combined with knowledge about the precursor mass, it is possible to construct a structure called a *peptide sequence tag* (PST), and this tag can then be used for searching. (See the comments in the bibliographic notes about the different use of the terms 'peptide sequence tag' and 'sequence tag'.)

Suppose that we have derived the subsequence $e_1 \ldots e_r$ from a stretch of peaks, using peaks at mass positions $x_0, x_1, \ldots, x_r$. This divides the precursor peptide into three parts, which are: (i) the mass before the derived subsequence (m_1), (ii) the derived subsequence (a sequence tag), and (iii) the mass after the derived subsequence (m_3). This construction

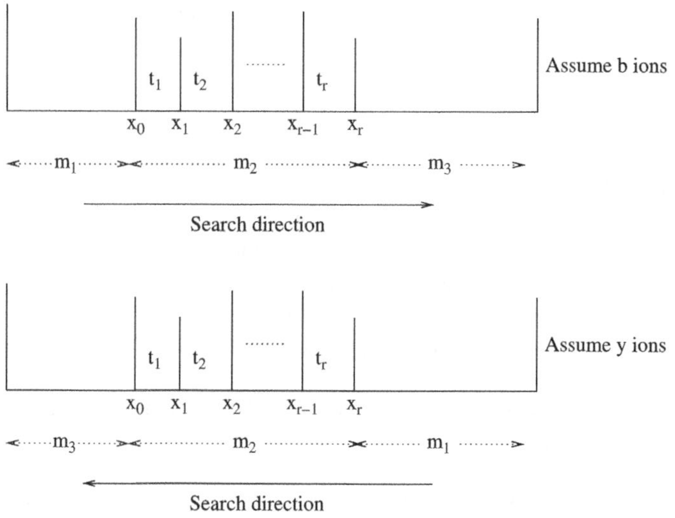

Figure 13.1 Illustration of how a peptide sequence tag is constructed. Note that the search in the database sequence moves from left (N-terminal) to right (C-terminal) in both cases. See text for a detailed explanation

is called a peptide sequence tag. Exactly what is meant by 'before' and 'after' depends on whether the subsequence is identified by b ions or y ions, which is unknown. The construction of a PST is illustrated in Figure 13.1. Note that the search direction (along the PST) depends on the ion type of the used peaks. Since this is unknown, both directions should be used.

A PST is then specified by $(m_1)e_1 \ldots e_r(m_3)$. The mass of the sequence tag, (m_2), equals the sum of the masses of the tag residues $(m_{e1} + \ldots + m_{er})$. Depending on whether the spectrum consists of b or y ions, (m_1) or (m_3) is read from the spectrum, and the remaining mass is calculated from $m_1 + m_2 + m_3 = M_P$. The database search locates a segment which consists of three parts, S_1, S_2, S_3, with residue masses (using the notation and equations from Section 8.6 and that Ψ means residue mass) $\Psi_{S_1}, \Psi_{S_2}, \Psi_{S_3}$, such that $m_1 = \Psi_{S_1} + \langle N \rangle, m_2 = \Psi_{S_2}, m_3 = \Psi_{S_3} + \langle C \rangle$. The calculations are done separately for b and y ions.

b ions: $m_1 = x_0$;
$m_3 = M_P - (m_1 + m_2) = M_P - x_r$.

y ions: From $x_r = \Psi_{S_2} + \Psi_{S_3} + \langle C \rangle + 2H$ we get $\Psi_{S_1} = \Psi_P - (\Psi_{S_2} + \Psi_{S_3}) = M_P - \langle C \rangle - \langle N \rangle - (x_r - \langle C \rangle - 2H) = M_P - x_r + H$. Then we have

$$m_1 = \Psi_{S_1} + H = M_P - x_r + 2H;$$
$$m_3 = \Psi_{S_3} + \langle C \rangle = x_o - \langle C \rangle - 2H + \langle C \rangle = x_0 - 2H.$$

Example We have a peptide with mass $M_P = 1624.7$ Da, and let the amino acids PS be derived from the peaks $x_0 = 819.3, x_1 = 916.4, x_2 = 1003.4$. Assume first b ions. Then $m_1 = x_0 = 819.3, m_3 = M_P - x_2 = 621.3$. Searching with this results in the database segment QTSESTGQ*PS*SEGLSM. The monoisotopic summed residue masses of the subsequence corresponding to m_1 is $818.3(= m_1 - \langle N \rangle)$. The mass of the other subsequence is 604.3 $(= m_3 - \langle C \rangle)$. If we assume y ions instead, we get $m_1 = M_P - x_2 + 2H = 1624.7 - 1003.4 + 2 = 623.3$ and $m_3 = x_0 - 2H = 817.3$.

△

In some programs it is not necessary to use all three values; in these only one of the masses and the derived sequence need be used (partial search). After a match is found between the peptide sequence tag and a segment, a theoretical spectrum can be created for the segment, and compared to the experimental spectrum.

13.3.1 A general model for PST search programs

In the last few years several programs that use PSTs have been developed. With intelligent methods for sequence tag extraction and their identification in the database, these programs work fairly fast. They also commonly include the possibility to allow identification of sequences with modifications.

Such programs typically consist of five steps:

1. Extract a set of (scored) sequence tags (25–50), where the tags typically are of a fixed length of three residues.
2. Search the database for matches to the sequence tags, *sequence tag hits*.
3. Extend the sequence tag hits with the flanking residues to investigate if there is a match to the PST.
4. Score the PST matches.
5. Calculate the statistical significance for the highest scoring segment.

The procedure described is for one spectrum, but several PSTs can be used in the searching, and more than one of these can match to the same segment.

There is a certain similarity between this approach and some of the programs described in the previous section, in that one first searches for hits between small amino acid tuples. Note also that this approach can be looked upon as a hybrid of the sequential and structural approach, since some *de novo* sequencing is used for finding the tags, and structural scorings are employed for scoring the PST matches.

We will mainly use two newly developed programs for discussing the listed steps, *GutenTag* and *InspecT*.

13.3.2 Automatic extraction and scoring of sequence tags

In extracting a set of sequence tags from a spectrum there are two important points: a correct sequence tag should be in the set, such that the correct segment can be identified in the database; and the number of sequence tags should not be too large, to prevent an unsatisfactory running time. A reasonable approach for the automatic extraction of tags relies on the use of spectrum graphs. A tag is identified by a path of $r+1$ nodes with uniquely labeled edges. (This labeling can include modified amino acids, if desired.) To extract the most informative ones, the tag candidates should be scored using node and edge scores, calculated as described in Chapter 12. The scoring components used by GutenTag and/or InspecT are the following:

- the observed intensities are compared to the expected intensities, where the latter are calculated based on a fragmentation model (the mass relative to the peptide mass, and one or both of the amino acids of the fragmentation site are known);
- the accuracy of the measured m/z value as compared to the mass of the derived amino acid;
- the supporting fragment types (b, y, a, including water and ammonia losses);
- observed isotopic patterns.

13.3.3 Database search

Effective methods should be used for finding sequence tag hits in databases. One way to do this is by indexing the database. An index table contains entries for all amino acid combinations of the sequence tag length (20^3 for tags of length 3). Each entry contains a list of database positions where the entry tag occurs. It could also be extended with the ability to index whole PSTs, if the flanking masses were included. The index organization has the drawback that it must be recomputed when the database is changed, and that the number of indices grows exponentially with the number of modifications allowed in the flanking masses. Another approach is to use a *trie*, as in InspecT. A trie is a tree data structure for storing a set of strings, where there is one node for every prefix. Instead of preprocessing the database, as is done when using an index table, the set of extracted tags is collected in a trie. The prefix of a node is recognized by a path from the root to the node. Figure 13.2 shows an example. The trie is then searched against the database, such that a search for all tags from several spectra can be performed in a single scan of the database.

13.3.4 Extending the sequence tag hits with flanking amino acids

For each of the sequence tag hits, we then must explore if the flanking residues can match the sequence tag masses m_1 and/or m_3. This may be performed by testing for m_1 and m_3

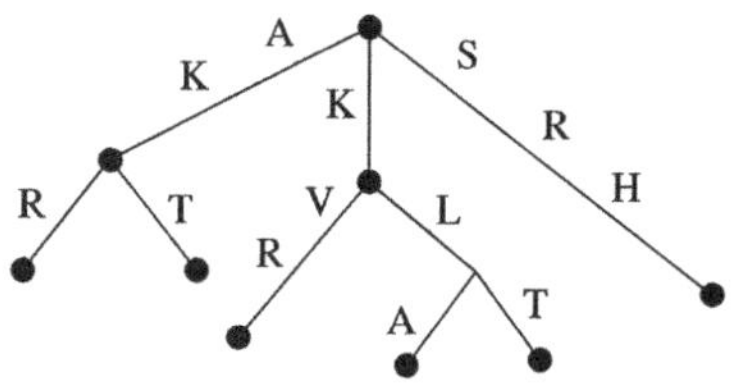

Figure 13.2 A trie for the set of tags {AKR, AKT, KVR, KLA, KLT, SRH}

separately. The challenge is to include possible substitutions and/or modifications. One simple strategy is to require that one of the masses matches, and assume modification if the other does not match. This can be explored further to see if the mass difference corresponds to a known modification/substitution. A more advanced technique works as follows. Let p be the first position in the database sequence S after the tag hit. A straightforward strategy is to search for a segment $S(p \ldots n)$ where n is found by

$$n = \arg\min_{n \in [1 \ldots]} |\Psi_{S(p \ldots n)} - m_3|$$

($\Psi_{S(p \ldots n)}$ is the residue mass of the subsequence $S(p \ldots n)$.) One can then test if the minimum mass difference is less than ϵ, corresponding to a match without substitution/modification, or that the minimum mass difference corresponds to a known substitution/modification.

Example Suppose that the sequence tag is `QPS`, with $m_3 = 385.7$, and that this tag is found in the database sequence `... TG`*QPS*`SEGLSM...` . The added masses of the residues after the tag are 87.03, 216.07, 273.09, 386.17, and 473.20. We conclude that the best match is given for $n = 4$, implying that m_3 corresponds to `SEGL`. Note that if tryptic digestion is assumed, the fact that the last amino acid is `L` and not `R/K` will reduce the likelihood that the segment correctly reflects the original peptide.

△

Another approach is to search with a defined set of modifications, as in InspecT. The masses of a set of modifications are ordered by increasing mass in a table t. One then searches for all segments $S(p \ldots n)$ and index i such that $|\Psi_{S(p \ldots n)} + t(i) - m_3| \leq \epsilon$. An algorithm with linear time in $|t|$ exists for this approach.

When a sequence tag hit is found with mass matches to both m_1 and m_3, it is important to verify that the assigned (attached) modifications are feasible. Also, if several attachments are possible, the most likely candidate should be found. In InspecT this is done by a dynamic programming procedure.

Example Suppose we have a sample where carboxyamidomethylation of cysteines has been performed. However, we are not sure that all cysteines have been modified. In the peptide analysis we then allow for up to one carboxyamidomethylation (57.02 Da mass difference) and one phosphorylation (79.96 Da mass difference). The allowed mass modifications are then {0, 57.02, 79.96, 136.98}. Let $m_3 = 513.1$, and the subsequence following the sequence tag hit in the database sequence be `SCVSH...` . The sum of residues masses of `SCVSH` is 513.20, and the mass of `SCVS` with the addition of one phosphorylation and one carboxyamidomethylation is 513.12. Phosphorylation is possible on one of the serines, and carboxyamidomethylation on the cysteine. Both alternatives are possible (to the required accuracy), but information about the protease specificity can be used here to resolve the dilemma. If the alternative with the modification is accepted, one should try to determine which serine the modification is on. One can of course also explore which modification suits the MS/MS spectrum best, the first or second serine.

△

13.3.5 Scoring the PST matches

Normally a set of segments is found that (more or less) matches the PSTs, and these should be scored in order to rank them. Two approaches are used. A score can be directly calculated from how well each of the three parts (S_1, S_2, S_3) of the PST match the segment. A more informative approach is to construct theoretical spectra for the segments, and perform scoring as described in Chapter 11. However, as mentioned in Tanner *et al.* (2005), the score functions must be changed if modifications are considered. This is due to the fact that the length of the retrieved segments may then vary more, and longer segments will have more putative fragments, which can either match or be missing. Thus several of the scoring schemes will have a length bias, and a length normalization has to be performed. This is done in the scoring scheme of InspecT, which uses a probabilistic model, with a log-odds score.

If several spectra are included, there can be several disjunct matches to the same protein sequence, and protein sequence scores must therefore be calculated.

13.3.6 Statistical significance

As mentioned earlier, the highest scoring segment need not be the correct match, and a statistical significance should be calculated. A method for calculating the **P**-value is included in InspecT. A linear combination of four factors is formed by (1) the score C, (2) how much of the total intensity of the spectrum can be explained by matched peaks, (3) how many of the peaks can be explained, and (4) b/y ion score as the fraction of the theoretical b/y ions found in the experimental spectrum. These four components are independent of the database size. A fifth factor, the score difference between the best and second best score, depends on the database size, and is not included, but reported separately.

13.4 Comparison by threading

Two methods that are different from the previous approaches (spectral and sequential) are classified as threading. One uses suffix trees, and one uses deterministic finite automata.

13.4.1 Use of suffix tree

A method using a suffix tree is described in Lu and Chen (2003b). The experimental spectrum is transformed to a spectrum graph, and the database is represented as a generalized suffix tree (see Gusfield (1997) for a description of suffix trees). Two different algorithms are used for the search, and these are outlined below. Intensities are not used in this program.

Searching the suffix tree against the spectrum graph

The suffix tree, ST, is exhaustively traversed in a depth-first order, checking against the spectrum graph G (threading the sequences on the spectrum). When a current path from the root to u in ST matches a path from the first node to a node g in G, the path in ST is extended to v. It is then examined if (the mass of) uv corresponds to an edge going out from g. Note that uv can represent several (successive) amino acids. The algorithm has space complexity $O(n+|V|)$ and time complexity $O(n)$, where n is the total length of the sequences, and $|V|$ is the number of vertexes in the spectrum graph.

Searching the spectrum graph against the suffix tree

The spectrum graph is exhaustively searched according to the topological order of the spectrum graph (in a breadth-first-like search). When visiting a node h in G, this node should contain all paths from the root in *ST* that correspond to any path from the root in G to h. To achieve this, each node g of the spectrum graph is extended by an *ST* field which is a collection of pointers. Each of them points to a position p in the suffix tree where the path from the root (of the suffix tree) to the position p corresponds to a path from the first node in G to g. Suppose we traverse the edge (g, h). Then for each pointer in the *ST* field of g, let u be the position that is pointed at. A search is then performed down the suffix tree, to see whether there exists a corresponding path (u, v) to (g, h). If such a path is found, a pointer to v will be included in the *ST* field of node h. When reaching the end node in G, the *ST* contains the paths in *ST* matching G. The algorithm has space complexity $O(n+|V|)$ and time complexity $O(n+|E|)$, where $|E|$ is the number of edges in the spectrum graph.

Modifications

Modifications can easily be included, by drawing edges in the spectrum graph for differences corresponding to defined modifications. This will increase the number of edges, but the execution time is still claimed to be linear.

13.4.2 *Use of deterministic finite automata*

A method using deterministic finite automata (DFA) is described in Falkner and Andrews (2005). The idea is to see if a segment (theoretical peptide) can match the spectrum, without constructing a theoretical spectrum or performing *de novo* sequencing. First a spectrum graph is constructed from the spectrum. A spectrum graph is (principally) equivalent to a non-deterministic finite automaton (NDFA) where the nodes are the states, and the edges represent the transitions. There is also a start state and an end state. The automaton is non-deterministic, since there may be several transitions from a state for the same amino acid, for example both a modified and unmodified amino acid, or a single residue and a part of an unresolved pair. A comparison between a segment and an automaton can be made by verifying if there is a path from the start state to the end state performing the transitions specified by the amino acids in the segment.

Example Assuming b ions, a spectrum graph is constructed with nodes at 71.0, 174.1, 244.1, 275.1, 372.2, 471.2, ... Considering the possibility for crotonaldehyde modification (70.04 Da mass difference), an NDFA can be obtained as shown in Figure 13.3.

△

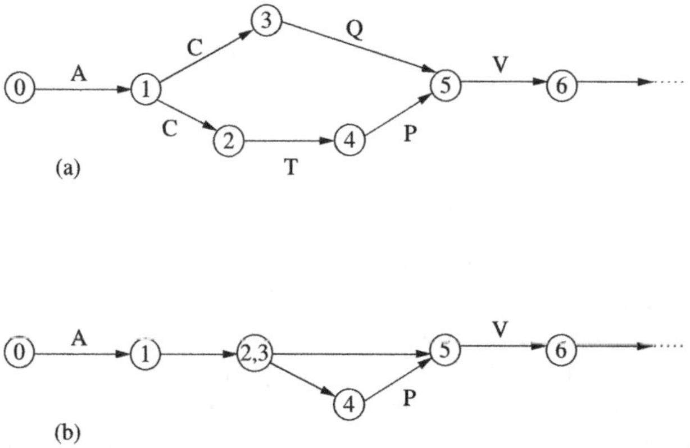

Figure 13.3 (a) An NDFA, from the peaks in the example. The states are numbered by increasing order of mass. State 3 is reached by crotonaldehyde modification of C. (b) A deterministic version of the automata

The search time for an NDFA can at worst be proportional to the sum of the number of states in every possible path, since every path may be searched. To reduce this time, the automaton can be transformed to a deterministic finite automaton (DFA), as described in general books about formal languages and automata. For a search in a DFA there is always only one transition for a given symbol, hence the search time becomes proportional to the number of amino acids in the segment.

A set of NDFAs from several spectra can be combined into one by combining the start states, and then transform it to a DFA. In this way one can compare a segment to several spectra by a single scan of the automata. Several segments will commonly be recognized by the automata, and these must afterwards be scored by a scoring scheme.

Exercises

13.1 Suppose a search with MS BLAST with 11 queries resulted in two HSPs. Estimate what the mininum total score should be to be significant at the 1 % level.

13.2 A sequence `T<140.2>LMAVT<250.2>K` is derived, and a database search is performed by OpenSea. Show how a search against the sequence `...YFTQLVMAVSHIKVG...` is done, when a requirement is that identical tags with a length at of least three residues should be found.

13.3 An MS/MS spectrum {72, 142, 232, 331, 402, 418, 565} from a peptide with mass $M_P = 692$ exists. Find a peptide sequence tag where the length of the derived subsequence is three. Calculate the flanking masses, assuming both b and y ions.

13.4 Develop an algorithm linear in $|t|$ which searches with a set of specified modifications, as explained in Section 13.3.4.

Bibliographic notes

MS BLAST
Altschul *et al.* (1997); Shevchenko *et al.* (2001)

FASTS
Mackey *et al.* (2002); Pearson (1990)

CIDentify
Taylor and Johnson (1997)

MS-Shotgun
Huang *et al.* (2001); Pegg and Babbitt (1999)

OpenSea
Searle *et al.* (2004, 2005)

SPIDER
Han *et al.* (2005)

Use of peptide sequence tag (PST)

The concept of a 'peptide sequence tag' was first introduced in Mann and Wilm (1994), where they simply called it 'sequence tag'. The actual term 'peptide sequence tag' was introduced in Frank *et al.* (2005). A sample of programs is listed below:

PeptideSearch
Mann and Wilm (1994)

GutenTag
Tabb *et al.* (2003b)

InspecT
Frank *et al.* (2005); Tanner *et al.* (2005)

MS-Seq
http://prospector.ucsf.edu/, works essentially as PeptideSearch

Mascot SequenceQuery
Pappin *et al.* (1996); Perkins *et al.* (1999)

MS-Tag
Clauser *et al.* (1999), http://prospector.ucsf.edu/

PepFrag
http://prowl.rockefeller.edu/PROWL/pepfragch.html

MultiTag
Sunyaev *et al.* (2003)

X!Tandem
Craig and Beavis (2003), http://thegpm.org/GPM/index.html

PPM-chain
Day *et al.* (2004)

P-Mod
Hansen *et al.* (2005), considers known protein

Other programs
Lie *et al.* (2006b), blind search for modifications

A method for tag extraction is presented in Lie *et al.* (2006a).

14 Large-scale proteomics

Large-scale proteomics usually means analyzing the whole proteome of an organelle, cell, tissue, or organism in one analysis. At the present time it mainly means identification, though some form of characterization and/or quantification can be included. Essentially, the goal is to analyze as many of the proteins in the sample as possible in a reasonable time.

The common (bottom-up) large-scale procedure is to start with a protein mixture, and perform all or some of the following steps, depending on the approach used:

1. Chemically treat the proteins for the succeeding steps. The exact kind of treatment depends on the specific method used for the further processing, see for example the description of COFRADIC in Section 14.2.
2. Digest the whole protein mixture into peptides (shotgun proteomics).
3. Select a representative sample from the peptide mixture. This step is performed in some large-scale approaches, and omitted in others. Possible protein treatments (Step 1) are often performed to enable or facilitate this selection step.
4. Separate the (selected) peptides into fractions. HPLC is typically used for this step, and often multiple separation steps are included.
5. Record MS/MS spectra. This step is usually performed simultaneously (on-line LC-MS) with the last separation step.
6. Filter out bad (for example, derived from non-biopolymer contaminants) spectra.
7. Recognize (and cluster) spectra coming from the same peptide.
8. Assign peptide identifications to the spectra.
9. Infer from which protein(s) these peptides may come.
10. Score the identified proteins.

14.1 Coverage and complexity

We have defined (protein) sequence coverage as the coverage of a specific protein sequence by the identified peptides. In large-scale proteomics we consider a large number

Computational Methods for Mass Spectrometry Proteomics I. Eidhammer, K. Flikka, L. Martens and S.-O. Mikalsen
© 2007 John Wiley & Sons, Ltd

of proteins, and the number of proteins correctly identified from the sample will be called *sample coverage*.

In large-scale analyses the peptide mixture is usually very complex: a sample of 10 000 proteins will produce about 300 000–400 000 tryptic peptides upon digest. Experiments have shown that it is difficult or even impossible to analyze all these peptides during a single analysis, as the mass spectrometer is essentially overwhelmed. One way to achieve high sample coverage is to try to reduce the peptide mixture complexity by reducing the number of different peptides, ideally maintaining at least one peptide from each protein. This approach, however, results in poor protein sequence coverage and as a result there is a trade-off between loss of protein sequence coverage and loss of sample coverage.

When trying to maintain high sequence coverage, the result is usually poor sample coverage. Some approaches are trying to alleviate this by performing many independent (protein or peptide) separation steps, such that the peptides are presented to the mass spectrometer over a longer time. If the peptides are roughly equally spread out over this time interval, the mass spectrometer has more time to analyze each peptide, resulting in good sample coverage. However, the many separation steps each incur a loss of sample (for example, through adsorption to columns, precipitation, incomplete chemical transformation, etc.) and this typically results in the disappearance of the more low-abundance peptides before the well-separated sample reaches the mass spectrometer. As shown in Section 1.7.3, if F_n is the number of fractions after separation step n, then the number of times the $(n+1)$th step needs to be applied is $\prod_{i=0}^{n} F_i$. Another way to increase sample coverage is to avoid choosing the same peptide several times for fragmentation analysis in the mass spectrometer. This is called *dynamic exclusion* and typically increases the number of different peptides analyzed.

14.2 Selecting a representative peptide sample – COFRADIC

A simple yet effective method for reducing the sample complexity, or for studies of specific peptides of interest, is to select only a subset of peptides from the mixture after digest. The challenge is to discard certain peptides without reducing the possibility for high sample coverage. The optimal reduction is to select just one unique peptide from each protein, preferably in such a way that high-quality MS/MS spectra are produced for all the selected peptides. We describe one such procedure, called COFRADIC (COmbined FRActional DIagonal Chromatography). Also, note that COFRADIC performs a separation in combination with the selection.

COFRADIC relies on two identical chromatographic runs for the selection:

1. The whole peptide mixture is separated into a set of chromatographic *fractions* in a *primary* run.
2. Each fraction is chemically treated, such that the peptides one wants to select will be modified and therefore change their chromatographic properties, while the properties of the unwanted peptides will remain unmodified and their chromatographic properties are maintained.

3. Each fraction is presented for a second separation in a *secondary* run, using the same column under exactly the same conditions. The unwanted, unmodified peptides will elute at the same time as in the primary run, but the modified peptides will elute either earlier or later, depending on the effect of the chemical modification. The peptides one wants can therefore be isolated by selecting only shifting peptides, and these can be subjected to subsequent MS/MS analysis, as illustrated in Figure 14.1.

The actual type of chromatography used is reverse phase HPLC, implying that the chemical modification step should alter the hydrophobicity of the peptides.

As described, the fractions are run separately in the secondary runs, meaning that a large number of runs have to be performed, resulting in a long duration time for the whole experiment. To reduce the overall time required, secondary runs can be combined to reduce the total number of runs. Suppose each fraction is collected in a time interval t. Let the maximum decrease in retention time for a modified peptide be Δt_M. Then peptides eluting from fraction f and fraction $f-i$ will not interfere if $i \cdot t > \Delta t_M$. Therefore several fractions can be combined, as shown in Figure 14.2. The figure also illustrates that the

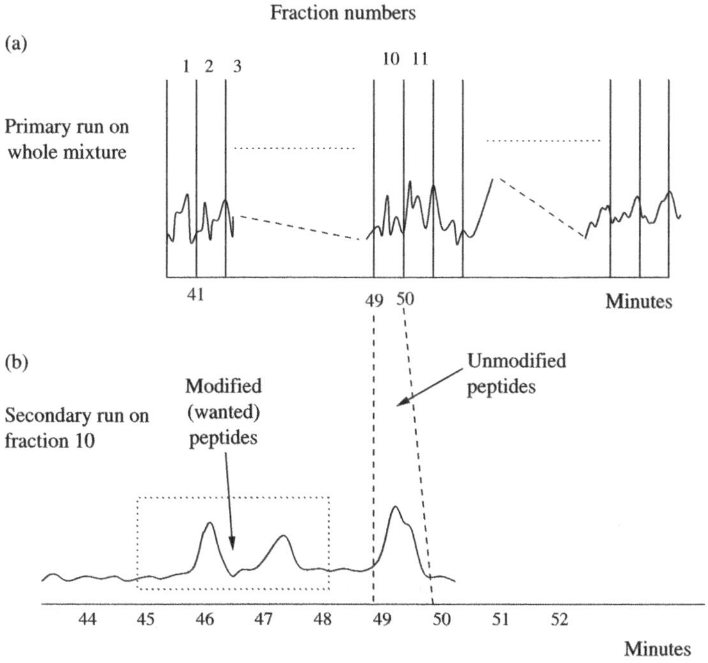

Figure 14.1 Illustration of the COFRADIC principle. (a) The first run. The separation starts after 40 minutes, and the whole peptide sample is separated in fractions, with a fraction consisting of peptides eluted in one minute interval. (b) The secondary run for fraction 10. The unmodified peptides are eluted at the same time as in the first run, but the modified peptides are (in this example) eluted at an earlier time. (After Gevaert *et al.* (2002))

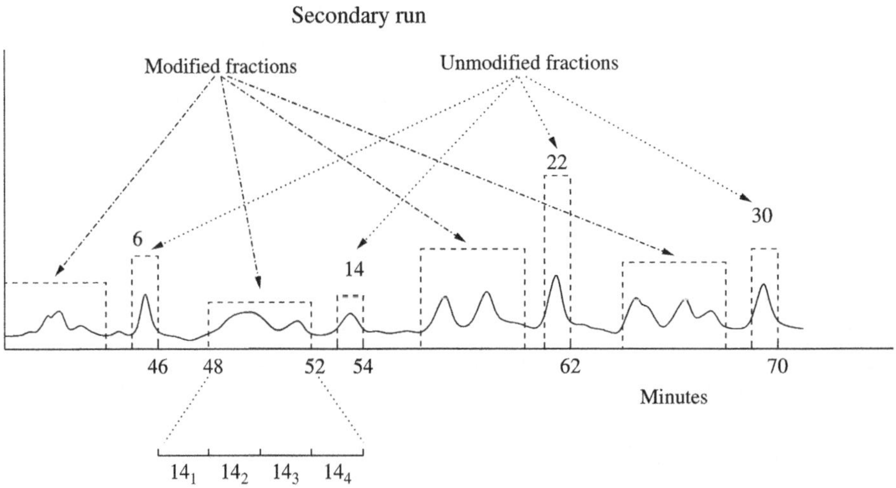

Figure 14.2 Illustration of the principle of COFRADIC. It is assumed that the decrease in retention time is such that fractions 6, 14, 22, 30, etc., can be run together. The figure also illustrates that each collected secondary fraction can be further separated into subfractions. (After Gevaert *et al.* (2002))

modified peptides retrieved from each secondary fraction can be further separated into subfractions, such that each subfraction can be analyzed in a subsequent LC-MS/MS experiment.

Example Suppose the first run results in 35 fractions, and each fraction contains the peptides eluted during 1 minute. Furthermore, let the reduction in retention time for the modified peptides be greater than 1 and less than 5 minutes. To have a 1 minute 'buffer time' between modified and unmodified peptides, the second runs should combine fractions that have a retention time difference of 7 minutes. The first secondary run can then combine five fractions {1, 8, 15, 22, 29}, thus resulting in seven secondary runs in total. Each of the five fractions collected from the seven secondary runs is further divided into four subfractions, thus yielding a total separation of the modified peptides into 140 fractions, which can be submitted to MS analyses.

△

Different methods for chemical modifications

Several chemical modification methods can be used to isolate specific peptides, of which some published examples are given below. Note that *identifiable* peptides here simply means peptides with a mass in the range of 600–4000 Da.

Methionine-containing peptides For the human subsection of Swiss-Prot of 24 January 2006, this results in a reduction in (tryptic) peptide complexity of 74 % (only 26 % of the identifiable peptides contain at least one methionine residue), with only 1.73 % of all proteins being lost due to the absence of identifiable tryptic methionine-containing peptides.

N-terminal peptides This presents a theoretically optimal selection, since each protein contains exactly one N-terminal peptide. The actual chemistry employed here alters the elution time of the non-N-terminal peptides while maintaining the elution characteristics of the N-terminal peptides. The secondary fractions of interest are therefore the ones that do not exhibit a shift. This approach is therefore also called *reverse COFRADIC*.

Cysteine-containing peptides Similar to the isolation of methionine-containing peptides, the reduction in (tryptic) peptide complexity here is 75 % (thus 25 % of the identifiable peptides contain cysteine), yet 4.22 % of the proteins in the human subset of Swiss-Prot of 24 January 2006 will not be represented in this case. COFRADIC can also be applied to selectively isolate modified amino acids, such as phosphorylated or glycosylated peptides.

14.3 Separating peptides into fractions

Another method to obtain high sequence and sample coverage consists of using several (orthogonal) separation techniques in series. The most popular of these methods is called *MudPIT* (for *Multidimensional Protein Identification Technology*), in which an SCX column is coupled to an RP column. The first separation dimension is therefore based on electrostatic interaction, and the second on hydrophobic interaction (thus employing orthogonal separation techniques). MudPIT can therefore be considered the high-throughput, peptide analog of 2D-PAGE separation for proteins.

This is illustrated in Figure 14.3. The peptide sample is injected into the SCX column under acidic conditions (typically a pH of 2–3), which will result in the protonation of most peptides. These charged peptides bind to the column, while those peptides that remain uncharged pass through. These uncharged peptides are then further separated by hydrophobicity when they enter the RP column. The charged peptides are subsequently

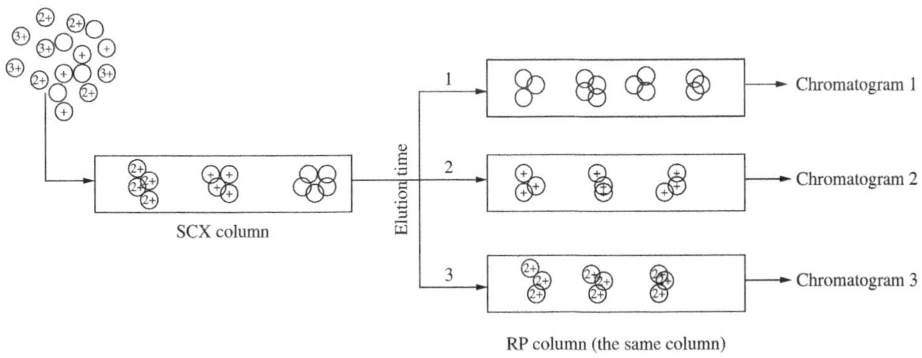

Figure 14.3 Illustration of tandem LC, where the first separation occurs on an SCX column and the second on an RP column. The numbers 1, 2, 3 indicate that fractions from the SCX column are trapped and separately run on the RP column. As the SCX column separates the peptides based on charge, the first SCX fraction will contain peptides with no charge (or negative charge, if any), and the subsequent SCX fractions will contain peptides with increasing numbers of charges

eluted from the SCX column, by using either a linear or a stepwise salt gradient, and each of the charged fractions is also further separated by the RP column.

The MudPIT approach can be set up in a variety of ways, resulting in both on-line and off-line separations. In the method most suited to high-throughput analysis, *biphasic* columns (containing SCX material on top of RP material) are used, and a voltage is applied over the end of the RP column. This allows the column to be directly coupled to the inlet of a mass spectrometer with an ESI source, requiring no further handling or transfer steps after the sample has been applied to the SCX end of the biphasic column.

14.4 Producing MS/MS spectra

In Figure 14.4 we repeat an illustration of the production of MS and MS/MS spectra. The spectra are obtained in time frames, in which the instrument alternates between MS and MS/MS operation modes. The time interval between two MS operations is often chosen as the average elution time of a peptide from the chromatography column. In this example the three most intense peaks are subsequently chosen for MS/MS analysis. During the first time slice after the initial MS scan, an MS/MS spectrum from peptide A is obtained by selectively fragmenting this mass only. In the next time slice a spectrum for peptide B is produced, followed by a third time slice for recording the MS/MS spectrum for peptide C. After these three fragmentation spectra have been obtained, a new MS scan is started. From this scan, three peptides A′, B′, C′ are selected for fragmentation and the cycle starts over again. This approach is often called *data-dependent acquisition*. Typical time slices for recording fragmentation spectra are 0.1–8 seconds.

This method of alternating production of MS/MS spectra involves some characteristics that one should be aware of (referring to Figure 14.4):

- Peptides enter into the mass spectrometer in a continuous manner, so that by the time peptide C is selected for MS/MS analysis, most of the peptide ions may have already eluted, leaving only a few (or even none) ions of that m/z to be analyzed in the instrument. If there are not enough ions to be analyzed, the spectrum quality and overall signal intensity will typically be low.
- When choosing the peptides for MS/MS analysis, some of them may have been chosen at an earlier analysis cycle. This will result in the creation of several MS/MS

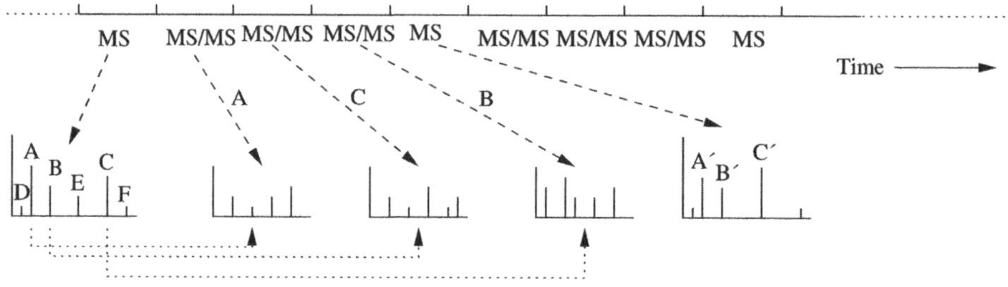

Figure 14.4 Illustration of the alternating production of MS and MS/MS spectra

spectra for the same peptide. To reduce this generation of redundant spectra, the instruments may remember the last several m/z values chosen, and will attempt to avoid fragmenting any of these again. The time during which these m/z values are remembered (and during which they will therefore be ignored for analysis) must be a trade-off between the desire to avoid redundant fragmentation, while minimizing the danger of ignoring a different peptide with a different elution time that happened to have the same m/z.

- Note that the selection of only three peptides for MS/MS analysis in the figure precludes the recording of fragmentation spectra for peptides D, E, and F.

To achieve 'optimal' efficiency in an experiment, one must fine-tune the following instrument parameters:

- The time period for acquiring the MS spectra.
- The time period for acquiring the MS/MS spectra. In Figure 14.4 the time periods for MS and MS/MS are chosen to be equal, but this does not need to be the case.
- The number of peptides that will be selected for MS/MS analysis from an MS scan, and the method by which to select them (for example, by intensity rank or a predefined inclusion mass list).
- The time period during which an m/z value will be excluded from being selected again.

14.5 Spectrum filtering

From the discussion above it is clear that many low-quality spectra can be produced, meaning that they do not contribute to the identification of peptide sequences. Removing these before starting a search could save computational time in the analysis. In large-scale proteomics, manual assessment of spectrum quality is not possible, and automatic methods are therefore desirable. It is also probable that automatic methods are better than manual ones as they are typically more consistent. One way of assessing quality relies on classifying the spectra into two classes: *good* and *bad*. Spectra classified as good are supposed to lead to an identification, and those classified as bad are not. Such a classification is then used as a filter to select against bad spectra.

14.5.1 Classifying good and bad spectra

In order to perform the classification, a classifier has to rely on a number of attributes (or features) whose values can be extracted from the spectra. The first challenge is to determine the attributes to use. Some attributes that correlate with spectrum quality have been discovered through manual inspection of spectra, including the number of peaks and the total intensity. By building a filter that utilizes these attributes, it has been shown to be possible to remove up to half of the bad spectra. The cumulative intensity normalization (Section 6.2.6) is also used as a filter, as it measures how well real peaks deviate from noise in terms of intensity. This system establishes the intensity rank of each peak (from 1 to $n+1$, with n the number of peaks and rank $n+1$ an artificial rank), and then creates a coordinate system in which the rank (in increasing order) forms the horizontal axis and the normalized intensities (0 to 1) the vertical axis. If the original intensities of all peaks are equal, the curve will be a straight line from $(1, 1)$ to $(n+1, 0)$. The more the real peaks deviate from the noise peaks in intensity, the more the curve will deviate from this line. This deviation can be measured as the size of the area between the curve and the straight line, and can be used as one attribute of a classifier.

A complicating factor in constructing classifiers (or filters) is that they must generally be different for each type of mass spectrometer and for each set of experimental conditions. It must also be considered whether there should be different classifiers for different types of spectra, for example as derived from precursors with different charges. The procedure for constructing classifiers should therefore be:

1. Collect a large set of spectra, the training set, for the specific instrument and set of experimental conditions one wants to construct a classifier for.
2. Divide the training set into good and bad spectra.
3. Define a set of attributes that potentially correlate with the quality.
4. Use the training set to investigate which of the proposed attributes actually correlate with the quality, and how this correlation is expressed. This can be accomplished by machine learning techniques.
5. Construct a classifier based on the suggested attributes.
6. Test the classifier on other spectra (that is, spectra not in the training set) to verify correct functioning.

The training set

One way of producing the training spectra is to collect a large number of spectra from a mixture of known proteins. Redundant (similar) spectra should be removed, since the inclusion of redundant spectra can easily bias the classifier towards the characteristics of highly abundant peptides. Next, the training spectra should be grouped into the *good* and *bad* categories. Since the training set should be large, this cannot be easily done manually, and some form of automatic grouping should be performed. One way to do this is to search the spectra against a sequence database with a reliable search program (for example, SEQUEST or Mascot). Spectra matched to proteins known to be in the mixture are denoted good, all others bad. Note that this approach may result in high-quality spectra that are considered bad, if they for instance carry modifications that are not considered in the search, or are derived from unexpected (contaminating) proteins in the mixture.

Attributes to consider

We can roughly divide the possible attributes into two groups:

- *General attributes* for which a numerical value can be calculated. Attributes that have been shown to have an impact on the quality are: the number of peaks in a spectrum, the number of peaks with a relative intensity greater than k, the average distance between peaks, the intensity difference between the top two peaks, etc.

- *m/z-related attributes* that specify which m/z peak values and m/z differences (delta values) are significant in the classification. To find these values one first needs to define what kind of values to extract, for example only integer mass values, and delta values in the range of 0–150. Next, the training set must be examined to find the relative importance of these attributes. This can for example be done by performing a chi-square test or by maximizing the entropy. Typical attributes that would be selected are for example m/z values corresponding to commonly observed ions or delta values corresponding to amino acid residue masses.

Constructing the classifier

Different machine learning techniques are used for constructing a classifier. The classifier is trained using the training set(s) and the proposed attributes. Examples of some commonly used techniques are (i) Bayesian classifiers; (ii) neural networks; (iii) support vector machines (SVM), which try to construct an $(n-1)$-dimensional hyperplane for the n attributes, such that the good and the bad spectra fall on different sides of the plane; (iv) quadratic discriminant analysis (QDA), which uses a quadratic classifier, trying to separate the good and the bad spectra by a quadric surface (hypersurface); (v) decision trees.

14.5.2 Use of the classifier

The classifier calculates two values, one related to the probability of the spectrum being good, and one to the spectrum being bad, or a value related to the relative probabilities of these two cases. The spectrum is assigned to the class with the highest score (probability). This behavior can, however, be changed if 'symmetry' between the classes is not desired, for example when it is considered much worse to classify good spectra as bad than vice versa.

Using a spectrum classifier can produce two types of error:

- Good spectra are classified as bad (which we will call false negatives here). This may result in a loss of peptide identifications, since identifiable spectra (good spectra) are not considered for identification (classified as bad). Note, however, that the end goal is usually to recognize proteins, not peptides. This means that if a protein can also be identified by other peptides, removal of a good spectrum for that protein is not such a big problem. How deleterious false negative classifications are can therefore vary based on the type of experiment and the desired result. Naturally, if a highly efficient peptide selection is performed (for instance, using N-terminal COFRADIC), the impact of a false negative classification is more severe, and therefore completely undesired.
- Bad spectra are classified as good (false positives). This will imply longer search time, but can also lead to false positive identification, albeit still to a lesser degree than when using unfiltered data. False positive classifications are therefore typically

less severe than false negatives, although in specific applications they might still be very annoying (for instance, in building spectral libraries).

Trying to avoid these two types of potential errors is often regarded as a trade-off situation, and the software should consider this trade-off as an input parameter. The user can then decide which error should be minimized and which can be tolerated depending on the relative importance of the two errors with regards to the desired results. Such a trade-off can be implemented by imposing a higher cost for one of the two error types, to bias the classifier towards the desired outcome.

14.6 Spectrum clustering

As explained in the introduction to this chapter, several spectra may have been recorded for a given peptide ion with a given charge (fragmentation of that same peptide but with different charge typically results in a different spectrum). This leads to some form of redundancy in the complete set of acquired MS/MS spectra. Such spectra are often called *sibling spectra*. Experience has shown that it is not unusual for over 20 % of the spectra to have siblings.

Sibling spectra are seldom exactly equal, due to different reasons:

- The abundance of the peptide ions may have been different each time the peptide was selected for MS/MS analysis, resulting in spectra of different quality (ions of higher abundance generally result in less noisy spectra).
- The fragmentation may have been different at different time points (for example, due to variations in the collision energy or the collision gas pressure).
- Random noise may influence which peaks are detected, as well as their intensities.

It has, however, been observed that sibling spectra typically result in differing intensities rather than in different m/z values.

There are several reasons why sibling spectra should be recognized:

- To reduce the computational processing time.
- Spectra from independent fragmentation (of the same peptide) may complement each other, increasing the total amount of information that is available during the identification process (for example, if two spectra each have one unique fragment ion).
- To avoid the assignment of sibling spectra to different subsequences or protein sequences.
- To remove noise, based on the assumption that noise occurs more randomly in the spectra than real peaks do. Peaks occurring in only one (or a few) sibling spectra can then be considered candidates for being noise.
- If a group of sibling spectra remains unidentified, the reasons for investigating them further are increased compared to single, unidentified spectra.

Often only spectra with a similar precursor mass are considered potential siblings. However, if the precursor mass equality criterion is relaxed, the clustering can be used to discover differently modified versions of the same peptide. If one for example discovers that a group of sibling spectra consists of two subgroups that are differentiated by a specific mass, one can suppose that the two groups correspond to both a modified and an unmodified version of the same peptide. One can then use a more comprehensive method to try to discover these modifications, as described in Section 14.6.4.

14.6.1 Recognizing sibling spectra

Sibling spectra are found by first performing pairwise spectrum comparisons. It is therefore essential to have a scoring function Ψ for scoring the similarity between two spectra. Ψ should satisfy some properties:

- Ψ should be large and have a small variance when two sibling spectra are compared;
- there should be a significant difference between the Ψ for sibling spectra and the Ψ for non-sibling spectra (that is, Ψ should yield good discrimination).

In Chapter 11 we described several methods for comparing spectra, with the slight nuance that the focus was to compare experimental and theoretical spectra. The situation is typically different for directly comparing two experimental spectra. Mass spectrometers may for instance have systematic deviations in their m/z measurements, which must be dealt with when comparing them to theoretical spectra. In the case where two spectra that originate from the same peptide and mass spectrometer are compared, however, there should be comparable systematic shifts in the two spectra. In the latter case one can therefore apply tighter constraints on m/z similarity for finding corresponding fragment peaks.

Spectrum preprocessing

As explained in Section 11.2, the comparison can be based either on the peaks themselves, or on the intensities in m/z bins (intervals). In the former case one must select the peaks that should take part in the comparison, a decision that is usually based on their intensity. In the latter case the intensities inside a bin are added up to compute the bin's intensity. In both cases the intensities should be normalized, however.

It has been observed that spectra from different peptides with a similar mass can erroneously be reported as siblings due to strong peaks in the fragmentation spectra that correspond to their unfragmented precursor mass, or neutral losses thereof. Such peaks should therefore be removed in the preprocessing.

Similarity scoring

The most often used scoring functions are spectral contrast angle or some form of (cross) correlation. In order to take advantage of the potentially higher accuracy due to the assumed similar systematic deviation in the measurement, the score can include a weight factor based on the m/z difference between corresponding peaks. Such a weight can for instance be implemented as a sigmoid function.

One general scoring function can be described as follows. Let there be m and n peaks in the two spectra (R_1, R_2), and let the peaks of the spectra be indexed by i and j respectively. Then the score between two peaks (i, j) is calculated as

$$s(i, j) = f(|m_i - m_j|)g(I_i, I_j)$$

where m_i, I_i are the mass and intensity of peak i, respectively, and f and g are two functions. f should be a (steeply) decreasing function, for example $f(x) = 1/(1+e^{\frac{x-a}{b}})$, in which constants a and b influence the location and the pivot of the sigmoidal curve, respectively. $g(x, y)$ could for example be the minimum, maximum, or average intensity of the matched peaks. A similarity score for two spectra can then be calculated as

$$S(R_1, R_2) = \sum_{i=1}^{m}\sum_{j=1}^{n} s(i, j)$$

In order to obtain comparable scores for sequences of different lengths, the score should be divided by the 'selfscore' for R_1 and R_2:

$$SC(R_1, R_2) = \frac{S(R_1, R_2)}{F(S(R_1, R_1), S(R_2, R_2))}$$

where F can for example be the average. Note that this score is symmetric.

MS/MS specific scoring The scoring method described above does not specifically utilize the information that the comparison is made for MS/MS spectra. This piece of information, however, is directly used in a method described in Bandeira *et al.* (2004).

Based on a spectrum R, a new spectrum is constructed as the union of R and the inverse of R. Each peak therefore corresponds to a node in a spectrum graph where b and y ions are considered, see Section 12.1. Each peak is given a weight, in the same way as the nodes are scored in the spectrum graph. The spectrum is called PRM (Prefix Residue Mass spectrum), indicating that one aims to construct a spectrum including many peaks corresponding to prefix masses (b ions). It is called sparse if no two peaks are less than 57 Da (the mass of glycine, the lightest amino acid) apart. The peaks are supposed to be b ions, so subsets of the sparse PRMs are constructed that contain no complementary peaks. Such a spectrum is called *antisymmetric*. A subset of a PRM spectrum R is optimal if it is antisymmetric and achieves the maximum total weight of all such subsets. Optimal subset(s) can be found by dynamic programming. A simple score between two spectra can then be defined as the score of the optimal sparse subset of the intersections of their PRMs. Note, however, that this does not need to be the optimal score between the two spectra, see Section 14.6.4.

Example Suppose we have two spectra R_1 and R_2, in which the peptide masses are equal to $M_P = 692$, namely $R_1 = \{129, 276, 331, 363, 519, 622\}$, $R_2 = \{129, 363, 418, 462, 519, 565\}$. The number of common peaks is three.

The inversed spectra are found as $\bar{R}_1 = \{72, 175, 331, 363, 418, 565\}$, $\bar{R}_2 = \{129, 175, 232, 276, 331, 565\}$. The PRM spectra become

$$R_1^{PRM} = \{72, 129, 175, 276, 331, 363, 418, 519, 565, 622\}$$

$$R_2^{PRM} = \{129, 175, 232, 276, 331, 363, 418, 462, 519, 565\}$$

Taking the intersection we get $\{129, 175, 276, 331, 363, 418, 519, 565\}$. This includes several masses less than 57 apart and also complementary masses.

The optimal sparse subset will contain four masses, which masses depend on the weight of the peaks. For an interpretation of the masses, see Exercise 14.1.

△

It was observed that the mass range of the intersection of sibling spectra was generally larger than the mass range of the intersection of non-sibling spectra with the same similarity scores. To take this observation into account, the original score was multiplied by a factor d/m, where d was the mass range of the intersected spectrum and m the precursor mass.

14.6.2 Clustering of sibling spectra

Clustering or grouping of objects, based on pairwise similarity measures, is a well-studied subject. Two objects are considered to be similar if their similarity score exceeds a given threshold T. Different approaches are used to decide if two objects should be in the same group. The strongest criterion is that all objects in a group should be pairwise similar, thus forming a *clique*. This is called *complete linkage*. The weakest criterion is that each object should be similar to at least one of the others, called *single linkage*. A drawback with the latter is that it can end up as a 'chain' in which many of the group pairs have similarity scores below the threshold. Contaminating spectra can typically contribute to such a chaining effect. On the other hand, single linkage has the advantage that not all pairwise scores between the spectra need to be accessible simultaneously, which is the case for most of the other techniques. Single linkage may therefore be the best alternative when a high number of spectra are to be clustered on a standard desktop computer with limited memory.

Many clustering techniques occupy a position somewhere between complete linkage and single linkage. An example of such a technique requires that each object must be similar to at least $k-1$ of the others. A further specialization of this approach is that each group must include a clique of size k and that all other spectra in the group must be similar to at least $k-1$ of the spectra in the clique. Such a group is called a *paraclique* (Tabb *et al.* (2005)), and a typical value for k is three. Another example consists of forming a star, with one spectrum as the 'kernel', and with all the other spectra in the star similar to the kernel.

Example Assume we have six spectra, with mutual scores as in Table 14.1. Assume that a score of five or more is required before two spectra are considered sibling spectra. If single linkage is used, all spectra will be collected in one group. If complete linkage is used, three groups form: {A,B,C}, {E,F}, {D}. If paracliques of minimal size 3 are used, there will be only one group satisfying the paraclique constraint, {A,B,C,D}.

△

Table 14.1 Mutual scores of six spectra as used in the example

	A	B	C	D	E	F
A		5	6	4	3	2
B			6	7	4	3
C				5	5	4
D					4	4
E						6

14.6.3 Representative spectra for the groups

When a group of (supposedly sibling) spectra has been formed, it can be useful to let one spectrum represent the group, and to use this spectrum for subsequent identification. There are two main methods used for obtaining such a representative spectrum:

1. Choose the best (usually the most intense) spectrum in the group.
2. Construct a synthetic spectrum representing all the spectra in the group.

Neither of these methods have been shown to hold significant advantages over the other, but they will generally lead to different spectra being identified. Synthetic spectra in some cases identify peptides which could not be identified using any of the grouped spectra separately. The synthetic spectrum can therefore increase the chances of identifying low-quality spectra by combining them. On the other hand, peptides identified by only a single spectrum in a group (with the rest of the group remaining unidentified) may not be found when using a representative spectrum. A method using both the 'best' spectrum in a group as well as a synthetic spectrum therefore seems to be a reasonable and exhaustive approach.

The synthetic spectrum can be constructed in different ways. The goal of this spectrum is to contain all the peaks that are found in spectra obtained after fragmenting the same precursor peptide. The task is thus to decide which peaks to include, and to determine an intensity for each of the included peaks. A reasonable way is to construct a peak at a mass that occurs in at least k of the spectra, where k is a value in the interval between one and n, with n the number of spectra in the group. If $k = 1$, we effectively take the union of the peaks found in the spectra, whereas if $k = n$ we take the intersection. To avoid inclusion of too many noise peaks in the synthetic spectrum, k should be chosen such that the probability that a noise peak will occur in k spectra should be less than a predefined value, for example 0.01.

Because some groups will consist of many sibling spectra, each one with its unique peak profile, the construction of good synthetic spectra is not a trivial task. Especially in the case of large-scale proteomics, the number of recorded spectra is very high, resulting in many redundant recordings of the same MS/MS spectrum. This leads to a situation in which some of the groups could best be represented by a synthetic spectrum with $k = 1$, others with $1 < k < n$, and yet others with $k = n$. Since k is typically chosen as a global parameter, however, the resulting synthetic spectrum will be suboptimal for a considerable number of groups. Many of these suboptimal synthetic spectra may subsequently elude identification, leading to a loss of peptide identifications.

14.6.4 *De novo* sequencing from representative PRM spectra

A motivation for the clustering of spectra is that a consensus spectrum should contain more explicit information from the original peptide than each spectrum separately. With respect to *de novo* sequencing this should imply a more reliable derived sequence. This is analyzed in Bandeira *et al.* (2006). They treat the case in which at least two spectra come from peptides that overlap, or in which one is a modified version of the other. Such pairs of spectra are called *spectral pairs*. The approach requires:

- an efficient method for discovering spectral pairs;
- an efficient method for *de novo* sequencing.

These two tasks are solved by a combined procedure. Let $\{r_i^{x,b}\}$, $\{r_i^{x,y}\}$ be the masses of the b and y ions of spectrum R_x. Let the number of peaks in the two spectra be m and n, respectively. Furthermore, let the suffix of length $m-i$ of R_1 overlap with the prefix of R_2, as shown in Figure 14.5(a). Then we have

$$r_{i+k}^{1,b} = r_k^{2,b} + r_i^{1,b}, i < k < m-i; \quad r_{j+k}^{2,y} = r_k^{1,y} + r_j^{2,y}, j < k < n-j$$

A *product matrix* of two spectra is a matrix with the masses of the two spectra along the horizontal and vertical axes respectively, and with a dot at all points $(r_i^{1,x}, r_j^{2,z})$, where x, z are from the set $\{b, y\}$. This means that there will be two subdiagonals in the product matrix of the b and y ion masses corresponding to the overlap, as shown in Figure 14.5(b). The presence of two such subdiagonals with many dots in a product matrix indicates a large overlap between the original peptides. The problem of finding siblings is therefore solved by exploring if the product matrix contains two subdiagonals with a sufficient number of dots. To increase the number of dots in the subdiagonals, both spectra are first transformed to the PRM representation. Note that if claiming $i = 0$ or $j = 0$ the search is for spectral pairs where one is a subspectrum of the other. This will drastically reduce the computational time, but it is shown that the reduction in identification is not reduced correspondingly.

An important aspect of the method is that the two subdiagonals divide the ions into b and y ions (without specifying which is which), hence making *de novo* sequencing much simpler.

The product matrix can also be used for examining if two spectra correspond to the same peptide that is differently modified. Let the spectra of the unmodified peptide be R_1, and that of the modified peptide be R_2. Then the masses corresponding to the 'unmodified' fragments are equal in R_1 and R_2, but those corresponding to modified fragments are increased by a mass δ in R_2. Finding the optimal path of the PRMs of R_1 and R_2 when including modifications can be done by spectral alignment, as described in Section 11.4.2. However, the alignment procedure must be changed such that the path found is both a sparse as well as an antisymmetric subpath of both R_1 and R_2. Without going into the details of the algorithm we provide an example next.

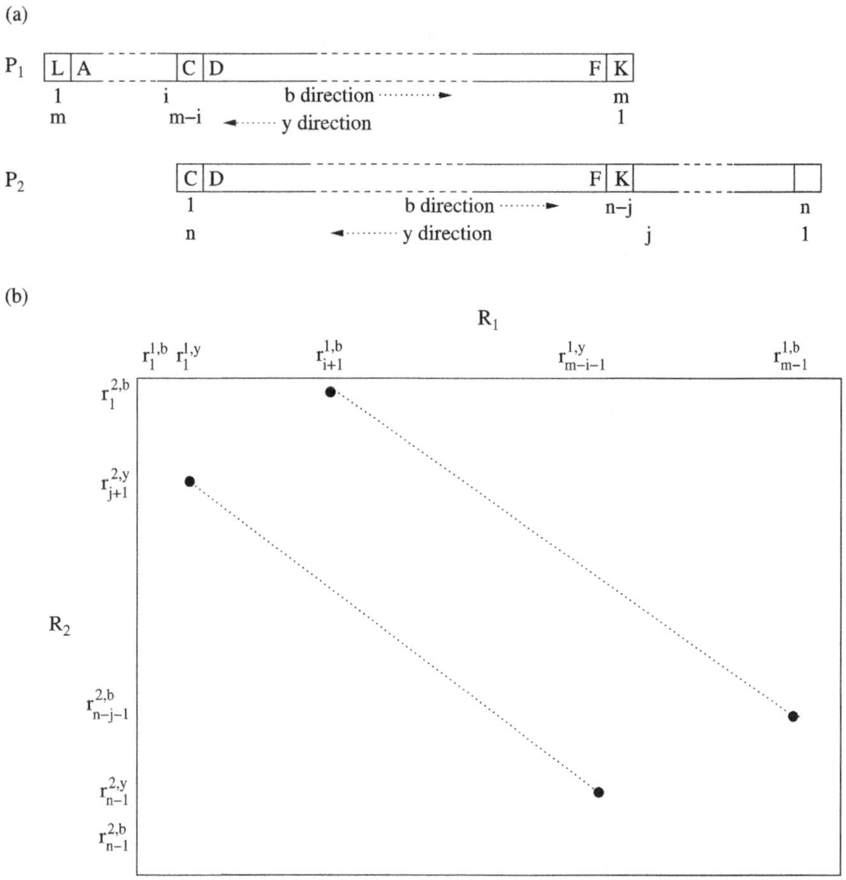

Figure 14.5 Illustration of two spectra R_1 and R_2 from overlapping peptides P_1 and P_2. The overlap will show up as subdiagonals in the product matrix, as explained in the text

Example Assume we have a spectrum $R_1 = \{129, 175, 276, 331, 519, 565\}$ with peptide mass $M_{P_1} = 692$. The inverse spectrum becomes $\bar{R}_1 = \{129, 175, 363, 418, 519, 565\}$, and the PRM spectrum $R_1^{PRM} = \{129, 175, 276, 331, 363, 418, 519, 565\}$. From this we can derive the sequence `(AG)(CV)STR`.

Assume another spectrum with $R_2 = \{345, 363, 476, 533\}$ with peptide mass $M_{P_2} = 706$. The inverse is $\bar{R}_2 = \{175, 232, 345, 363\}$, and $R_2^{PRM} = \{175, 232, 345, 363, 476, 533\}$. One should then try to find an optimal path through the combined graph, such that it is sparse and that the projection on both of the spectra is antisymmetric. One such path is $(129, 232, 331, 345, 418, 519, 533)$. The projection on R_1^{PRM} is $(129, 331, 418, 519)$, corresponding to the sequence `(AG)(CV)STR`. The projection on R_2^{PRM} is $(232, 345, 533)$, corresponding to the sequence `(AGC)V+(ST)+R+`, where $+$ means that a mass of 14 Da is added to the amino acid masses. Combining these two, we derive the common sequence `(AG)CVSTR`, where the sequence for R_2 has an addition of 14 Da at amino acid `V`.

Since modifications on V are extremely rare, the R_2 sequence could probably represent a polymorphism where V has been changed to L.

△

The example shows that the use of sibling spectra can increase the quality of the derived sequences. This is utilized in Bandeira *et al.* (2006) by constructing star clusters. The kernel spectrum is pairwise compared to each of the other spectra, and the common peaks in the PRMs of such a pairwise comparison are recognized. The representative spectrum is the union of the results from all the pairwise comparisons. *De novo* sequencing is then performed on that representative spectrum.

To increase the number of sibling spectra containing modifications Bandeira *et al.* (2006) propose to perform chemical modifications of peptides.

14.7 Searching the database

The methods that have been described in the preceding chapters can also be used for the identification of spectra. In this application of the algorithms, the experimental spectra are compared against theoretical spectra obtained from a sequence database rather than against other experimental spectra.

14.8 LIMS

It is important to realize that the large amount of data generated in high-throughput experiments presents both a challenge with regard to data management, as well as a treasure trove of information to mine for optimizing algorithms or improving experimental design. Many approaches are described throughout this book that highlight the importance of having access to learning and testing sets for training machine learning algorithms, so we will briefly discuss the data management challenge in this section.

Since large-scale, high-throughput experiments can generate several tens of thousands of spectra per run, and several runs can be performed in a day, integration of the acquired data into a robust data management infrastructure has become increasingly important for proteomics labs. The software systems that are developed for this purpose are called *Laboratory Information Management Systems* (LIMS). LIMS packages are available from commercial companies, and freely available and open-source systems produced by academia also exist. LIMS can vary from the extremely complex, detailed, and comprehensive systems used in pharmaceutical companies, to the relatively simple and focused versions developed to support the specific needs of a single academic lab. The data management backbone of typical LIMS consists of a *Relational Database Management System* (RDBMS). The relational storage used in an RDBMS is particularly suited to the fast and efficient retrieval of specific linked information amongst vast amounts of data. Software applications are provided to manage the import and retrieval of the various results obtained during a typical proteomics workflow: importing spectra from the mass spectrometer, constructing clusters of sibling spectra and creating representative spectra from these, submitting spectra to a search engine, interpreting and storing the results of the search, filtering for suspect identifications, etc.

Exercises

14.1 The spectra in the example in Section 14.6.1 are supposed to come from the peptide `AGCVSTR`. Assign the spectra masses to different *b* and *y* ions.

Bibliographic notes

General

Bandeira *et al.* (2004); Colinge *et al.* (2003); Craig and Beavis (2003); Sunyaev *et al.* (2003)
Nesvizhskii and Aebersold (2004, 2005)

Covering and complexity

Martens (2006)

COFRADIC

Gevaert *et al.* (2002, 2004, 2003)

MudPIT

Washburn *et al.* (2001); Peng *et al.* (2003)

Spectra quality (filtering)

Flikka *et al.* (2006); Purvine *et al.* (2004); Salmi *et al.* (2006)
Bern *et al.* (2004); Na and Paek (2006); Savitski *et al.* (2005a)

Clustering spectra

Beer *et al.* (2004); Gentzel *et al.* (2003); Tabb *et al.* (2003a, 2005)
Bandeira *et al.* (2004, 2006); Wong *et al.* (2005)

15 Quantitative MS-based proteomics

The main topics of this book address the use of mass spectrometry to deal with the qualitative aspects of proteomics, mainly the identification and characterization of proteins. It is, however, useful to add a quantitative dimension to proteomics experiments. The reason for this is that some proteins can be up- or down-regulated in certain cell states or disease states. Other proteins might be translocated to different subcellular compartments, or modified. In order to utilize the full potential of proteomics, quantitative techniques are therefore needed as part of the analytical toolbox.

In this chapter, the main techniques for obtaining quantitative measurements of peptides and proteins are described, together with certain computational challenges associated with these techniques.

15.1 Defining the quantification task

We have a set of n samples, each containing a set of molecular components. The majority of the components are the same in each sample, and in our context the components are mainly peptides, but in some cases they are proteins. The task is to explore how the abundance of the corresponding peptides varies from sample to sample. Due to the way the instruments detect the ions, *relative quantification* is the easiest form of quantification, meaning that, instead of the absolute concentration, we measure *fold changes* in the molecules between samples. The relative measurements are performed either within a sample or across the samples, and both label-based methods and label-free methods exist.

Example Let the amount of three components in sample S_1 be (10, 20, 18) (of an undefined unit), and in sample S_2 be (30, 20, 17). The relative increase in the first component between S_1 and S_2 is then 300 %, corresponding to a three-fold increase.

△

Relative quantification is the main topic of this chapter. However, *absolute quantification* is briefly described at the end.

The challenge in quantitative proteomics is to perform the experiments (sample gathering and preparation, mass measurement, etc.) such that the resulting outcomes from the

Computational Methods for Mass Spectrometry Proteomics I. Eidhammer, K. Flikka, L. Martens and S.-O. Mikalsen
© 2007 John Wiley & Sons, Ltd

different samples can be compared. This implicitly requires that the samples are similar, for example having approximately the same total amount of protein content, and that only a few of the molecular components vary.

Two-dimensional gels can be used for quantitative proteomics, as described in Chapter 2. The quantitation aspect is reflected in the size and intensity of a protein spot, where a big and intense spot usually implies an abundant protein. The large majority of the quantitative methods, however, rely on mass spectrometry, in which comparing abundances means comparing peaks in spectra. We will concentrate on these methods in this chapter.

15.2 mRNA and protein quantification

The quantification and comparison of mRNA (or cDNA) expression levels has already been extensively studied. Since mRNA is the source of protein production, a reasonable question is to ask whether there is a correlation between the levels of an mRNA and the protein it encodes. Such a (general) correlation exists, but it is not strong enough to be used on individuals; it is therefore not possible to predict the protein amount from the corresponding mRNA amount. This can partly be explained by the fact that RNA levels depend on the transcription efficiency and degradation rates of mRNA, while protein level also depends on translational and posttranslational mechanisms (including degradation of proteins).

However, several investigations have shown that there is some correlation when the *codon bias* is high (> 0.5). Codon bias is a measure of the propensity of an organism to selectively utilize certain nucleotide codons over others that encode the same amino acid residue. Several measures for codon bias exist, and the result is often a value between zero and one.

Since proteins are the executive molecules in the cells, the main focus should be on the protein level. However, quantitative analyses on mRNA level are easier to perform, and they also help in exploring the mechanism for protein production. The end result is that studies on these two levels are complementary rather than mutually exclusive or redundant.

15.3 Quantification of peaks

In nearly all the MS-based methods, the expression levels of the peptides or proteins are measured by observing the signal intensity detected by the mass spectrometer. As shown in Section 6.2.4, this can essentially be done in two different ways, either by using the maximum height of a given peak, or by calculating the area under the curve representing the peak.

In Figure 15.1 a zoomed view of two peptide peaks is shown, and the two ways of describing their intensities are depicted. It is not obvious which method gives the most reliable results, since some reports indicate that peak area is best, but has limited value if the signal intensity is too low. In this example, it can be observed that the ratio between the highest and lowest peak is 2.15 when calculated using the peak height, and 2.00 using the area. Which approach to choose is largely dependent on the signal intensity, the shape of the peaks, the instrument type, and other, experiment-specific conditions.

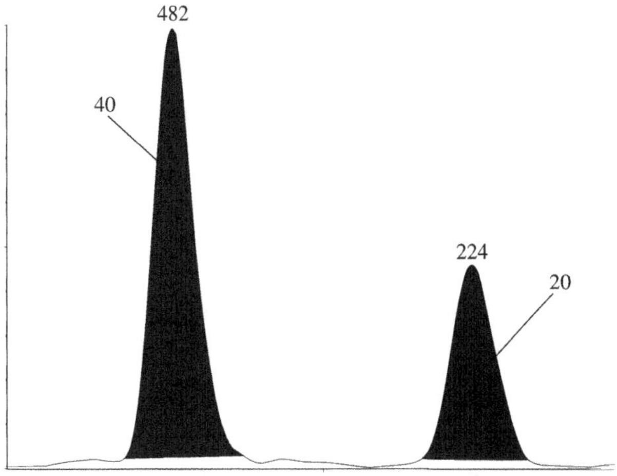

Figure 15.1 Two MS peaks, where numbers indicate intensities, based either on height (number above each peak) or on area (number connected to line)

15.4 Normalization

Usually, the signal intensities from run to run are not directly comparable. This is caused by a systematic intensity shift, which causes the total ion count between runs to differ, even though the samples are identical. In order to compare samples we must therefore first correct for such shifts. This is typically done by normalizing the signal intensities. Normalization is briefly treated in Sections 6.2.6 and 9.2.1, but is especially important for quantitative comparisons. After the normalization, statistical analyses are often performed to discover the peptide differences between samples. The normalization should obviously be such that the change in peptide amounts in the samples is reflected in the normalized values.

Several procedures exist for signal intensity normalization, of which we mention three:

- Normalization using internal standards by adding known components (peptides/proteins) to the sample. The intensity of each peak can now be normalized relative to that of the internal standard closest in m/z.
- Normalization to constant sum implies using the total ion intensity as the normalization factor, expressed as the normalized intensity $x_N = x/\sum_{i=1}^{n} x_i$.
- Normalization to unit length; that is, if the vector of intensities is denoted x, $x_N = x/||x||$, where $||x|| = \sqrt{\sum_{i=1}^{n} x_i^2}$.

The normalization used depends on the spectra and the experimental conditions. One issue to consider is the potential presence of high peaks. These can sometimes be problematic, because the intense peaks typically show a higher variability than less intense peaks, and this variation affects the normalized values of all data points. This is a major drawback of normalization to unit length, since it gives more impact to intense peaks as compared to normalization to constant sum, because the denominator will be more affected by the most intense peaks.

Example Given two vectors of intensities, for example $x = 1, 3, 2$ and $y = 1, 1, 4$, normalization to constant sum will yield the vectors $x_N = \frac{1}{6}, \frac{3}{6}, \frac{2}{6}$ and $y_N = \frac{1}{6}, \frac{1}{6}, \frac{4}{6}$, whereas normalization to unit length gives $x_N = 1/\sqrt{14}, 3/\sqrt{14}, 2/\sqrt{14}$ and $y_N = 1/\sqrt{18}, 1/\sqrt{18}, 4/\sqrt{18}$.

△

In this example, when using normalization to constant sum, both x and y are normalized using the same normalization factor, whereas when unit length is used, the denominator is higher for y, because of the presence of a higher peak.

15.5 Different methods for quantification

Several options exist for turning a proteomics experiment into a quantitative protein measurement setup. We will divide the discussion into methods related to label-free peptide methods and label-based peptide methods. In this context, a label is simply something attached to the peptides of a sample to enable the distinction of this sample from a differently labeled or unlabeled sample.

15.6 Label-free quantification

The general procedure for performing label-free quantification using peptides is described below, but note that not all steps are included in all methods:

1. Separate the samples into subsamples (or fractions). This must be done if the samples contain too many peptides for being handled in one mass analyzing procedure. The separation should ideally ensure that the subsamples will contain the same proteins or peptides for all the original samples. In practice, however, the extensive pre-analysis fractionation required to resolve all the main peptide components easily introduces differences between samples.
2. Record the mass spectra and obtain the intensities for the peptide ions. The number of detected peptides depends on the overall method and instrument used.
3. Normalize the intensities inside each subsample, such that the intensities of peptides in corresponding subsamples can be compared.
4. Find corresponding peptides over the subsamples.
5. Explore how the intensities of corresponding peptides vary over the samples.
6. Identify the proteins from which the interesting peptides are cleavage products.

Each sample or subsample is performed in one analysis. The spectra from corresponding subsamples are then compared. Ideally, the same peptide in corresponding subsamples

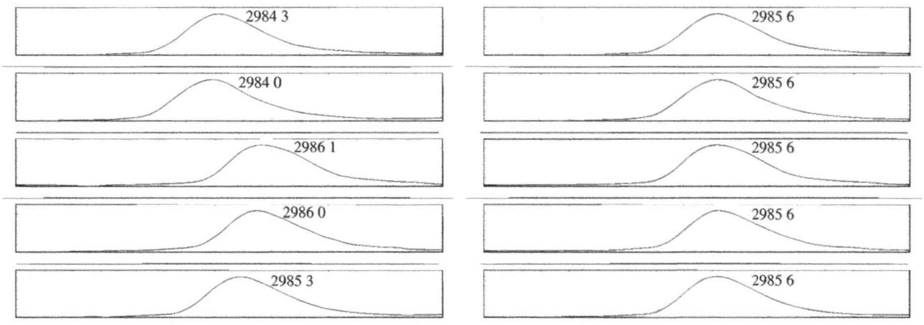

Figure 15.2 A zoomed view of a set of MALDI-TOF spectra before (left) and after (right) alignment in SpecAlign

should occur at the same m/z value in the produced spectra. In practice, imperfect precision usually results in slightly different m/z values for the same peptide, and this problem is discussed next. In the following subsections we then describe three different quantification methods.

Using label-free MS methods to quantify peptide samples will not give any indication of the identity of the components under analysis, but has a greater potential for discovering low-abundance molecules compared to MS/MS-based methods, because the instrument does not need to spend time in MS/MS mode. If interesting candidates are discovered, these can be selected for subsequent identification by MS/MS if suitable spectrometers are used.

15.6.1 *Comparing spectra*

The main reason for the observation that the same peptide may occur at different m/z values in different spectra lies in the lack of sufficient instrument precision. This is particularly noticeable in the higher mass regions. In order to correlate peptides over a set of (sub)samples, the spectra therefore often need to be aligned first. See Figure 15.2 for an illustration of the effect of an alignment of a set of raw MALDI-TOF spectra using the SpecAlign software.

When spectra are aligned and normalized, differences between various samples can be discovered. Traditionally, peak lists have been a common way to represent a spectrum, with the advantage that only the relevant parts of the spectra are retained for subsequent analysis. The disadvantage is that the algorithms that generate such peak lists sometimes fail to detect certain peaks. This can lead to situations where a given peak can only be found in some of the analyzed spectra, which can potentially be erroneously interpreted as signaling the absence of that peptide or protein in the samples where it is missing (a false negative error). An alternative to peak lists is to use the raw data spectrum. In that way no information is lost due to flawed peak detection, but at the same time large amounts of noise and background data remain present, which may confound analysis. In some cases this may lead to a complicated data analysis, simply because the number of data points or variables will be high (typically up to 50 000 points for one MALDI-TOF spectrum). The binning procedure provides a compromise between peak detection and retaining all points, by reducing the complexity by collapsing adjacent mass values into bins (intervals) of for example 1 m/z unit. This has the advantage of avoiding peak-detection errors while still reducing the number of data points. Where the peak-detection algorithm can result in completely discarding potential peptide or protein peaks, a binning procedure will only gradually degrade the information content, depending on the width chosen for the bins. If bins are used, the comparison between different samples is also made easier, because one can directly compare the bins, and avoid the need for matching peaks.

15.6.2 MALDI-TOF-based methods

MALDI-based methods do not include on-line separation during the MS analysis, but an off-line separation into subsamples can be performed beforehand if the samples are very complex. Since each subsample is spotted on one spot, only one spectrum is produced for each subsample. Therefore the analysis becomes simple, only comparing spectra.

15.6.3 SELDI-TOF-based methods

In SELDI-TOF (Surface-Enhanced Laser Desorption Ionization-TOF) analysis, separation is obtained by using different chromatographic surfaces on the different spots on the ionization plate. Such a plate is often called a SELDI chip. Surfaces with different properties and affinity bind to different classes of proteins and peptides. In this way different classes of molecules will be analyzed on each spot, with a single spectrum produced from each spot. Figure 15.3 illustrates a SELDI chip with five different surfaces, and three of the resulting spectra. In a multi-sample setup, each sample will (in this case) generate five different spectra. In this way the spectrum from a subsample can be identified by the spot-surface identifier, and corresponding spectra from different (sub)samples can be compared.

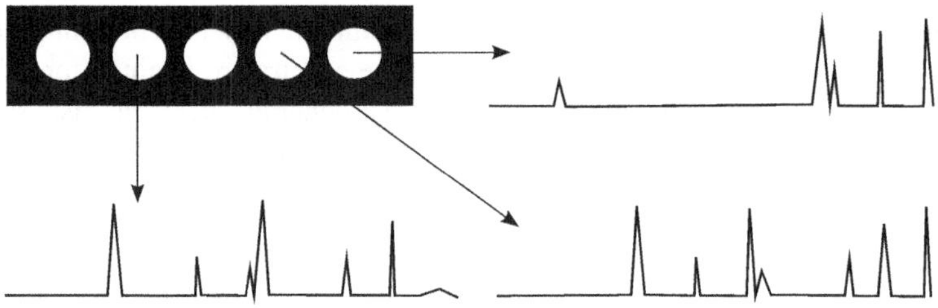

Figure 15.3 Illustration of a SELDI chip and some of the resulting spectra. The sample is applied to all spots on the chip, but after washing, different peptides are stuck on the different surfaces. Each spot therefore corresponds to a subsample

15.6.4 LC-MS quantification

Chromatographic separation systems (gas chromatography, liquid chromatography) coupled to mass spectrometers have been used to directly quantify components in samples such as urine or air pollution samples for a long time. This way of quantification was not applied to quantitative LC-MS analysis of peptides and proteins, however, mainly due to the poor reproducibility obtained for these complex samples. Because of the availability of techniques that have increased reproducibility in recent years, this approach has now become more commonly used in proteomics quantification as well.

The way these methods work is simply to load a peptide sample onto the LC column coupled to an MS instrument. For simplicity, assume that every sample (a patient, a given experimental condition, a time point in a time series, etc.) is run as only one LC-MS experiment, thus no pre-analysis separation is performed. The mass of a peptide (peak) is usually found during several consecutive MS scans, depending on how much time it takes the analyte (peptides) to elute from the column (corresponding to the width of its chromatographic peak). See Figure 15.4 for an example. Some of the peaks in scan 1 are clearly present in all three scans, whereas analyte B is practically gone in the last scan. This example shows that in order to quantify peaks from an LC-MS experiment, a total intensity for each analyte should be obtained by combining the signal contributions for that analyte over several scans.

Numerous spectra are produced over the course of an LC-MS run. Ideally, it should be easy to identify corresponding spectra from different (sub)samples based on their retention time. However, the exact retention time of an analyte (and thus the occurrence of its chromatographic peak) may shift from run to run. An analyte can also occur in

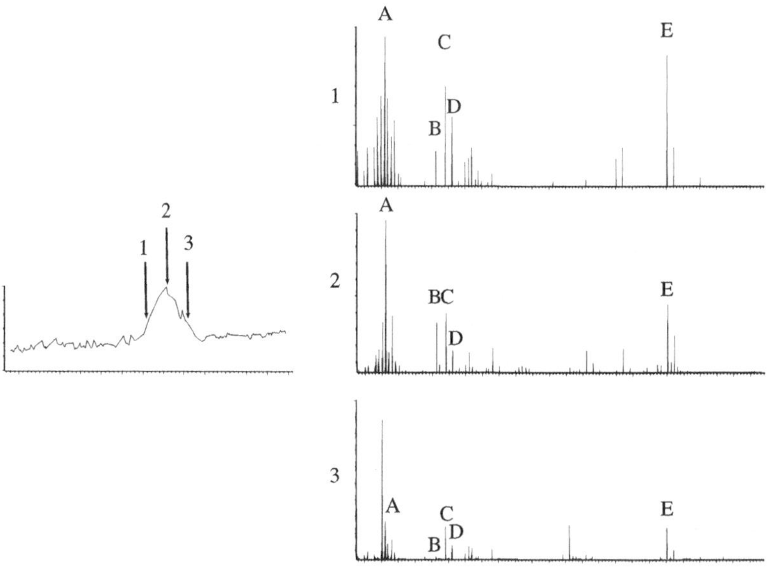

Figure 15.4 Example showing the elution profile of a chromatographic peak (left), and three consecutive MS scans taken at the highlighted time points (1, 2, and 3, right)

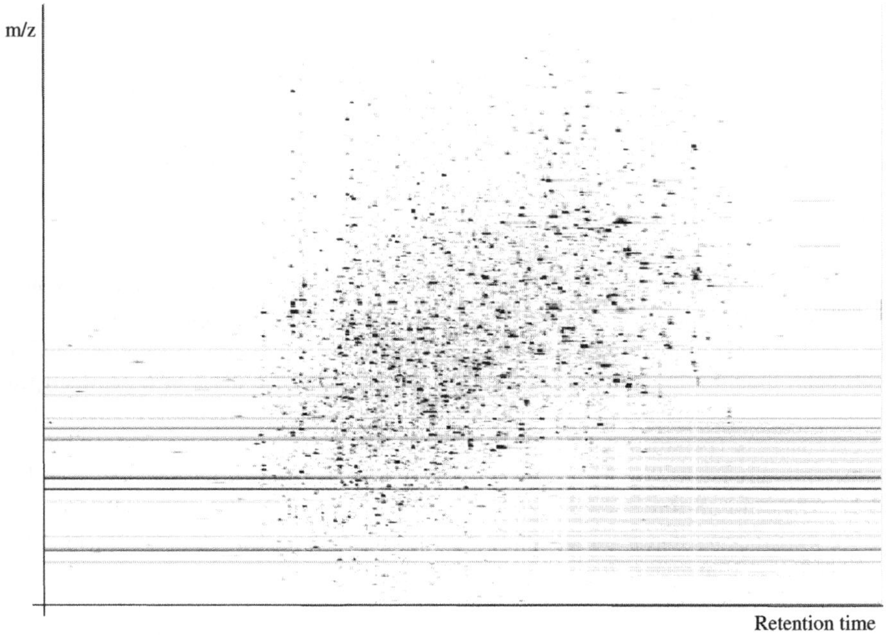

Figure 15.5 Example of an LC-MS run converted to a two-dimensional image, where each spot represents a measured ion. Horizontal lines correspond to background noise caused by non-biopolymers that consistently elute from the LC column over the entire run

several spectra from the same (sub)sample as outlined above. The relative simplicity of comparing only one spectrum from each sample is therefore lost in this approach.

To correct for the variation in retention time from run to run, linear or nonlinear transformations of the retention times must be performed, to ensure comparable times from sample to sample. Such a computational alignment often takes into account user-supplied input about the location of some common peaks across the different samples, which are used as 'landmarks'.

A useful way to visualize a quantitative LC-MS experiment is to represent it as an image, as seen in Figure 15.5. Most of the peptides appear as spots, and the representation resembles that of traditional 2D gels. Notice that some of the peptides appear 'smeared out' over the retention time, an effect due to the fact that some peptides elute over a broad retention time range on the LC column. Other molecules are 'background contaminants', and result in bands that cover the entire LC run. It is possible to use such images directly for analysis, by comparing them with methods similar to those used for 2D gel images.

15.7 Label-based quantification

Together with 2D gel methods (with or without DIGE labeling) and SELDI-TOF, quantification by MS labeling is by far the most frequently applied technique. Roughly, the concept is to make originally identical peptides from a limited number of different samples distinguishable by their masses. The exact way this is accomplished is described in the following subsections for each technique. Quantification methods based on signals from both MS and MS/MS analyses are discussed. Common to the methods described is that they simultaneously analyze more than one sample in each experiment, removing the difficulties associated with between-run variability, as outlined above.

15.7.1 MS-based labeled quantification

The peptide molecules are labeled differently for each sample, with each label contributing a different mass delta (see the next example for a detailed explanation). The samples are then mixed and analyzed as a single sample. The same peptide from different samples will turn up as different m/z peaks in a mass spectrum due to their different labels, and their intensities can be compared. In order to do this, the peaks of corresponding peptides in the spectrum must be found. If the masses of the labels are known, this is a straightforward exercise.

Example The SILAC (Stable Isotope Labeling by Amino acids in Cell culture) method for comparing two samples relies on feeding cells either normal ('light' label), or heavy (essential) amino acids ('heavy' label) in culture. Heavy arginine and lysine (or only one of these) are typically used for this. One way to make these amino acids heavy is to substitute the normal ^{12}C atoms with ^{13}C atoms. Both these amino acids have six carbon atoms, thus the heavy forms will be 6 Da heavier than the normal, light forms. A tryptic peptide normally contains one of these amino acids, so a typical MS spectrum of a peptide from such a light/heavy two-sample mixture will look as depicted in Figure 15.6. This doubly charged peptide is identified as `SYELPDGQVITIGNER`. In this example, each peptide consists of five visible peaks that compose the isotopic envelope of the peptide. The first peak is the one originating from those peptide molecules that are uniquely composed of light isotopes, while the second peak contains peptides with one ^{13}C (or deuterium, or ^{15}N, etc.) atom, and the third contains peptides with (for instance) two ^{13}C atoms etc. Typically, at least three naturally occurring isotopes can be observed if the

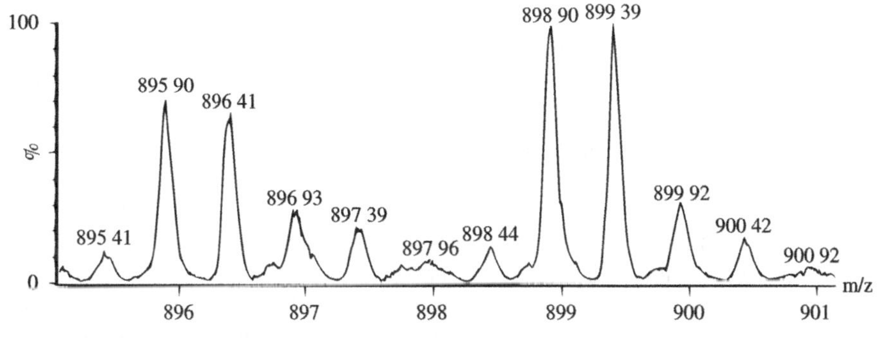

Figure 15.6 MS spectrum of the doubly charged peptide `SYELPDGQVITIGNER` (from human actin). The monoisotopic m/z value of the normal peptide is 895.9, whereas the peptide with the heavy (^{13}C-containing) arginine has an m/z value of 898.9

mass spectrometer has sufficient resolution. From the isotope peaks in the figure, one can easily determine that the peptides are doubly charged. The difference between the heavy and light versions of a tryptic, doubly charged peptide (assuming it contains only one residue of arginine/lysine) is then expected to be 3 units, as can indeed be seen in the spectrum (898.90 for the heavy peptide versus 895.90 for the original light version). △

When representing this couple as a quantitative entity, it is often expressed as a ratio, for example *heavy/light*, or as a percentage $heavy/(heavy + light) * 100\ \%$. In Figure 15.6 the light version is less abundant than the heavy one, thus the ratio as calculated above will be greater than 1, with the percentage greater than 50 %.

When using such labeling techniques for tryptic peptides the mass difference between the light and heavy peptides is normally 6 Da, and the corresponding peptide peaks can easily be found if their charge is known. However, uncertainties in the measurement and about the charge state usually necessitate a successful identification of MS/MS spectra in order to correctly couple the light and heavy versions of a peptide. The introduction of MS/MS analyses means that the sample coverage will be limited compared to methods that only require LC-MS to function (less peptide ions will be recorded due to the time lost during MS/MS analysis of a single precursor). The obvious advantage of the MS/MS-based methods, on the other hand, is that the identity of a quantified peptide is known instantly.

In addition to SILAC (which only works for samples that can be fed a specific diet of (essential) heavy amino acids), some of the more generally applicable post-isolation labeling strategies include ICAT (Isotope-Coded Affinity Tags), $^{16}O/^{18}O$ labeling, and ICPL (Isotope-Coded Protein Label).

15.7.2 MS/MS-based quantification

In contrast to the MS-based methods described above, these methods use special labeling techniques that only allow sample differentiation for a specific peptide during MS/MS mode. In MS mode, a peptide ion from different samples will appear with exactly the same (modified) mass. This is accomplished by incorporating *isobaric* labels (having equal mass) that result in charged fragments with different masses (so-called *marker ions* or *reporter ions*) upon fragmentation. iTRAQ (isobaric Tags for Relative and Absolute Quantification) is the most commonly used MS/MS-based labeling strategy, typically consisting of up to four different labels, marking up to four different samples.

As already outlined above, the label consists of a marker ion and a neutral balancing part. The masses of the four marker ions are different and they detach under MS/MS conditions, allowing sample differentiation upon fragmentation. In order to achieve the same overall mass delta for the intact peptides from different samples, the balancing parts consequently need to have masses as well. Simply put, for an introduced peptide mass delta of x, with four marker ions a_1, a_2, a_3, and a_4, and with four different masses m_{a1}, m_{a2}, m_{a3}, and m_{a4}, four balancer parts b_1, b_2, b_3, and b_4 are required that satisfy $m_{bi} = x - m_{ai}$, where m_{bi} is the mass of balancer part i.

When a peptide (peak) is selected for MS/MS processing, it contains occurrences from all the four samples. In MS/MS mode the labels then fragment, yielding marker ions that occur in the fragmentation spectrum. Since the marker ions will occur at their different m/z values, the relative abundances of that peptide in the four different samples can be deduced from the relative intensities of the marker ions.

Figure 15.7 shows an example of an MS/MS spectrum from four different samples. The masses of the four marker ions (m_{ai}) for iTRAQ are 114, 115, 116, and 117 Da, respectively. In this example, the sample labeled with iTRAQ marker ion 115 has the highest expression of the fragmented peptide among the four samples.

Quantification by isobaric tags that are distinguishable in MS/MS often enable a higher number of samples to be analyzed simultaneously, but the fact that MS/MS spectra are the basis of quantitation reduces the accuracy of the results. The reason for this is that

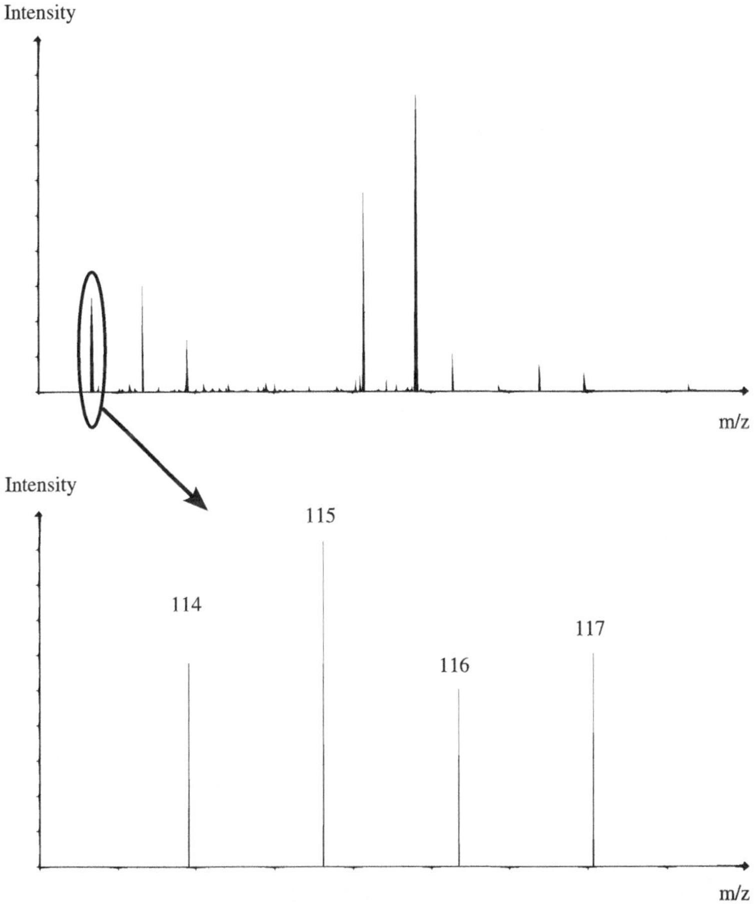

Figure 15.7 Illustration of MS/MS-based quantification using iTRAQ. The MS/MS spectrum of the peptide KVPQVSTPTLVEVSR is shown (top spectrum), with the m/z interval containing the four highlighted iTRAQ marker ions zoomed in (bottom spectrum)

the acquisition of MS/MS spectra often results in a limited amount of fragments, thus reducing the actual ion count. The causes for this effect are described in Chapter 8. Some MS/MS spectra will contain high-intensity fragments, while others will only have low-intensity fragment ions. This strongly influences the accuracy of the measured ratios, because strong signals give more reliable ratios, as outlined below.

15.8 Variance-stabilizing transformations

Common to the above-mentioned methods is that the measurement variance is often dependent on signal intensity. This means that if the variance is calculated between identical samples for all variables (mass points), it will not be equal for masses with high and low intensity. In such cases there will be a correlation between average signal intensity and variance, resulting in too much emphasis put on intense signals. It also hampers most statistical methods that assume constant variance for all variables. This is often referred to as heteroscedastic noise.

Various methods have been explored to remedy this effect. Common transformations include taking the logarithm of the intensity or relying on *n*th roots. The method of choice will depend on the actual level of the noise; if there is a small increase of variance with increasing intensities, taking the square root may be sufficient.

15.9 Dynamic range

All of the above methods rely on a correlation between the signal intensity and the actual number of molecules in the original sample. Ideally, this correlation would always be linear (two times the signal means two times the number of molecules), and constant across many orders of magnitude (one-tenth the signal still means one-tenth the number of molecules). In reality, however, the most commonly used detectors in mass spectrometers can only resolve a limited dynamic range (typically four orders of magnitude), with an even smaller section of that resolvable range following a linear correlation. The reason for this is that the response to very intense signals levels off as the detector reaches its maximum output capacity; stronger input signals will either result in a smaller relative detector output (when approaching the maximum output) or remain at the maximum detector output level (if the maximum output capacity has been reached). For weak signals, on the other hand, detector output is only observed when a minimal amount of ions is present. When plotting ion intensities against a range of fold changes for known calibrants, the curve will follow a sigmoidal path, leveling off at both ends. It is therefore important to experimentally determine the dynamic range (and the linear part of this range) for an instrument prior to using it for quantification purposes.

15.10 Inferring relative quantity from peptide identification scores

An indirect method for deducing the relative quantity of a given peptide or protein within a sample involves using peptide identification scores. For a hint about a given protein's abundance in a sample, the sum of constituent peptide scores has been shown to yield useful results (Colinge *et al.* (2005)). With this method, 2.5–5-fold changes are reported to be detected with 90–95 % confidence, using an *abundance indicator*. The abundance indicator of a protein p is calculated using the following formula:

$$E(p) = \sum_{\rho \in T(p)} \sum_{\sigma \in I(\rho)} s(\sigma, \rho) \tag{15.1}$$

$T(p)$ is the set of tryptic peptides in p, ρ is a tryptic peptide, $I(\rho)$ is the set of experimental spectra identifying ρ, and $s(\sigma, \rho)$ is a normalized score of the match between σ and ρ (Colinge *et al.* (2003c)).

15.11 Absolute quantification methods

Some methods have also been developed to supply absolute quantities of analytes, usually expressed as the molar concentration (M), rather than the relative quantification discussed so far. One example where the absolute concentration of a given peptide can be of interest is when looking for biomarkers, where the absolute concentration of a peptide biomarker will provide useful information about the suitability of different assays to detect this peptide in a subsequent diagnostic procedure.

The simplest methods for the determination of absolute concentration levels rely on utilizing a feature of the relationship between the number of molecules in the sample and a mass spectrometer's signal. For the three most intense tryptic peptides, the signal per mole of protein was shown (Silva *et al.* (2006)) to be constant within a coefficient of variation of $\pm 10\,\%$. When this is known, the protein concentrations in a sample can be calculated by establishing the relationship between the signal and the known concentration of an internal standard. To do this, a number of samples spiked with increasing concentrations of known test proteins were analyzed. After the concentrations of these proteins were plotted against the raw signal generated for these proteins, a regression function was obtained. This calibration function was then used to quantify compounds of unknown concentration. In this approach, the signal for a protein was defined as the average signal of the three most intense peptides in the protein.

A different approach to absolute quantification involves the synthesis of an isotopically labeled partner for each analyzed peptide. In practice, this is both expensive and time consuming. As an alternative to chemical synthesis, one tryptic peptide from each monitored protein is chosen to represent this protein. Then, DNA is synthesized that represents the polypeptide chain obtained after concatenating each of these peptides. When this nucleotide sequence is added into an expression vector and thereafter transfected into a bacterium, the resulting protein (denoted QCAT protein) will consist of the necessary tryptic peptides. By growing these bacteria on a medium that contains one or more heavy amino acids, the peptides can be labeled. Each of these peptides can then serve as a basis for finding the absolute quantities of the corresponding proteins, since they can be spiked in known amounts into the sample.

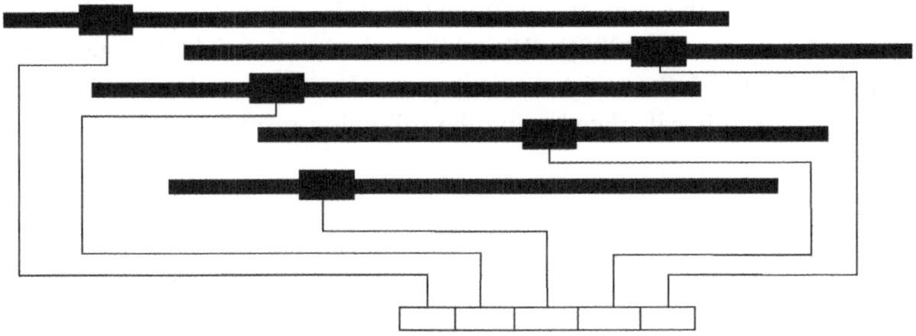

Figure 15.8 Schematic overview of the generation of a synthetic protein designed to monitor five different proteins. One peptide from each protein is selected, and concatenated into the QCAT expression module

Figure 15.8 illustrates the process of generating the QCAT protein. The output for each peptide in this analysis is a ratio between the sample peptide intensity and the QCAT peptide intensity. Since the absolute quantity of the QCAT protein is known, the absolute intensity of the sample peptide can now be inferred.

Bibliographic notes

Different forms of alignments

Fischer *et al.* (2006); Nordstrom *et al.* (2006); Wong *et al.* (2005)

Quantification by identification scores

Colinge *et al.* (2005, 2003c)

Absolute quantification

Beynon *et al.* (2005); Silva *et al.* (2006)

Software

VEMS	http://personal.cicbiogune.es/rmatthiesen/
MSQuant	http://msquant.sourceforge.net/
SpecAlign	http://physchem.ox.ac.uk/ jwong/specalign/
I-Tracker	http://www.dasi.org.uk/download/itracker.htm

16 Peptides to proteins

As has been highlighted previously in this book, the fundamental unit of proteomics is the protein. It has also become clear, however, that information about protein identity is usually based on indirect information as derived from the sequence of peptides, obtained after proteolytic digest of one or more proteins. It is therefore important to investigate the issues that may arise when one attempts to retrace the origin of the identified peptide sequences to their precursor proteins.

16.1 Peptides and proteins

As we have already seen, peptides are the short stretches of amino acids that are obtained after the proteolytic cleavage of proteins. Peptides are usually around 10–15 amino acids long, and a single protein yields approximately 35 peptides on average. While it would be convenient to uniquely identify proteins directly based on their physicochemical properties, there are several reasons why this is almost impossible in practice, as discussed in Chapter 1. Since peptides are more readily identifiable, processing of the proteins into peptides at some stage during the sample preparation is used to overcome this problem.

As explained in the previous chapters, the different techniques used for peptide-centric protein identification fall into two broad categories: the first approach relies only on the masses of the peptides obtained after proteolytic cleavage, while the second approach attempts to retrieve actual sequence information from one or more of the peptides. It is clear that the first approach, relying solely on the masses of the peptides, has less overall distinguishing power than the second approach, which relies on mass as well as (partial) sequence information. Finally, both approaches are also influenced by the amount of additional information that may be collected during sample preparation and separation and both methods may also be confounded by biological variation.

16.2 Protein identification using peptide masses: an example revisited

In order to get a more realistic picture of the problems encountered in matching peptide masses to unique proteins, let us revisit a revised version of the example given in Section 1.7.1. This example originally assumed that there are five proteins in a sample

Computational Methods for Mass Spectrometry Proteomics I. Eidhammer, K. Flikka, L. Martens and S.-O. Mikalsen
© 2007 John Wiley & Sons, Ltd

(A, B, C, D, E), and that they are separated by their isoelectric point, resulting in the separated groups $\{A, D\}$, $\{B\}$, $\{C, E\}$. The following ideal digestion table was produced for the five proteins:

Protein	Peptides
A	A_1, A_2, A_3, A_4
B	B_1, B_2, B_3
C	C_1, C_2, C_3, C_4, C_5
D	D_1, D_2, D_3, D_4
E	E_1, E_2, E_3

There were substantial simplifications involved, however: (i) all peptide masses were considered unique, (ii) no (partial) missed cleavages were considered, (iii) only five proteins were considered to be in the sample, rather than the thousands readily encountered in real samples, and (iv) each protein resulted in only 3–5 peptides rather than the average of 35 peptides. However, even a slight revision of the example will adequately serve to explain the complexity typically encountered in real samples. Indeed, since in practice the cleavage of the protein may not occur at every theoretical cleavage site, we can assume that digestion of the first group resulted in the peptides A_1, $A_{2,3}$, A_4, $D_{1,2}$, D_3, D_4, which means that there is one missed cleavage in each of the proteins. The second revision is to allow indistinguishable masses for some of these peptide, along with a failure to detect certain other peptides. Suppose, therefore, that:

- A_4 and $D_{1,2}$ have a similar mass, m_1.
- $A_{2,3}$ has a similar mass to C_2, m_2.
- A_1 and D_4 are not recognized in the experiment, hence these masses are not available.

This means that three masses are ultimately observed in the experiment: m_1 (for peptides A_4 and $D_{1,2}$), m_2 (for $A_{2,3}$), and m_3 (for D_3). Assuming that proteins A, C, and D are all present in the database, a search with one allowed missed cleavage will yield the following hit table:

Protein	Peptide mass
A	m_1, m_2
C	m_2
D	m_1, m_3

We can exclude protein C from the identification list since we have found that the calculated pI of protein C does not correspond to the experimentally obtained pI for this fraction, which does correspond to the calculated values for both A and D. This illustrates the usefulness of additional information obtained from the sample processing or separation in resolving identification ambiguity. We are still confronted with our inability to distinguish between proteins *A* and *D*, however. Since there is unambiguous evidence for the presence of protein *D* through the detection of m_3 (which uniquely matches peptide D_3), it is reasonable to consider *D* as identified. Protein *A* can only be considered identified because we eliminated the possibility that m_2 was derived from peptide C_2. Indeed, the occurrence of mass m_1 by itself would not have allowed the identification of protein *A* as we already positively identified protein *D* through mass m_3, and mass m_1 can therefore be explained without invoking the presence of the additional protein *A*.

We can, however, remove yet another idealization from our example by supposing that 50 % of all protein *C* is modified by phosphorylation *in vivo*, and that this modification occurs on a serine residue in peptide C_4. Let us further postulate that the pI of the modified protein C' is very similar to the pI of proteins *A* and *D*. It is clear that we are now no longer certain whether protein *C* is present in the fraction or not. This problem extends to the identification of protein *A* that is now suspect as well. Indeed, only one of the two proteins *A* and *C* is sufficient to account for mass m_2, with mass m_1 already explained by protein *D*. We are now confronted with a situation in which the following identification sets are indistinguishable based on the peptide mass measurements: $\{A, D\}$, $\{C, D\}$, and $\{A, C, D\}$. Although it is possible to argue in favor of the $\{A, D\}$ set as the most reasonable answer (by invoking the principle of parsimony, also known as Occam's razor, over the extra modification required for sets $\{C, D\}$ and $\{A, C, D\}$, for instance), the conclusion is necessarily subjective and unverifiable based on the available evidence. Further experiments are clearly required to come to definitive conclusions.

16.2.1 Extension to MS/MS-derived peptide sequences instead of masses

The example developed above can easily be adapted to peptide sequences instead of masses. Indeed, we can compose a hit table between proteins and identified peptide sequences, assuming that s_x represents a unique peptide sequence, that sequence s_1 is present in both protein *A* and *D*, and that s_2 is shared between protein *A* and *C*:

Protein	Peptide sequence
A	s_1, s_2
C	s_2
D	s_1, s_3

We are again faced with the same situation as for the peptide masses, yielding exactly the same problems. It can be noted in this context that many proteins belong to large protein families, in which the members share a substantial sequence similarity. It is therefore not a rare occurrence to see one peptide match to multiple proteins. Since the same problems with regard to unambiguous protein identification occur in both peptide mass and peptide-sequence-based inference, we will treat both cases together in the subsequent sections.

16.3 Minimal and maximal explanatory sets

As shown in the previous example, the identification of proteins can sometimes lead to several potential sets of identified proteins. When these datasets cannot be distinguished using the available evidence, one can wield both Occam's razor as well as an anti-razor in an attempt to resolve the issue. The former dictates that one should not invoke novel entities for phenomena that can readily be explained without them. In our previous example, a first iteration based on this principle will reduce the number of possible sets to two: $\{A, D\}$ and $\{C, D\}$. The second iteration will then settle on set $\{A, D\}$ since set $\{C, D\}$ necessarily invokes the presence of a modification for which no other evidence is found. An anti-razor would instead choose set $\{A, C, D\}$ as the correct answer, since these three proteins could potentially all be there, and any other choice would throw away potential identifications without proper empirical justification.

These two views are reflected in what is commonly referred to as *minimal* and *maximal* explanatory sets. The actual protein composition of the sample can then be any set that is equal to or broader than the minimal set, and smaller than or equal to the maximal set. Usually, a distinction between equally valid minimal sets (such as between $\{A, D\}$ and $\{C, D\}$ in our example) is not made, resulting in indistinguishable subsets of proteins in minimal datasets.

16.3.1 Minimal and maximal sets in peptide-centric proteomics

In Section 1.7.2, we discussed the general outline of peptide-centric proteomics, in which proteolytic digest of the proteins in the sample is performed prior to separation. As such, no protein properties will have been recorded in these analyses. This can increase the difficulty involved in retracing identified peptides to their parent proteins, yet it is at the same time free of the risk of false negative associations due to differences between calculated and observed protein properties. We have seen that such discrepancies can occur when unexpected posttranslation modifications take place on (at least a subset of) the protein. This is in essence the benefit of the undirected peptide-centric proteomics experiments, and it is therefore particularly important to calculate and report both minimal and maximal explanatory sets for these types of experiment.

The problems encountered with regard to ambiguous protein identification in peptide-centric proteomics essentially led to the investigation of the problem as well as to the development of several tools to help construct meaningful protein identification lists from the amassed peptide identifications. This has been well summarized in Nesvizhskii and Aebersold (2005). It is important to note that, although mostly studied in peptide-centric proteomics, these problems also exist in more traditional gel-based proteomics and similar protein-based proteomic methods.

16.3.2 Determining maximal explanatory sets

A maximal explanatory set is necessarily dependent on the database used, as well as on the optional search parameters such as allowed missed cleavages and potential modifications. Yet once these have been decided upon, the maximal set can easily be determined by matching each recorded peptide mass or identified sequence against all theoretically derived peptides. For each matching peptide, the precursor protein is added to the set (note that we consider a 'set' to automatically exclude redundant additions). The resulting set composes the maximal explanatory set of proteins for the database used and the search settings employed.

It should be noted that the maximal explanatory set does not necessarily encompass the complete protein composition of the original mixture. Indeed, a substantial number of proteins may have eluded detection altogether. The old adage therefore holds true here: that *absence of evidence is not evidence of absence*.

16.3.3 Determining minimal explanatory sets

Contrary to the maximal explanatory set, construction of a minimal explanatory set of proteins for a given list of peptide masses or peptide sequences is a non-trivial task. Usually a probability is calculated for each matching protein to be present in the sample. These probabilities are typically derived from the confidence one has in the underlying peptide identifications and as such are mostly designed to work with the results of MS/MS identification algorithms, incorporating for instance the scores attributed to the identifications. It is clear that this strategy will render the construction of a minimal set dependent on external factors such as the underlying search algorithm and any search and/or validation parameters used. Examples of such algorithms are given in the bibliographic notes.

These different algorithms all attempt to solve what is essentially an *ill-posed problem*; there is no way to accurately reconstruct the original protein composition of the sample based on the peptide properties measured. The output delivered by the algorithms therefore necessarily has its limitations and it should be stressed again that follow-up experiments are often the best way to proceed when validation of protein presence (or absence) is required.

Bibliographic notes

Protein inference	Nesvizhskii and Aebersold (2005)
Search algorithms	
Protein Prophet	Nesvizhskii *et al.* (2003)
Experimental Peptide	Identification Repository (EPIR)
	Kristensen *et al.* (2004)
Isoform Resolver	Resing *et al.* (2004)
DBParser	Yang *et al.* (2004)

17 Top-down proteomics

The main reason for digesting proteins into peptides, as performed in traditional bottom-up proteomics, is that peptides are much more suitable to analysis by mass spectrometry. Also, since many (independent) peptides are created for each protein, a certain redundancy is introduced making protein identification more reliable. However, as explained in Section 6.1.3, not all peptides are ionized and detected in a mass spectrometer, resulting in a lack of coverage for (sometimes a large part of) the sequences. This makes full characterization (primarily PTM detection) difficult and haphazard.

An alternative approach is to perform mass spectrometric analysis on intact proteins, without any digestion into peptides. Such an approach has the potential to yield full protein characterization. On the other hand, the requirements for the resolution and accuracy of the instruments are much higher. Since such instruments are now available, this approach (called top-down proteomics) is gaining in popularity. It must be noted that top-down proteomics is not new; analysis by 2D gel only is for example a form of top-down proteomics. The novelty lies in the considerable gain in efficiency by using mass spectrometry.

Figure 17.1 illustrates the main differences and similarities between bottom-up and top-down proteomics using MS/MS.

17.1 Separation of intact proteins

2D gel electrophoresis is the most popular technique for the separation of proteins. In Meng *et al.* (2002), however, an alternative 2D separation for top-down proteomics is described. The protein sample is first separated by size using ALS-PAGE, a 1D gel system in which the familiar SDS denaturing agent is replaced by ALS (Acid-Labile Surfactant). Each resulting fraction is then further separated by reverse phase LC. The number of proteins in each fraction in the top-down experiments typically performed today varies from one up to six or eight.

17.2 Ionization of intact proteins

Both MALDI and ESI are used as ionization sources, but ESI is more popular. If ESI is used as a source, the proteins may get a large number of charges: up to 30 charges for a

Computational Methods for Mass Spectrometry Proteomics I. Eidhammer, K. Flikka, L. Martens and S.-O. Mikalsen
© 2007 John Wiley & Sons, Ltd

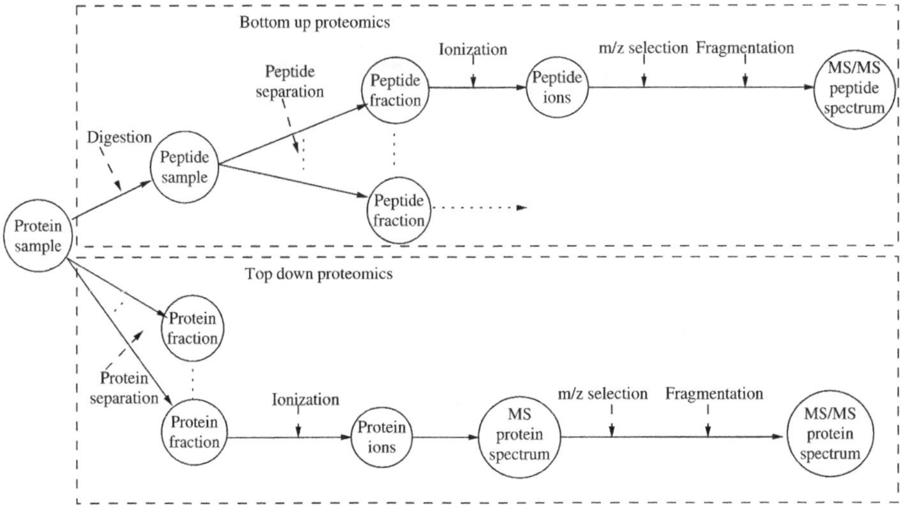

Figure 17.1 Illustration of the bottom-up and top-down paradigms for proteomics

20 kDa protein is not unusual. Since some ions of the same protein will end up carrying more charges than others, a protein typically shows up in an MS spectrum as a collection of several peaks, each corresponding to a different charge state.

Liu *et al.* (2007) describe a simple method to estimate the possible charge state distribution of proteins ionized by ESI. This method was based on constraints derived from empirical observations of ESI of proteins. The highest possible charge number was estimated by dividing the protein mass value by 700 and rounding down to the next integer value, and the lowest charge number was estimated by dividing the protein mass value by 2000 and rounding down to the next integer value. Furthermore, it is important to realize that the average protein charge state is influenced by the electrospray conditions, as described by Sze *et al.* (2002).

This parallel occurrence of different charge states does not need to confound analysis, as a single peak can be selected for MS/MS analysis. It is important, however, that the charge of the protein corresponding to the selected peak is determined.

17.3 Resolution and accuracy requirements for charge state determination and mass calculation

As explained earlier, the most common method for determining the charge of an ion relies on the analysis of the isotopic distribution of the peak in the mass spectrum. Thus the spectrometer must be able to resolve the individual isotopes, and for a large number of charges this requires a high resolution. The minimal required resolution R_m can be calculated for an ion with an m/z of m_z and a charge of z as $R_m = m_z/\frac{1}{z} = m_z \cdot z$.

Once the charge is determined, the original mass can be calculated from the measured m/z. The absolute error on this calculated mass will be the absolute error on the m/z multiplied by the charge, however. It is therefore important that the m/z measurement is sufficiently accurate. We can calculate the allowed error x (in ppm) for an observed m/z value m_z with charge z to achieve a maximal predefined error of ϵ (in Da) on the calculated mass as $x \leq \epsilon/(m_z \cdot z) \cdot 10^6$.

Example Suppose we have a protein ion of $m/z = 1600$ and $z = 25$. Then the distance between the isotopic peaks is 0.04. For the peaks to be resolved at FWHM, the width of the peaks Δm must satisfy $\Delta m < 0.04$. This means that the resolution must satisfy $m_z/\Delta m > 1600/0.04 = 40\,000$.

The minimum required accuracy in ppm to obtain the original mass of this protein to within ± 0.5 Da is $0.5/(1600 \cdot 25) \cdot 10^6$, which works out to 12.5 ppm. △

17.4 Fragmentation of intact proteins

The only instruments that routinely achieve the specific requirements for resolving power and accuracy are FT-ICR analyzers. Recently, Macek *et al.* (2006) have also demonstrated the suitability of the new hybrid linear ion trap–Orbitrap mass spectrometer for top-down proteomics. An additional requirement is an efficient fragmentation of large ions, with many methods developed specifically for FT-ICR instruments. The most common fragmentation mechanism is ECD (Electron Capture Dissociation; used in FT-ICR instruments), but other methods such as electron transfer dissociation (ETD; used in ion trap instruments), infrared multiphoton dissociation (IRMPD), and sustained off-resonance irradiation are also in use. In the top-down experiment described by Macek *et al.* (2006), CID is used for fragmentation. They show that CID for large proteins mostly fragments at the terminal areas. ECD, on the other hand, can fragment nearly anywhere along the backbone, but has low overall efficiency (resulting in a lower sensitivity).

The goal of achieving a total sequence coverage of 100 % can be attained by this fragmentation of intact proteins, as shown in Figure 17.2. The figure illustrates that complementary pairs (for example, b and y ions) can be found for intact proteins as well.

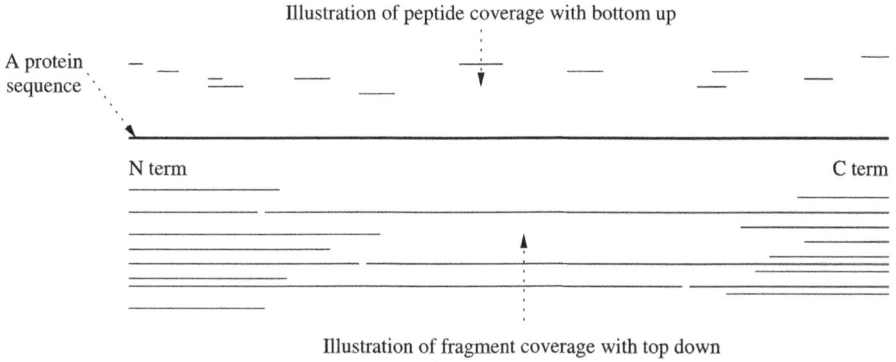

Figure 17.2 Illustration of the difference between typical sequence coverage for bottom-up and top-down approaches. In bottom-up experiments, coverage is achieved through small peptides, whereas in top-down studies the increased coverage results from much larger fragments

Analyzing efficiency can be further increased by combining CID and ECD, as explained in Section 12.6.1.

17.5 Charges of the fragments

As the precursor is typically highly charged, the product ions will also carry a large number of charges, which must be determined as well.

Because of the difficulties in determining the charge state of highly charged ions on lower resolution instruments, an alternative approach has been developed, which incorporates chemical steps to reduce the charge of the product ions to one. This method is applied for top-down proteomics in the relatively low-resolution quadrupole ion trap instruments, by using gas phase ion chemistry as described by Reid and McLuckey (2002).

17.6 Protein identification

Basically, the intact protein mass (measured with high accuracy) can be used to identify possible protein sequences in a database. A defined modification mass can be taken into account by allowing the experimental and theoretical protein masses to differ by the mass of that modification. If complementary fragment masses are found in the spectrum, these can be used to further verify a potential sequence. The test is then to simply find backbone fragmentation positions in the candidate protein sequence that result in the formation of the two observed complementary fragment ions. Note that the presence of fragment ions also allows more precise localization of a modification, if present. Non-complementary peaks can also be matched against predicted b or y ions of the proposed proteins to enhance confidence in the identification.

When one cannot rely on a match in protein mass (for example, due to many modifications occurring on the protein, if a splice variant of the database protein is measured, or if incomplete protein sequence databases for the organism under study are available) one can search for sequence tags in the spectrum instead. As CID fragmentation mostly occurs at the terminal parts of the sequence, sequence tags can often be found in these areas. Note that complete *de novo* sequencing cannot be expected from such studies, and that spectral comparison is not used due to the large number of theoretical fragment ions.

17.7 Protein characterization – detecting modifications

When a protein is identified, one can try to localize possible modifications. Consider an offset mass δm between the experimental and theoretical protein masses, corresponding to a possible modification (or an amino acid substitution) with mass difference δm. Upon protein fragmentation, a set of b and y ions, and possibly some internal product ions, will be observed. When the MS/MS spectrum is deconvoluted (to a singly charged spectrum), peaks corresponding to (either modified or unmodified) b and y ions can be found. Let b_i and y_j be the largest of the b and y ions that have the same mass as their theoretical ions. Furthermore, let b_k and y_l be the smallest of these ions with modified masses. With the length of the protein given as n, the modification is then localized between residue $\max(i, n-l-1)$ and residue $\min(k+1, n-j)$. If ions b_k or y_l are not observed, k or l is set to n respectively.

Example Suppose $n = 93$ and b ions of length $\{23, 45, 52, 63\}$ are identified as unmodified, but b_{78} is modified. Suppose further that the y ions of length $\{6, 11, 12\}$ are identified as not modified, and y_{28} as modified. Then the modified residue is localized between residue $\max(63, 93-28-1)$ and residue $\min(79, 93-12)$, which means between residues 64 and 79.

△

17.8 Problems with top-down approach

Although top-down proteomics studies have some benefits over bottom-up experiments, several reasons contribute to their relatively slow adoption. As we have seen, efficient top-down proteomics relies on specialized and expensive instrumentation to achieve the required resolution and accuracy. Indeed, until recently only the FT-ICR could routinely be used for such studies. The high price and corresponding maintenance costs, as well as the high demands on infrastructure that come with these instruments (they use liquid helium to supercool magnets that deliver immense magnetic fields), have greatly limited their adoption. Additionally, proteins need to be separated prior to introduction in the mass spectrometer, limiting the high-throughput potential of top-down proteomics. Bottom-up proteomics is also more suited to identify proteins, and is more readily amenable to high-throughput approaches. Finally, top-down proteomics requires more sophisticated data processing software, which typically has to be developed in-house. The availability of the much simpler and more affordable Orbitrap mass analyzer and the proof that it can be used for top-down proteomics might help boost the popularity of this approach, although it is likely that it will remain a more specialized tool which deals with the thorough characterization of one or a few proteins rather than the mapping of an entire proteome. The realization that many of the most interesting studies in top-down proteomics to date have been performed on highly customized top-of-the-line instruments further illustrates that some time will probably be required before entire proteomes can be run through affordable standard production instruments.

Exercises

17.1 Derive the formula for the maximal allowed error in ppm, given a maximal absolute error on the calculated precursor mass as at the end of Section 17.3.

17.2 Suppose that the following values are found for the variables in Section 17.7: $i = 5, k = 9, j = 9, l = 15$. Derive limits for where the modification can be.

Bibliographic notes

Kelleher *et al.* (1999); Meng *et al.* (2002); Reid and McLuckey (2002); Sze *et al.* (2002) Liu *et al.* (2007); Macek *et al.* (2006); Meng *et al.* (2004)

Tools for identification and characterization
ProSight PTM https://prosightptm.scs.uiuc.edu/

18 Standards

Standards are a very important part of our daily lives. Although we may not always be aware of their influence, commonplace things such as postcodes and the geometry of electric power sockets are dictated by standards. The standardized postcodes make sure your letters arrive quickly and the standardized power sockets will allow your newly bought appliance to connect to the electric power grid.

18.1 Standard creation

A standard represents a model item that is established as the consensus through one of three means: by authority, by general consent, or by custom. It is worthwhile to briefly discuss these three mechanisms of standard creation, along with an example from proteomics for each.

Standard creation by authority can be interpreted as a standard that is derived from a single source yet subsequently adopted by others through either enforcement (relatively infrequent in science) or general consent. An example of the latter case in proteomics is the nomenclature of the fragment ions obtained after fragmentation of a precursor peptide. The *a*, *b*, *c*, *x*, *y* and *z* ions (amongst others) have been named by Roepstorff and Fohlman (1984) and Biemann (1988) and these names are commonly referred to as 'Biemann nomenclature'. The component of general consent can be found in the inherent logic of the fragment ion naming. Indeed, since *a*, *b* and *c* ions are derived from the N-terminus and *x*, *y* and *z* ions are derived from the C-terminus, the proposed standard appealed to scientists familiar with the Roman alphabet, which eased its widespread adoption.

Standard creation by general consent relies on representative committees working out a standard. Usually the proposed standard is then communicated to a wider audience and can be corrected or revised according to the public comments received. After final approval, the standard is released and should therefore be adhered to. An example in proteomics is the mzData standard for MS-derived data.

Standard creation by custom occurs when the usage of a particular model has been so widely adopted that a *de facto* standard can be said to have developed. An example in proteomics (and the life sciences in general) is the usage of the FASTA format for sequence databases. Originally developed as the input format for the FASTP/FASTA

Computational Methods for Mass Spectrometry Proteomics I. Eidhammer, K. Flikka, L. Martens and S.-O. Mikalsen
© 2007 John Wiley & Sons, Ltd

program family (Lipman and Pearson (1985)), the simplicity of the format appealed to many programmers, eventually making it popular enough to be supported by default by almost all sequence-related software tools.

18.1.1 *Types of standards*

Standards can take a variety of forms, ranging from the mass defined by 1 kilogram or the length of 1 meter, to the way an mp3 file is formatted. A possible way to divide the many different types of standards follows:

- measurement standards, for example the SI standards for mass, length, and temperature;
- dimensional standards, for example the size of a battery or the diameter of the central hole in a compact disc;
- design standards, for example the maximum electric current that a particular wire can carry or the way to conduct a clinical trial;
- test standards, for example calibrants or a protocol for measuring the toxicity of a compound;
- terminology standards, for example the sign for toxic substances or the E numbers for food additives;
- formatting standards, for example MIME, XML, CSS.

18.2 Standards from a proteomics perspective

In this section we will be discussing some of the properties of standards that are related to proteomics. In particular, properties that are relevant for computational access to proteomics data will be covered in more detail. There are three broad areas of proteomics where standards come into play:

Standard operating procedures are design standards and belong squarely in the laboratory, as they detail the exact conditions for sample preparation, separation, and mass analysis.

Test samples are test standards and are located halfway between the laboratory and the computer.

Data standards are composed of both formatting and terminology standards and are the domain of computational approaches.

The latter two will be dealt with in subsequent subsections.

18.2.1 *Creation of test samples*

Test samples are often used in analytical chemistry, where they serve as a quality control for the performance of a laboratory. They can be thought of as 'blind calibrants' in that the experimentalist does not know the analyte, but the issuing body does. As such, the laboratory as a whole is 'calibrated' through test samples rather than the instrument. In practice, failure to reproduce the correct measurements from a test sample can result in severe consequences for the analyzing laboratory, including a loss of reputation. It is therefore important that any test sample is judiciously constructed, that it is stable over time, and that it does not introduce a bias by favoring one analytical approach over another.

Computational methods can be applied to achieve a standard test sample that corresponds to these requirements. Potential candidate proteins can be analyzed for hydrophobicity, pI range, and mass in order to construct a sample that either covers as much of the parameter space as possible, or is aimed at testing performance in a particular range of these parameters. The same type of analyses can be performed on the peptides that are obtained after proteolytic digest. Important factors here can be sequence diversity, peptide length, number of charged residues, and uniqueness of the peptide sequence in the overall peptide mixture.

The creation of suitable test samples for proteomics is currently undertaken by the Human Proteome Organization (HUPO).

18.2.2 Data standards in proteomics

Data standards concern both the structure and content of the information. The relevant types of standards are formatting and terminology standards, respectively. It should be noted that data standards are usually about data communication rather than data storage and that the field of proteomics is no exception. The objective of data standards in computational proteomics therefore is to allow different software tools to exchange information.

Terminology standards in proteomics We have already discussed a terminology example when explaining the Biemann nomenclature above, but there are more complex examples as well. For instance, the different types of mass spectrometers are all composed of a number of parts (an ion source, one or more m/z analyzers, and a detector) which can be named. A particular analyzer can be called 'TOF', 'TOF tube', 'Time-of-flight', etc., depending on the person reporting. It would be much easier if the field were to settle on a unique name for this part. Even better would be an abstraction of such a concept to a number or accession string, since we could then tie an unlimited number of synonyms to the concept. One example is shown in Figure 18.1.

Such a structure of unambiguous concepts linked to synonyms is called a *controlled vocabulary*, or *CV* for short. Any concept in a CV is called a *term*. Although it is already a good thing to have such a CV, another improvement can be made by linking terms by their relationships. The result ideally represents a model of the entire domain, also called an *ontology*. Figure 18.2 shows a section of a possible ontology.

Formatting standards in proteomics Apart from a common, unambiguous language to describe information, the structure of this information also needs to be standardized. Fortunately, several data formatting standards already exist, including ASN.1, MIME, and XML. Since XML provides a number of interesting advantages (both human and

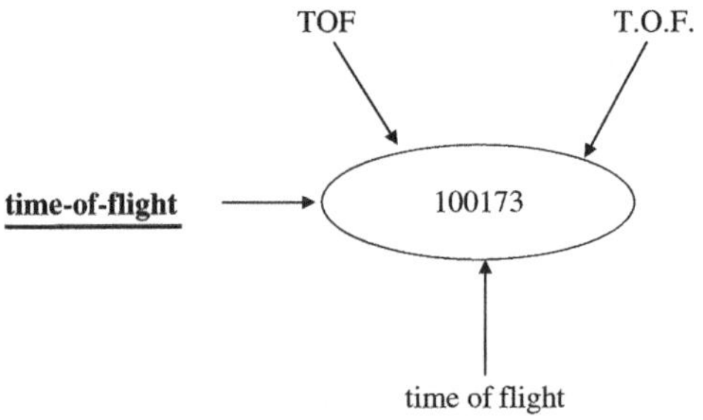

Figure 18.1 A particular concept (the time-of-flight mass analyzer) is abstracted to a number. All relevant synonyms reference this number. Note that one of the synonyms is selected to be the preferred name, indicated here by the bold face and underscoring

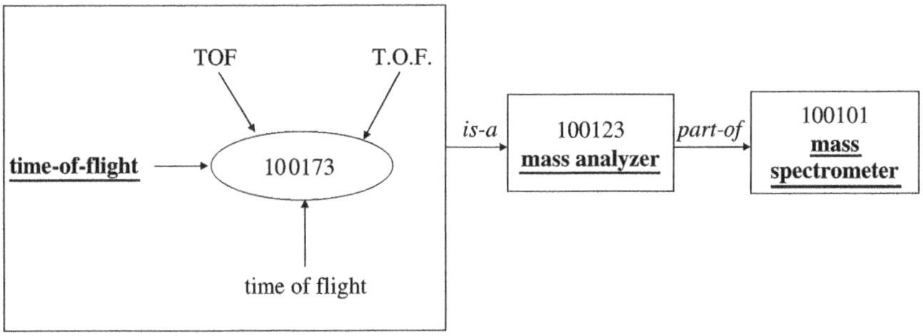

Figure 18.2 A representation of a small part of an ontology, illustrating some of the relationships by which terms can be linked. Note that the relationships are annotated in italics next to the arrows linking the boxes

machine readable, validatable, readily transformable through XSL transformation), most proposed data standards in proteomics are written in it. One of the difficulties in tying a format to an ontology lies in the overlap between the two in terms of structure. For instance, we can model the XML to show the hierarchical relationship of a time-of-flight analyzer to a mass spectrometer, or we can rely on the ontology to implicitly model this relationship. The resulting XML structure in the latter case will obviously be relatively flat. A balance between the modeled structure in the format and the implicit structure in the ontology can usually be found, however, yielding a very flexible data standard that can cope with possible future modifications.

18.2.3 Requirements for data standards

When data standards are used for the communication of information between software programs, they should cater to certain requirements. The most important of these are as follows:

- adequately model at least the minimal information required to perform basic functionalities on the data;
- enforce the presence of this minimal information in the standard, so that any data consumer knows what information can always be reliably retrieved;
- allow as much as possible (ideally, any) additional detail to be communicated along with the core information in optional fields;
- accommodate future extensions and additions while ideally remaining fully backwards compatible (new files should be readable by older software, while new software should be able to read older files);
- adoption of the standard should be relatively straightforward (for example, by providing model implementations for popular programming languages or by enlisting support from commercial instrument and software vendors to support input and output of their data in the standard format).

It is not trivial to adhere to this list of requirements, and even when a standard does adhere to all of these points, some problems can still occur.

18.2.4 Problems with data standards

There are many different problems that can interfere with the successful adoption of a data standard. We summarize here several of the most common problems:

- Competing standards for the same purposes have the tendency to divide the field, necessitating a multiplication of efforts for each software programmer who wants to read and/or write both standards. Multiple coexisting standards also lead to confusion and often even erroneous interpretation of information. An unfortunate example of problems due to conflicting standards is the failed NASA Mars Climate Orbiter mission, where confusion about the exact units of measurement used (imperial units were reported, metric units were expected) resulted in the loss of a spacecraft worth well over 100 million dollars.
- Relatively flexible standards can sometimes be achieved through a simple structure, such as in the case of the FASTA format highlighted earlier. An important caveat is that the simplicity of the format usually results in a lack of detailed modeling. As such, the actual fine-grained structure of the information in the format is up to the implementer and therefore can no longer be considered standard. In the FASTA database format, for instance, it is impossible to automatically extract accession numbers from the most popular formats in a consistent way.
- Standards that are very complex, on the other hand, often resist adoption due to the large overhead their implementation incurs. This is one reason why the ASN.1 format never caught on as an interchange format for sequence databases.
- When a standard needs to be revised often in order to keep in step with the developments of the field it tries to standardize information for, the overhead of

implementing the standard quickly rises. While not necessarily presenting a threshold for implementation in the first iteration of the standard, support for the standard during subsequent iterations might rapidly decline as the overhead of reimplementing the changes increases.

18.3 The Proteomics Standards Initiative

The Proteomics Standards Initiative (*PSI*) was founded in April 2002, at the HUPO meeting in Washington. It was founded to define community standards for data representation in proteomics to facilitate data comparison, exchange, and verification. The PSI consists of several working groups, each of which covers a specific area of interest within the field of proteomics. The current working groups are: *Molecular Interactions (MI), Mass Spectrometry (MS), Sample Processing (SP), Gel Electrophoresis (Gel), Proteomics Informatics (PI),* and *Protein Modifications (Mod).*

The PSI welcomes contributions from people in the field and allows anybody to join any of the working groups. Two meetings are held each year (one in spring and one in autumn) where the individual working groups present their progress to the participants. These meetings are free and are meant to allow interested parties to comment on the work of the PSI. Working groups stay in touch through phone conferences that are also open to interested parties, and sometimes organize workshops where working group members convene to work on their standard.

Through the PSI website, several open mailing lists, and publications in the scientific literature, the PSI attempts to reach as wide an audience as possible.

18.3.1 Minimal reporting requirements

The PSI also requires each working group to develop minimal reporting requirements for its specific field. A minimal reporting requirement consists of a document that specifies the minimal amount of information about the experiment or analysis that should be provided upon publication. The combined documents form the *Minimal Information About a Proteomics Experiment (MIAPE)* specification. The MIAPE documents should capture *sufficient* information for the understanding and reproducing of the work, while remaining *practical* for the submitter at the same time. It is important to emphasize that MIAPE specifications are not aimed at prescribing how an experiment should be done, but instead informs researchers how best to describe what they have done.

18.4 Mass spectrometry standards

Mass spectrometers accumulate substantial amounts of information during a data acquisition period as raw data. The primary storage format of this data typically consists of one or more binary files with a proprietary layout. These files are difficult to read directly, yet may sometimes be accessed programmatically through software libraries provided by the instrument vendor. These files can also be very large, easily comprising several gigabytes per LC run for modern LC-MS instruments. As such, these formats are seldom used for further analysis. Instead, highly processed digests of these files are exported as text-based *peak lists*, which are more convenient for submission to search engines.

Peak lists can be formatted differently depending on the type of analysis performed. For MS analysis (peptide mass fingerprinting), a simple list of recorded m/z values is the minimal information in such a peak list. Sometimes a two-column table of values is reported, in which the first column holds m/z values while the second column holds the corresponding intensity of these signals. In MS/MS studies (fragmentation spectra), a peak list will also contain the selected precursor m/z and its charge (if known) along with a table of m/z values and their intensities.

There are many formats for peak lists in existence, the most prominent of which are DTA, PKL, and MGF. Examples of these three formats are shown in Figure 18.3. Historically, instrument vendors settled on slightly different peak list formats, while search engine vendors typically preferred yet other formats as input for their software. It is obvious that these many competing formats cause some unnecessary overheads and potential misinterpretations,[1] yet another flaw in the peak list formats is the oversimplification of the data structure. For instance, finding out from which instrument the peak list is derived is impossible. Another perceived problem with peak lists is the lack of peak-specific metadata that can be stored. This makes it impossible to record potentially important pieces of information such as fragment peak charge state or precursor peak elution time.

```
1823.78 2
113.34 654
125.10 345
238.91 1120
315.67 619
450.73 8415
517.33 10765
599.12 312
610.08 9371
780.76 6541
980.36 3201
1100.95 1201
```

DTA file

```
912.39 12304 2
113.34 654
125.10 345
238.91 1120
315.67 619
450.73 8415
517.33 10765
599.12 312
610.08 9371
780.76 6541
980.36 3201
110.95 1201
```

PKL file

```
# A comment line.
BEGIN IONS
TITLE=Spectrum title
PEPMASS=912.39 12304
CHARGE=2+
113.34 654
125.10 345
238.91 1120
315.67 619
450.73 8415
517.33 10765
599.12 312
780.76 6541
980.36 3201
110.95 1201
END IONS
```

MGF file

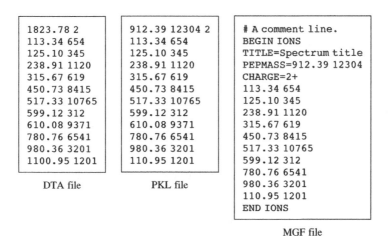

Figure 18.3 Examples of three popular peak list formats. The same fragmentation spectrum has been formatted in DTA, PKL, and MGF format. Note the use of comments in the MGF format, as well as the more verbose structural elements. Also note that the PKL file contains the precursor m/z value and the DTA file holds the singly charged precursor mass

[1] For instance, DTA and PKL files are very similar in format, but where PKL files contain the precursor m/z value, DTA files hold the singly charged precursor mass.

In order to solve the problems with existing peak list formats, the Institute for Systems Biology (ISB) in Seattle developed the *mzXML* format (Pedrioli *et al.* (2004)). As an XML-based format, mzXML benefits from all the underlying properties of XML: that is, it is human and machine readable, validatable, and architecture independent. The ISB also provided convenient tools that could extract mzXML files from several vendor-specific raw data formats, making it relatively easy for others to generate mzXML formatted data. Finally, a collection of freely available tools developed by the ISB provided powerful data analysis and processing functionality based on mzXML input files.[2]

At the same time, the HUPO PSI Mass Spectrometry (*PSI-MS*) working group developed the mzData standard as another standard for peak list replacement. Also XML based, mzData inherits the same underlying advantages of XML as mzXML. Contrary to mzXML, which was developed by a single institute, mzData was developed by a committee of interested people from academia, industry, and mass spectrometer vendors. mzData also differs from mzXML in that it relies on a controlled vocabulary to annotate data elements, rather than on the free text descriptions found in mzXML.

Currently, both teams are working on a joint standard to supersede both mzXML and mzData under the PSI-MS umbrella.

18.5 Modification standards

The PSI effort on modifications (*PSI-Mod*) is building an ontology that will contain all modifications currently held by the RESID, DeltaMass, and UniMod databases. This ontology will therefore contain posttranslational modifications (effected by the cellular machinery *in vivo*) as well as chemical modifications (due to sample handling and processing in the lab). For more information on modifications, see Section 1.4.

18.6 Identification standards

As with the peak lists, the format of protein or peptide identifications is specific to the software that produced the output. Some of these are based on existing formatting standards, such as the MIME files reported by MASCOT (also known as *datfiles* after their '.dat' extension) and the different XML-based formats (for example, from X!Tandem and OMSSA), while others define custom formats, such as the text-based SEQUEST '.dta' files.

Additionally, Ruedi Aebersold's group at ISB wrote its own post-analysis software to refine peptide and protein identification, which came with its own XML-based output formats as well. These are known as pepXML and protXML.

The PSI Proteomics Informatics (*PSI-PI*) working group is currently developing the analysisXML standard which will standardize search engine output and possible postprocessing steps. The format is XML based and will rely on terms from a corresponding ontology.

[2] The need to exchange information from different instruments between different software applications was one of the driving forces behind the development of mzXML.

Bibliographic notes

FASTA Lipman and Pearson (1985)

Standards Pedrioli *et al.* (2004)

HUPO http://www.hupo.org

PSI http://psidev.info

Bibliography

Altschul SF, Madden TL, Schffer AA, Zhang J, Zhang Z and Lipman DJ 1997 Gapped BLAST and PSI-BLAST: a new generation of protein database search programs. *Nucl. Acids Res.* **25**(17), 3389–3402.

Bafna V and Edwards N 2001 SCOPE: a probabilistic model for scoring tandem mass spectra against a peptide database. *Bioinformatics* **17**(Suppl. 1), S13–S21.

Bandeira N, Tang H, Bafna V and Pevzner P 2004 Shotgun protein sequencing by tandem mass spectra assembly. *Anal. Chem.* **76**, 7221–7223.

Bandeira N, Tsur D, Frank A and Pevzner P 2006 A new approach to protein identification *Lecture Notes in Computer Science, volume 3909, Proceedings of Research in Computational Molecular Biology: 10th Annual International Conference, RECOMB 2006*. Springer.

Barrett A, Rawlings ND and Woessner JF 1998 *Handbook of proteolytic enzymes*. Academic Press.

Barsnes H, Mikalsen SO and Eidhammer I 2006 MassSorter: a tool for administrating and analyzing data from mass spectrometry experiments on proteins with known amino acid sequences. *BMC Bioinformatics* **7**, 42.

Bartels C 1990 Fast algorithm for peptide sequencing by mass spectroscopy. *Biomed. Mass Spectrom.* **19**, 363–368.

Baumgart S, Lindner Y, Kühne R, Oberemm A, Wenschuh H and Krause E 2004 The contribution of specific amino acid side chains to signal intensities of peptides in matrix-assisted laser desorption/ionization mass spectrometry. *Rapid Commun. Mass Spectrom* **18**, 863–868.

Beer I, Barnea E, Ziv T and Admon A 2004 Improving large-scale proteomics by clustering of mass spectrometry data. *Proteomics* **4**, 950–960.

Bern M and Goldberg D 2005 EigenMS: *de novo* analysis of peptide tandem mass spectra by spectral graph partitioning. In *Ninth International Conference on Research in Computational Molecular Biology (RECOMB 2005)* (ed. Miyano S), pp. 357–372.

Bern M, Goldberg D, McDonald WH and Yates III, JR 2004 Automatic quality assessment of peptide tandem mass spectra. *Bioinformatics* **20**(Suppl. 1), i49–i54.

Berndt P, Hobohm U and Langen H 1999 Reliable automatic protein identification from matrix-assisted laser desorption/ionization mass spectrometric peptide fingerprints. *Electrophoresis* **20**, 3521–2526.

Beynon R, Doherty M, Pratt J and Gaskell S 2005 Multiplexed absolute quantification in proteomics using artificial QCAT proteins of concatenated signature peptides. *Nat. Methods* **2**, 587–589.

Computational Methods for Mass Spectrometry Proteomics I. Eidhammer, K. Flikka, L. Martens and S.-O. Mikalsen

© 2007 John Wiley & Sons, Ltd

Biemann K 1988 Contributions of mass-spectrometry to peptide and protein-structure. *Biomed. Environ. Mass Spectrom.* **16**, 99–111.

Biemann K 1990 Nomenclature for peptide fragment ions. *Methods Enzymol.* **193**, Academic Press, pp. 886–887, Appendix 5.

Bjellqvist B 1993 The focusing positions of polypeptides in immobilized pH gradients can be predicted from their amino acid sequences. *Electrophoresis* **14**, 1023–1031.

Blom K 2001 Estimating the precision of exact mass measurements on an orthogonal time-of-flight mass spectrometer. *Anal. Chem.* **73**, 715–719.

Boyd SE, Garcia de la Banda M, Pike RN, Whisstock JC and Rudy GB 2005 PoPS: A computational tool for modeling and predicting protease specificity. *J. Bioinformatics Comput. Biol.* **3**, 551–585.

Bruni R, Gianfrancesci G and Koch G 2004 On peptide *de novo* sequencing: a new approach. *J. Pept. Sci.* **11**, 225–234.

Cagney G, Amiri S, Premawaradana T, Lindo M and Emili A 2003 In silico proteome analysis to facilitate proteomics experiments using mass spectrometry. *Protein Science Open Access*, **1**(1), 5.

Ceasy DM and Cottrell J 2004 UniMod: protein modifications for mass spectrometry. *Proteomics* **4**, 1534–1536.

Chamrad DC, Körting G, Gobom J, Thiele H, Klose J, Meyer HE and Blüggel M 2003 Interpretation of mass spectrometry data through high-throughput proteomics. *Anal. Bioanal. Chem.* **376**, 1014–1022.

Chamrad DC, Körting G, Stühler K, Meyer HE, Klose J and Blüggel M 2004 Evaluation of algorithms for protein identification from sequence databases using mass spectrometry data. *Proteomics* **4**, 619–628.

Chen SS, Deutsch EW, Yi EC, Li XJ, Goodlett DR and Aebersold R 2005 Improving mass and liquid chromatography based identification of proteins using Bayesian scoring. *J. Proteome Res.* **4**, 2174–2184.

Chen T, Kao MY, Tepel M, Rush J and Church GM 2001 A dynamic programming approach to *de novo* peptide sequencing via tandem mass spectrometry. *J. Comput. Biol.* **8**, 325–337.

Clauser KR, Baker P and Burlingame AL 1999 Role of accurate mass measurement in protein identification strategies employing MS or MS/MS and database searching. *Anal. Chem.* **71**, 2871–2882.

Cohen SL and Chait BT 1996 Influence of matrix solution conditions on the MALDI-MS analysis of peptides and proteins. *Anal. Chem.* **68**, 31–37.

Colinge J, Chiappe D, Lagache S, Moniatte M and Bougueleret. L 2005 Differential proteomics via probabilistic peptide identification scores. *Anal. Chem.* **77**, 596–606.

Colinge J, Magnin J, Dessingy T, Giron M and Masselot A 2003a Improved peptide charge state assignment. *Proteomics* **3**, 1434–1440.

Colinge J, Masselot A and Magning J 2003b A systematic statistical analysis of ion trap tandem mass spectra in view of peptide scoring. In *Algorithms in bioinformatics-WABI* (ed. Benson G and Page R), pp. 25–38. Springer.

Colinge J, Masselot A, Cusin I, Mahé E, Niknejad A, Argoud-Puy G, Reffas S, Bederr N, Gleizes A, Rey PA and Bougueleret L 2004 High-performance peptide identification by tandem mass spectrometry allows reliable automatic data processing in proteomics. *Proteomics* **3**, 1977–1984.

Colinge J, Masselot A, Giron M, Dessingy T and Magnin J 2003c OLAV: towards high-throughput tandem mass spectrometry data identification. *Proteomics* **3**, 1454–1463.

Craig R and Beavis RC 2003 A method for reducing the time required to match protein sequences with tandem mass spectra. *Rapid Commun. Mass Spectrom.* **17**, 2310–2316.

Dancik V, Addona TA, Clauser KR, Vath JE and Pevzner PA 1999 *De novo* peptide sequencing via tandem mass spectrometry. *J. Comput. Biol.* **6**(3/4), 327–342.

Day RM, Borziak A and Gorin A 2004 PPM-chain *de novo* peptide identification program comparable in performance to SEQUEST *IEEE CSB'04*, pp. 505–508, IEEE.

de Hoffmann E and Stroobant V 2002 *Mass Spectrometry: Principles and Applications*. John Wiley & Sons, Ltd.

Domon B and Aebersold R 2006 Mass spectrometry and protein analysis. *Science* **312**, 212–217.

Du P, Kibbe WA and Lin SM 2006 Improved peak detection in mass spectrum by incorporation of continuous wavelet transform-based pattern matching. *Bioinformatics* **22**, 2059–2065.

Duda RO and Hart PE 1972 A generalized Hough transformation to detecting lines in pictures. *Commun. ACM* **15**, 1–15.

Earnshaw WC, Martins LM and Kaufmann SH 1999 Mammalian caspases: structure, activation, substrates, and functions during apoptosis. *Anal. Rev. Biochem.* **68**, 383–424.

Egelhofer V, Bussow K, Luebbert C, Lehrach H and Nordhoff E 2000 Improvements in protein identification by MALDI-TOF-MS peptide mapping. *Anal. Chem.* **72**, 2741–2750.

Egelhofer V, Gobom J, Seitz H, Giavalisco P, Lehrach H and Nordhoff E 2002 Protein identification by MALDI-TOF-MS peptide mapping; a new strategy. *Anal. Chem.* **74**, 1760–1771.

Elias JE, Gibbons FD, King OD, Roth FP and Gygi SP 2004 Intensity-based protein identification by machine learning from a library of tandem mass spectra. *Nat. Biotechnol.* **22**(2), 214–219.

Eng JK, McCormack AL and Yates III, JR 1994 An approach to correlate tandem mass spectral data of peptides with amino acid sequences in a protein database. *J. Am. Soc. Mass Spectrom.* **5**(11), 976–989.

Eriksson J and Fenyö D 2002 A model of random mass-matching and its use for automated significance testing in mass spectrometric proteome analysis. *Proteomics* **2**, 262–270.

Eriksson J and Fenyö D 2003 Probity: a protein identification algorithm with accurate assignment of the statistical significance of the results. *J. Proteome Res.* **3**, 32–36.

Eriksson J, Chait BT and Fenyö D 2000 A statistical basis for testing the significance of mass spectrometric protein identification results. *Anal. Chem.* **72**, 999–1005.

Falkner J and Andrews P 2005 Fast tandem mass spectra-based identification regardless of the number of spectra or potential modifications examined. *Bioinformatics* **21**, 2177–2184.

Fenyö D and Beavis RC 2003 A method for assessing the statistical significance of mass spectrometry-based protein identification using general scoring schemes. *Anal. Chem.* **75**, 768–774.

Fernández-de-Cossío J, Gonzalez J and Besada V 1995 A computer program to aid the sequencing of peptides in collision-activated decomposition experiments. *CABIOS* **11**(4), 427–434.

Fernández-de-Cossío J, Gonzalez J, Betancourt L, Besada V, Padron G, Shimonishi Y and Takao T 1998 Automated interpretation of high-energy collision-induced dissociation spectra of singly protonated peptides by 'SeqMS', a software aid for *de novo* sequencing by tandem mass spectrometry. *Rapid Commun. Mass Spectrom.* **12**(23), 1867–1878.

Fernández-de-Cossío J, Gonzalez J, Satomi Y, Shima T, Okumura N, Besada V, Betancourt L, Padron G, Shimonishi Y and Takao T 2000 Automated interpretation of low-energy collision-induced dissociation spectra by SeqMS, a software aid for *de novo* sequencing by tandem mass spectrometry. *Electrophoresis* **21**(9), 1694–1699.

Fischer B, Grossmann J, Roth V, Gruissem W, Baginsky S and Buhmann JM 2006 Semi-supervised LC/MS alignment for differential proteomics. *Bioinformatics* **22**, 132–140.

Flikka K, Martens L, Vandekerckhove J, Gevaert K and Eidhammer I 2006 Improving the reliability and throughput of mass spectrometry-based proteomics by spectrum quality filtering. *Proteomics* **6**, 2086–2094.

Frank A and Pevzner P 2005 PepNovo: *de novo* peptide sequencing via probabilistic network modeling. *Anal. Chem.* **77**, 964–973.

Frank A, Tanner S, Bafna V and Pevzner P 2005 Peptide sequence tags for fast database search in mass-spectrometry. *J. Proteome Res.* **4**, 1287–1295.

Gasteiger E, Hoogland C, Gattiker A, Duvaud S, Wilkins M, Appel R and Bairoch A 2005 Protein identification and analysis tools on the ExPasy server. In *The Proteomics Protocols Handbook* (ed. Walker JM). Humana Press.

Gatlin CL, Eng JK, Cross ST, Detter JC and Yates III, JR 2000 Automated identification of amino acid sequence variations in proteins by HPLC/microspray tandem mass spectrometry. *Anal. Chem.* **72**, 757–763.

Gattiker A, Bienvenut WV, Bairoch A and Gasteiger E 2002 FindPept, a tool to identify unmatched masses in peptide mass fingerprinting protein identification. *Proteomics* **2**, 1435–1444.

Gay S, Binz PA, Hochstrasser DF and Appel RD 1999 Modeling peptide mass fingerprinting data using the atomic composition of peptides. *Electrophoresis* **20**, 3527–3534.

Gentzel M, Köcher T, Ponnusamy S and Wilm M 2003 Preprocessing of tandem mass spectrometric data to support automatic protein identification. *Proteomics* **3**, 1597–1610.

Gevaert K, Damme JV, Goethals M, Thomas GR, Hoorelbeke B, Demol H, Martens L, Puype M, Staes A and Vandekerckhove J 2002 Chromatographic isolation of methionine-containing peptides for gel-free proteome analysis. *Mol. Cell Proteomics* **1**, 896–903.

Gevaert K, Ghesquiere B, Staes A, Martens L, Damme JV, Thomas GR and Vandekerckhove J 2004 Reversible labeling of cysteine-containing peptides allows their specific chromatographic isolation for non-gel proteome studies. *Proteomics* **4**, 897–908.

Gevaert K, Goethals M, Martens L, Damme JV, States A, Thomas GR and Vandekerckhove J 2003 Exploring proteomes and analyzing protein processing by mass spectrometric identification of sorted N-terminal peptides. *Nat. Biotechnol.* **21**, 566–569.

Gobom J, Müller M, Egelhofer V, Theiss D, Lehrach H and Nordhoff E 2002 A calibration method that simplifies and improves accurate determination of peptide molecular masses by MALDI-TOF MS. *Anal. Chem.* **74**, 3915–3923.

Gras R, Müller W, Gasteiger E, Gay S, Binz PA, Bienvenut W, Hoogland C, Sanchez JC, Bairoch A, Hochstrasser D and Appel R 1999 Improving protein identification from peptide mass fingerprinting through a parameterized multi-level scoring algorithm and an optimized peak detection. *Electrophoresis* **20**, 3535–3550.

Grossmann J, Roos FF, Cieliebak M, Lipták Z, Mathis LK, Müller M and Gruissem W 2005 AUDENS: a tool for automated peptide *de novo* sequencing. *J. Proteome Res.* **4**, 1768–1774.

Guo D, Mant CT, Taneja AK and Rodges R 1986a Prediction of peptide retention times in reversed-phase high-performance liquid chromatography II. Correlation of observed and predicted peptide retention times factors and influencing the retention times of peptides. *J. Chromatogr.* **359**, 519–532.

Guo D, Mant CT, Taneja AK, Parker JR and Rodges R 1986b Prediction of peptide retention times in reversed-phase high-performance liquid chromatography I. Determination of retention coefficients of amino acid residues of model synthetic peptides. *J. Chromatogr.* **359**, 499–518.

Gusfield D 1997 *Algorithms on Strings, Trees, and Sequences*. Cambridge University Press.

Halligan BD, Ruotti V, Jin W, Laffoon S, Twigger SN and Dratz EA 2004 ProMoST (protein modification screening tool): a web-based tool for mapping protein modifications on two-dimensional gels. *Nucl. Acids Res.* **32**, W638–W644.

Han Y, Ma B and Zhang K 2005 SPIDER: software for protein identification from sequence tags with *de novo* sequencing error. *J. Bioinformatics Comput. Biol.* **3**, 697–716.

Hansen BT, Davey SW, Ham AJL and Liebler DC 2005 P-Mod: an algorithm and software to map modifications to peptide sequences using tandem MS data. *J. Proteome Res.* **4**, 358–368.

Hansen BT, Jones JA, Mason DE and Liebler DC 2001 Salsa: a pattern recognition algorithm to detect electrophile-adducted peptides by automated evolution of CID spectra in LC-MS-MS analyses. *Anal. Chem.* **73**(8), 1676–1683.

Hao P, He WZ, Huang Y, Ma LX, Xu Y, Xi H, Wang C, Liu BS, Wang JM, Li YX and Zhong Y 2005 MPSS: an integrated database system for surveying a set of proteins. *Bioinformatics* **21**, 2142–2143.

Havilio M, Haddad Y and Smilansky Z 2003 Intensity-based statistical scorer for tandem mass spectrometry. *Anal. Chem.* **75**, 435–444.

Heredia-Langner A, Cannon WR, Jarman KD and Jarman KH 2004 Sequence optimization as an alternative to *de novo* analysis of tandem mass spectrometry data. *Bioinformatics* **20**(14), 2296–2304.

Hines WM, Falick AM, Burlingame AL and Gibson BW 1992 Pattern-based algorithm for peptide sequencing from tandem high energy collision-induced dissociation mass spectra. *J. Am. Soc. Mass Spectrom.* **3**, 326–336.

Horn DM, Zubarev RA and McLafferty FW 2000 Automated *de novo* sequencing of proteins by tandem high-resolution mass spectrometry. *Proc. Natl Acad. Sci. USA* **97**, 10313–10317.

Huang L, Jacob RJ, Pegg SCH, Baldwin MA, Wang CC, Burlingame AL and Babbitt PC 2001 Functional assignment of the 20 s proteasome from *Trypanosoma brucei* using mass spectrometry and new bioinformatics approaches. *J. Biol. Chem.* **276**, 28327–28339.

Huang Y, Triscari JM, Tseng GC, Pasa-Tolic L, Lipton MS, Smith RD and Wysocki VH 2005 Statistical characterization of the charge state and residue dependence of low-energy CID peptide dissociation patterns. *Anal. Chem.* **77**, 5800–5813.

Huang Y, Wysocki VH, Tabb DL and Yates III, JR 2002 The influence of histidine on cleavage C-terminal to acidic residues in double protonated tryptic peptides. *Int. J. Mass Spectrom.* **219**, 233–244.

James P, Quadroni M, Carafoli E and Gonnet G 1994 Protein identifiction in DNA databases by peptide mass fingerprinting. *Protein Sci.* **3**, 1347–1350.

Jensen ON, Podtelejnikov AV and Mann M 1997 Identification of the components of simple protein mixtures by high-accuracy peptide mass mapping and database searching. *Anal. Chem.* **69**, 4741–4750.

Johnson RS, Martin SA, Biemann K, Stults K and Watson JT 1987 Novel fragmentation process of peptides by collision-induced decomposition in a tandem mass spectrometer: differentiation of leucine and isoleucine. *Anal. Chem.* **59**(21), 2621–2625.

Kapp EA, Schutz F, Reid GE, Eddes JS, Moritz RL, O'Hair RAJ, Speed TP and Simpson RJ 2003 Mining a tandem mass spectrometry database to determine the trends and global factors influencing peptide fragmentation. *Anal. Chem.* **22**, 6251–6264.

Karty JA, Ireland MME, Brun YV and Reilly JP 2002 Artifacts and unassigned masses encountered in peptide mass mapping. *J. Chromatogr. B* **782**, 363–383.

Keil B 1992 *Specificity of proteolysis*. Springer.

Kelleher NL, Lin HY, Valaskovid GA, Aaserud DJ, Fridriksson EK and McLafferty FW 1999 Top down versus bottom up protein characterization by tandem high-resolution mass spectrometry. *J. Am. Chem. Soc.* **121**, 806–812.

Keller A, Nesvizhskij AL, Kolker E and Aebersold R 2002a Empirical statistical model to estimate the accuracy of peptide identifications made by MS/MS and database search. *Anal. Chem.* **74**, 5383–5392.

Keller A, Purvine S, Nesvizhskii AI, Stolyar S, Goodlett DR and Kolker E 2002b Experimental protein mixture for validating tandem mass spectral analysis. *OMICS* **6**(2), 207–212.

Klammer AA, Wu CC, MacCoss MJ and Noble WS 2005 Peptide charge state determination for low-resolution tandem mass spectra. *Proceedings of the 2005 Computational Systems Bioinformatics Conference*, pp. 175–185.

Kristensen DB, Brønd JC, Nielsen PA, Andersen JR, Sørensen OT, Jørgensen V, Budin K, Matthiesen J, Venø P, Jespersen HM, Ahrens CH, Schandorff S, Ruhoff PT, Wisniewski JR, Bennett KL and Podtelejnikov AV 2004 Experimental peptide identification repository (EPIR): an integrated peptide-centric platform for validation and mining of tandem mass spectrometry data. *Mol. Cell. Proteomics* **3**(3), 1023–1038.

Krokhin OV 2006 Sequence-specific retention calculator. Algorithm for peptide retention prediction in ion-pair RP-HPLC: application to 300- and 100-pore size C18 sorbents. *Anal. Chem.* **78**, 7785–7795.

Krokhin OV, Craig R, Spicert V, Enst W, Standing KG, Beavis RC and Wilkins JA 2004 An improved model for prediction of retention times of tryptic peptides in ion pair reversed-phase HPLC. *Mol. Cell. Proteomics* **3**, 908–919.

Kyte J and Doolittle RF 1982 A simple method for displaying the hydropathic character of a protein. *J. Mol. Biol.* **157**, 105–132.

Lehmann WD, Bohne A and von der Lieth CW 2000 The information encrypted in accurate peptide masses-improved protein identification and assistance in glycopeptide identification and characterization. *J. Mass Spectrom.* **35**, 1335–1341.

Levander F, Rögnvaldsson T, Samuelsson J and James P 2004 Automated methods for improving protein identification by peptide mass fingerprinting. *Proteomics* **4**, 2594–2601.

Lie C, Yan B, Song Y, Xu Y and Cai L 2006a Peptide sequence tag-based blind identification of post-translational modification with point process model. *ISMB06 Conference*.

Lie C, Yan B, Song Y, Xu Y and Cai L 2006b Peptide sequence tag-based blind identification of post-translational modifications with point process model. *Bioinformatics* **22**, 307–313.

Lipman DJ and Pearson WR 1985 Rapid and sensitive protein similarity searches. *Science* **227**, 1435–1441.

Liu J, Chrisman PA, Erickson DE and McLuckey SA 2007 Relative information content and top-down proteomics by mass spectrometry: utility of ion/ion proton-transfer reactions in electrospray-based approaches. *Anal. Chem.* **79**(3), 1073–1081.

Lu B and Chen T 2003a A suboptimal algorithm for *de novo* peptide sequencing via tandem mass spectrometry. *J. Comput. Biol.* **10**(1), 1–12.

Lu B and Chen T 2003b A suffix tree approach to the interpretation of tandem mass spectra: application to peptides of non-specific digestion and post-translational modifications. *Bioinformatics* **19** (Suppl. 2), ii113–ii121.

Lubeck O, Sewell C, Gu S, Chen X and Ca M 2002 New computational approaches for *de novo* peptide sequencing from MS/MS experiments. *Proc. IEEE* **90**, 1868–1874.

Ma B, Zhang K, Hendrie C, Liang C, Li M, Doherty-Kirby A and Lajoie G 2003 PEAKS: powerful software for peptide *de novo* sequencing by tandem mass spectrometry. *Rapid Commun. Mass Spectrom.* **17**, 2337–2342.

Macek B, Waanders LF, Olsen JV and Mann M 2006 Top-down protein sequencing and MS^3 on a hybrid linear quadrupole ion trap-Orbitrap mass spectrometer. *Mol. Cell. Proteomics* **5**, 949–958.

Mackey AJ, Haystead TAJ and Pearson WR 2002 Getting more from less. *Mol. Cell. Proteomics* **1**, 139–147.

Magnin J, Masselot A, Menzel C and Colinge J 2004 OLAV-PMF: a novel scoring scheme for high-throughput peptide mass fingerprinting. *J. Proteome Res.* **3**, 55–60.

Magrane M, Martin MJ, O'Donovan C and Apweiler R 2005 Protein sequence databases *The Proteomics Protocols Handbook*. Humana Press.

Mann M and Wilm M 1994 Error-tolerant identification of peptides in sequence databases by peptide sequence tags. *Anal. Chem.* **66**, 4390–4399.

Mann M, Hendrickson RC and Pandey A 2001 Analysis of proteins and proteomes by mass spectrometry. *Annu. Rev. Biochem.* **70**, 437–73.

Mann M, Meng CK and Fenn JB 1989 Interpreting mass spectra of multiple charge ions. *Anal. Chem.* **61**, 1702–1708.

Marouga R, David S and Hawkins E 2005 The development of the DIGE system: 2D fluorescence difference gel analysis technology. *Anal. Bioanal. Chem.* **382**, 669–678.

Martens L 2006 Novel bioinformatics tools assisting targeted peptide-centric proteomics and global proteomics data dissemination. PhD thesis, Universiteit Gent.

Matthiesen R, Bunkenborg J, Stensballe A, Jensen ON, Welinder KG and Bauw G 2004 Database-independent, database-dependent, and extended interpretation of peptide mass spectra in VEMS v2.0. *Proteomics* **4**, 2583–2593.

Meek JL 1980 Prediction of peptide retention times in high-pressure liquid chromatography on the basis of amino acid composition. *Proc. Natl Acad. Sci. USA* **77**, 1632–1636.

Meng F, Cargile BJ, Patrie SM, Johnson JR, McLoughlin SM and Kelleher NL 2002 Processing complex mixtures of intact proteins for direct analysis by mass spectrometry. *Anal. Chem.* **74**, 2923–2929.

Meng F, Yi D, Miller LM, Patrie SM, Robinson DE and Kelleher NL 2004 Molecular-level description of proteins from *Saccharomyces cerevisiae* using quadrupole FT hybrid mass spectrometry for top down proteomics. *Anal. Chem.* **76**, 2852–2858.

Na S and Paek E 2006 Quality assessment of tandem mass spectra based on cumulative intensity normalization. *J. Proteome Res.* **5**, 3241–3248.

Nelson DL and Cox MM 2004 *Lehninger Principles of Biochemistry*. Worth.

Nesvizhskii AI and Aebersold R 2004 Analysis, statistical validation and dissemination of large-scale proteomics datasets generated by tandem MS. *Drug Discov. Today* **9**, 391–202.

Nesvizhskii AI and Aebersold R 2005 Interpretation of shotgun proteomic data. *Mol. Cell. Proteomics* **4**, 1419–1440.

Nesvizhskii AI, Keller A, Kolker E and Aebersold R 2003 A statistical model for identifying proteins by tandem mass spectrometry. *Anal. Chem.* **75**(17), 4646–4658.

Nordstrom A, O'Maille G, Qin C and Siuzdak G 2006 Nonlinear data alignment for UPLC-MS and HPLC-MS based metabolomics: quantitative analysis of endogenous and exogenous metabolites in human serum. *Anal. Chem.* **78**, 3289–3295.

Olsen JV, Ong SE and Mann M 2004 Trypsin cleaves exclusively C-terminal to arginine and lysine residues. *Mol. Cell. Proteomics* **3**, 608–614.

Orchard S, Hermjakob H, Binz PA, Hoogland C, Taylor CF, Zhu W, Julian RK and Apweiler R 2005 Further steps towards data standardisation: The Proteomic Standards Initiative HUPO 3(rd) Annual Congress, Beijing 25–27 October 2004. *Proteomics* **5**, 337–339.

Owens KG 1992 Application of correlation analysis techniques to mass spectral data. *Appl. Spectrosc. Rev.* **27**(1), 1–49.

Paizs B and Suhai S 2005 Fragmentation pathways of protonated peptides. *Mass Spectrom. Rev.* **12**, 508–548.

Palmblad M, Ramstrm M, Markides KE, Håkonsson P and Bergquist J 2002 Prediction of chromatographic retention and protein identification in liquid chromatography/mass spectrometry. *Anal. Chem.* **74**, 5826–5830.

Papayannopoulos IA 1995 The interpretation of collision-induced dissociation tandem mass spectra of peptides. *Mass Spectrom. Rev.* **14**(1), 49–73.

Pappin DJC, Rahman D, Hansen D, Bartlet-Jones HF, Jeffery M and Bleasby W 1996 Chemistry, mass spectrometry and peptide-mass databases: evolution of methods for the rapid identification and mapping of cellular proteins. In *Mass Spectrometry in the Biological Sciences* (ed. Burlington A and Carr S). Humana Press, pp. 135–150.

Parker KC 2002 Scoring methods in MALDI peptide mass fingerprinting: Chemscore, and the ChemApplex program. *J. Am. Soc. Mass Spectrom.* **13**, 22–39.

Pearson W 1990 Rapid and sensitive sequence comparison with FASTP and FASTA. *Methods Enzymol.* **266**, 227–258.

Pedrioli PGA, Eng JK, Hubley R, Vogelzang M, Deutsch EW, Raught B, Pratt B, Nilsson E, Angelettiand RH, Apweiler R, Cheung K, Costello CE, Hermjakob H, Huang S, Julian RK, Jr, Kapp E, McComb ME, Oliver SG, Omenn G, Paton NW, Simpson R, Smith R, Taylor CF, Zhu W and Aebersold R 2004 A common open representation of mass spectrometry data and its application to proteomics research. *Nat. Biotechnol.* **22**, 1459–1466.

Pegg SC and Babbitt PC 1999 Shotgun: getting more from sequence similarity searches. *Bioinformatics* **15**, 729–740.

Peng J, Elias JE, Thoreen CC, Licklider LJ and Gygi SP 2003 Evaluation of multidimensional chromatography coupled with tandem mass spectrometry (LC/LC-MS/MS) for large-scale protein analysis: the yeast proteome. *J. Prot. Res.* **2**, 43–50.

Peri S, Steen H and Pandey A 2001 GPMAW - a software tool for analyzing proteins and peptides. *Trends Biochem. Sci.* **11**, 687–689.

Perkins DN, Pappin DJC, Creasy DM and Cottrell JS 1999 Probability-based protein identification by searching sequence databases using mass spectrometry data. *Electrophoresis* **20**, 3551–3567.

Petritis K, Kangas LJ, Ferguson PL, Anderson GA, Pasa-Tolic L, Lipton MS, Auberry KJ, Strittmatter EF, Shen Y, Zhao R and Smith RD 2003 Use of artificial neural networks for the accurate prediction of peptide liquid chromatography elution times in proteome analysis. *Anal. Chem.* **75**, 1039–1048.

Pevzner PA, Dančík V and Tang CL 2000 Mutation-tolerant protein identification by mass spectrometry. *J. Comput. Biol.* **7**(6), 777–787.

Pevzner PA, Mulyukov Z, Dančík V and Tang CL 2001 Efficiency of database search for identification of mutated and modified proteins via mass spectrometry. *Genet. Res.* **11**, 290–299.

Powell LA and Hieftje GM 1978 Computer identification of infrared spectra by correlation-based file searching. *Anal. Chim. Acta* **100**, 313–327.

Purvine S, Kolker N and Kolker E 2004 Spectral quality assessment for high-throughput tandem mass spectrometry proteomics. *OMICS* **8**, 255–265.

Rawlings ND, Tolle DP and Barrett AJ 2004 MEROPS: the peptidase database. *Nucl. Acids Res.* **32**, D160–D164.

Reid GE and McLuckey SA 2002 Top down protein characterization via tandem mass spectrometry. *J. Mass Spectrom.* **37**, 663–675.

Resing KA, Meyer-Arendt K, Mendoza AM, Aveline-Wolf LD, Jonscher KR, Pierce KG, Old WM, Cheung HT, Russell S, Wattawa JL, Goehle GR, Knight RD and Ahn NG 2004 Improving reproducibility and sensitivity in identifying human proteins by shotgun proteomics. *Anal. Chem.* **76**(13), 3556–3568.

Roepstorff P and Fohlman J 1984 Proposal for a common nomenclature for sequence ions in mass spectra of peptides. *Biomed. Mass Spectrom.* **11**, 601.

Sadygov RG, Eng J, Durr E, Saraf A, McDonald H, MacCoss MJ and Yates III, JR 2001 Code developments to improve the efficiency of automated MS/MS spectra interpretation. *J. Proteome Res.* **1**, 211–215.

Salmi J, Moulder R, Filen JJ, Nevalainen OS, Nyman TA, Lahesmaa R and Aittokallio T 2006 Quality classification of tandem mass spectrometry data. *Bioinformatics* **22**, 400–406.

Samuelsson J, Dalevi D, Levander F and Rögnvaldsson T 2004 Modular, scriptable and automatic analysis tools for high-throughput peptide mass fingerprinting. *Bioinformatics* **20**, 3628–3635.

Sasagawa T, Okuyarna T and Teller DC 1982 Prediction of peptide retention times in reversed-phases high-performance liquid chromatography during linear gradient elution. *J. Chromatogr.* **240**, 329–340.

Sauve AC and Speed TP 2004 Normalization, baseline correction and alignment of high-throughtput mass spectrometry data. *Proceedings of Gensips (Workshop on Genomic Signal Processing and Statistics).*

Savitski MM, Nielsen ML and Zubarev RA 2005a New data base-independent, sequence tag-based scoring of peptide MS/MS data validates Mowse scores, recovers below threshold data, singles out modified peptides, and assesses the quality of MS/MS techniques. *Mol. Cell. Proteomics* **4**, 1180–1188.

Savitski MM, Nielsen ML, Kjeldsen F and Zubarev RA 2005b Proteomics-grade *de novo* sequencing approach. *J. Proteomic Res.* **4**, 2348–2354.

Scarberry RE, Zhang Z and Knapp DR 1995 Peptide sequence determination from high-energy collision-induced dissociation spectra using artificial neural networks. *J. Am. Soc. Mass Spectrom.* **6**, 947–961.

Schmidt F, Schmid M, Jungblut PR, Mattow J, Facius A and Pleissner KP 2003 Iterative data analysis is the key for exhaustive analysis of peptide mass finger-prints from proteins separated by two-dimensional electrophoresis. *J. Am. Soc. Mass Spectrom.* **14**, 943–956.

Schwartz R, Ting CS and King J 2001 Whole proteome pI values correlate with subcellular localizations of proteins for organisms within the three domains of life. *Genet. Res.* **11**, 703–709.

Searle BC, Dasari S, Turner M, Reddy AP, Choi D, Wilmarth PA, McCormack AL, David LL and Nagalla SR 2004 High-throughput identification of protein and unanticipated sequence modifications using a mass-based alignment algorithm for MS/MS *de novo* sequencing results. *Anal. Chem.* **76**, 2220–2230.

Searle BC, Dasari S, Wilmarth PA, Turner M, Reddy AP, David LL and Nagalla SR 2005 Identification of protein modifications using MS/MS *de novo* sequencing and the OpenSea alignment algorithm. *J. Proteome Res.* **4**, 546–554.

Shevchenko A, Sunyaev S, Loboda A, Shevchenko A, Boork P, Ens W and Standing KG 2001 Charting the proteomes of organisms with unsequenced genomes by MALDI-quadrupole time-of-flight mass spectrometry and BLAST homology searching. *Anal. Chem.* **73**, 1917–1926.

Silva JC, Gorenstein MV, Li GZ, Vissers JPC and Geromanos SJ 2006 Absolute quantification of proteins by LCMSE. *Cell. Proteomics* **5**, 144–156.

Steen H and Mann M 2004 The abc's (and xyz's) of peptide sequencing. *Nat. Rev. Mol. Cell. Biol.* **5**, 699–711.

Sunyaev S, Liska AJ, Golod A, Shevchenko A and Shevchenko A 2003 MultiTag: multiple error-tolerant sequence tag search for the sequence-similarity identification of proteins by mass spectrometry. *Anal. Chem.* **75**, 1307–1315.

Sze SK, Ge Y, Oh H and McLafferty FW 2002 Top-down mass spectrometry of a 29-kDa protein for characterization of any posttranslational modification to within one residue. *Proc. Natl Acad. Sci. USA* **99**, 1774–1779.

Tabb DL, MacCoss MJ, Wu CC, Anderson SD and Yates III, JR 2003a Similarity among tandem mass spectra from proteomic experiments: detection, significance, and utility. *Anal. Chem.* **75**, 2470–2477.

Tabb DL, Saraf A and Yates III, JR 2003b GutenTag: high-throughput sequence tagging via an empirically derived fragmentation model. *Anal. Chem.* **75**, 6415–6421.

Tabb DL, Smith LL, Breci LA, Wysocki VH and Yates III, JR 2003c Statistical characterization of ion trap tandem mass spectra from double charged tryptic peptides. *Anal. Chem.* **75**, 1155–1163.

Tabb DL, Thompson MR, Khalsa-Moyers G, VerBerkmoes NC and McDonald WH 2005 MS2Grouper: group assessment and synthetic replacement of duplicate proteomic tandem mass spectra. *J. Am. Soc. Mass Spectrom.* **16**, 1250–1261.

Tang WH, Halpern BR, Shilov IV, Seymour SL, Keating SP, Loboda A, Patel AA, Schaeffer DA and Nuwaysir LM 2005 Discovering known and unanticipated protein modifications using MS/MS database searching. *Anal. Chem.* **77**, 3931–3946.

Tanner S, Shu H, Frank A, Wang LI, Zandi E, Mumby M, Pevzner PA and Bafna V 2005 InspecT: identification of posttranslationally modified peptides from tandem mass spectra. *Anal. Chem.* **77**, 4626–4639.

Taylor JA and Johnson RS 1997 Sequence database searches via *de novo* peptide sequencing by tandem mass spectrometry. *Rapid Commun. Mass Spectrom.* **11**, 1067–1075.

Taylor JA and Johnson RS 2001 Implementation and use of automated *de novo* peptide sequencing by tandem mass spectrometry. *Anal. Chem.* **73**, 2594–2604.

Thiede B, Lamer S, Mattow J, Siejak F, Dimmler C, Rudel T and Jungblut PR 2000 Analysis of missed cleavage sites, trytophan oxidation and N-terminal pyroglutamylation after in-gel tryptid digestion. *Rapid Commun. Mass Spectrom.* **14**, 496–502.

Thornberry NA, Rano TA, Peterson EP, Rasper DM, Timkey T, Garcia-Calvo M, Houtzager VM, Nordstrom PA, Roy S, Vaillancourt JP, Chapman KT and Nicholson DW 1997 A combinatorial approach defines specificities of members of the caspase family and granzyme b. Functional relationships established for key mediators of apoptosis. *Biol. Chem.* **272**, 9677–9682.

Tonge R, Shaw J, Middleton B, Rowlinson R, Rayner S, Young J, Pognan F, Hawkins E, Currie L and Davison M 2001 Validation and development of fluorescence two-dimensional differential gel electrophoresis proteomics technology. *Proteomics* **1**, 377–396.

Tsur D, Tanner S, Zandi E, Bafna V and Pevzner PA 2005 Identification of post-translational modifications by blind search of mass spectra. *Nat. Biotechnol.* **23**, 1562–1567.

Tuloup M, Hemandez C, Coro I, Hoogland C, Binz PA and Appel RD 2003 Aldente and Biograph: an improved peptide mass fingerprinting protein identification environment. In *Understanding Biological Systems through Proteomics, Basle, Switzerland* (ed. FontisMedia), pp. 174–176. Swiss Proteomics Society.

Walker JM 2005 *The Proteomics Protocols Handbook*. Humana Press.

Wang Y, Zhang J, Gu X and Zhang XM 2005 Protein identification assisted by the prediction of retention time in liquid chromatography/tandem mass spectrometry. *J. Chromatogr. B* **826**, 122–128.

Washburn MP, Wolters D and Yates III JR 2001 Large-scale analysis of the yeast proteome by multidimensional protein indentification technology. *Nat. Biotechnol.* **19**, 242–247.

Weiller GF, Djordjevic MJ, Caraux G, Chen H and Weinman JJ 2001 A specialised proteomic database for comparing matrix-assisted laser desorption/ionization-time of flight mass spectrometry data of tryptic peptides with corresponding sequence database segments. *Proteomics* **1**, 1489–1494.

Wilkins MR, Gasteiger E, Bairoch A, Sanchez JC, Williams KL, Appel RD and Hochstrasser DF 1999a Protein identification and analysis tools in the ExPasy server. *Methods Mol. Biol.* **112**, 531–552.

Wilkins MR, Gasteiger E, Herbert B, Molloy MP, Binz PA, Ou K, Sanchez JC, Bairoch A, Williams KL and Hochstrasser DF 1999b High-throughput mass spectrometric discovery of protein post-translational modifications. *J. Mol. Biol.* **289**, 645–657.

Wilkins MR, Gasteiger E, Wheeler CH, Lindskog I, Sanchez JC, Bairoch A, Appel RD, Duun MJ and Hochstrasser DF 1998 Multiple parameter cross-species protein identification using MultiIdent - a World-Wide Web accessible tool. *Electrophoresis* **19**, 3199–3206.

Wilkins MR, Lindskog I, Gasteiger E, Bairoch A, Sanchez JC, Hochstrasser DF and Appel RD 1997 Detailed peptide characterisation using PeptideMass - a World-Wide Web accessible tool. *Electrophoresis* **18**, 403–408.

Wolff JC, Fuentes TR and Taylor J 2003 Letter to the editor. *Rapid Commun. Mass Spectrom.* **17**, 1216–1219.

Wolski WE, Lalowski M, Jungblut P and Reinert K 2005 Calibration of mass spectrometric peptide mass fingerprint data without specific external or internal calibrants. *BMC Bioinformatics* **6**, 203.

Wong JWH, Cagney G and Cartwright HM 2005 SpecAlign-processing and alignment of mass spectra datasets. *Bioinformatics* **9**, 2089–2090.

Wool A and Smilansky Z 2002 Precalibration of matrix-assisted laser desorption/ionization-time of flight spectra for peptide mass fingerprinting. *Proteomics* **2**, 1365–1373.

Wu CC, MacCoss MJ, Howell KE and Yates III, JR 2003 A method for the comprehensive proteomic analysis of membrane proteins. *Nat. Biotechnol.* **21**, 532–538.

Wysocki VH, Tsaprailis G, Smith LL and Breci LA 2000 Mobile and localized protons: a framework for understanding peptide dissociation. *J. Mass Spectrom.* **35**, 1399–1406.

Yan B, Zhou T, Wang P, Liu Z, Emanuele II, VA, Olman V and Xu Y 2006 A point-process model for rapid identification of post-translational modifications. *Proceedings of 2006 Pacific Symposium on Biocomputing*, pp. 327–338.

Yang X, Dondeti V, Dezube R, Maynard DM, Geer LY, Epstein J, Chen X, Markey SP and Kowalak JA 2004 DBParser: web-based software for shotgun proteomic data analyses. *J. Proteome Res.* **3**(5), 1002–1008.

Yates III, JR, Eng JK, McCormack AL and Schieltz D 1995 Method to correlate tandem mass spectra of modified peptides to amino acid sequences in the protein database. *Anal. Chem.* **67**, 1426–1436.

Yates III, JR, Morgan SF, Gatlin CL, Griffin PR and Eng JK 1998 Method to compare collision-induced dissociation spectra to peptides: potential for library searching and subtractive analysis. *Anal. Chem.* **70**, 3557–3565.

Zhang N, Aebersold R and Schwikowski B 2002 ProbID: a probabilistic algorithm to identify peptides through sequence database searching using tandem mass spectral data. *Proteomics* **2**, 1406–1412.

Zhang W and Chait BT 2000 ProFound - an expert system for protein identification using mass spectrometric peptide mapping information. *Anal. Chem.* **72**, 2482–2489.

Zhang X, Asara JM, Adamec J, Ouzzani M and Elmagarmid AK 2005 Data preprocessing in liquid chromatography-mass spectrometry-based proteomics. *Bioinformatics* **21**, 4054–4059.

Zhang Z 2004 *De novo* peptide sequencing based on a divide-and-conquer algorithm and peptide tandem spectrum-simulation. *Anal. Chem.* **76**, 6374–6383.

Zhong H and Li L 2005 An algorithm for interpretation of low-energy collision-induced dissociation product ion spectra for *de novo* sequencing of peptides. *Rapid Commun. Mass Spectrom.* **19**, 1084–1096.

Index

Computational Methods for Mass Spectrometry Proteomics I. Eidhammer, K. Flikka, L. Martens and S.-O. Mikalsen
© 2007 John Wiley & Sons, Ltd

Printed and bound in the UK by
CPI Antony Rowe, Eastbourne

Printed and bound by CPI Group (UK) Ltd, Croydon, CR0 4YY
16/04/2025
14658554-0002